Manias, Panics

Examine the record of history, recollect what has happened within the circle of your own experience, consider with attention what has been the conduct of almost all the great unfortunate, either in private or public life, whom you may have either read of, or heard of, or remember; and you will find that the misfortunes of by far the greater part of them have arisen from their not knowing when they were well, when it was proper for them to sit still and be contented.

– Adam Smith
The Theory of Moral Sentiments

Much has been written about panics and manias, much more than with the most outstretched intellect we are able to follow or conceive; but one thing is certain, that at particular times a great deal of stupid people have a great deal of stupid money.... At intervals, from causes which are not to the present purpose, the money of these people – the blind capital, as we call it, of the country – is particularly large and craving; it seeks for someone to devour it, and there is a 'plethora'; it finds someone, and there is 'speculation'; it is devoured, and there is 'panic.'

– Walter Bagehot
'Essay on Edward Gibbon'

Manias, Panics and Crashes

A History of Financial Crises

Sixth Edition

Charles P. Kindleberger

and

Robert Z. Aliber

palgrave
macmillan

First edition published 1978
Second edition published 1989
Third edition published 1996
Fourth edition published 2000
Fifth edition published 2005
Sixth edition published 2011 by
PALGRAVE MACMILLAN

Palgrave Macmillan in the UK is an imprint of Macmillan Publishers Limited,
registered in England, company number 785998, of Houndmills, Basingstoke,
Hampshire RG21 6XS.

Palgrave Macmillan in the US is a division of St Martin's Press LLC,
175 Fifth Avenue, New York, NY 10010.

Palgrave Macmillan is the global academic imprint of the above companies
and has companies and representatives throughout the world.

Palgrave® and Macmillan® are registered trademarks in the United States,
the United Kingdom, Europe and other countries.

ISBN: 978–0–230–36535–3 paperback

This book is printed on paper suitable for recycling and made from fully
managed and sustained forest sources. Logging, pulping and manufacturing
processes are expected to conform to the environmental regulations of the
country of origin.

A catalogue record for this book is available from the British Library.

A catalog record for this book is available from the Library of Congress.

10 9 8 7 6 5 4 3 2 1
20 19 18 17 16 15 14 13 12 11

Printed and bound by CPI Group (UK) Ltd, Croydon, CR0 4YY

Contents

Tables

Foreword

Charlie Kindleberger (CPK from now on) was a delightful colleague: perceptive, responsive, curious about everything, full of character, and, above all, lively. Those same qualities are everywhere evident in *Manias, Panics and Crashes*. I think that CPK began to work on the book in the spirit of writing a natural history, rather as Darwin must have done at the stage of the *Beagle* – collecting, examining and classifying interesting specimens. Manias, panics, and crashes had the advantage over rodents, birds, and beetles that they were accompanied by the rhetoric of contemporaries, sometimes with insight, sometimes just blather. It was CPK's style as an economic historian to hunt for interesting things to learn, not to pursue a systematic agenda.

Of course, he was an economist by training and experience, and he soon found patterns and regularities, and causes and effects. What caught his eye especially were the irrationalities that seemed so often to enmesh those directly or indirectly enmeshed in the events themselves. By itself that would have been merely entertaining. The story got interesting for CPK with the inter-action of behavior and institutions. The occurrence of manias, panics, and crashes, and their ultimate scope, also depended very much on the monetary and capital-market institutions of the time.

CPK could not have known at the start just how hardy a perennial financial crises would turn out to be. The quarter-century *after* the publication of the first edition featured a whole new level of turbulence in national banking systems, exchange-rate volatility and asset-price bubbles. There was always new material to be digested in successive editions. This history cannot have been merely the result of increasing human irrationality, though CPK would have been charmed by what a German friend of ours called 'Das Gesetz der Verschlechtigung aller Dinge' (the Law of the Deterioration of Everything). Increasing wealth, faster and cheaper communication, and the evolution of national and international financial systems also played an indispensable role, as sketched in Chapter 13, added to this edition by Robert Aliber. CPK's effort at economic history found a subject that does not appear to be going out of style.

The shape of a 'new financial architecture' and the possible utility of a lender of last resort – national and/or international – along with the guidelines that ought to govern it were also among CPK's preoccupations. Those who are engaged in reforming (or at least changing) the system would do well to ponder the lessons that emerge from this book.

One of those lessons is very general, and is most applicable in contexts where irrationality may trump sober calculation. CPK was a skeptic by nature, just the

opposite of doctrinaire. He mistrusted iron-clad intellectual systems, whether their proponents were free marketeers or social engineers. In fact, he considered clinging to rigid beliefs in the face of disconcerting evidence to be one of the more dangerous forms of irrationality, especially when it is practiced by those in charge. The international economy would be a safer place if CPK's tolerant skepticism were more common among the powers that be. I am thinking, in particular, about current discussions of the so-called 'Washington consensus', and the pros and cons of both freely floating exchange rates and unfettered capital markets.

Any reader of this book will come away with the distinct notion that large quantities of liquid capital sloshing around the world should raise the possibility that they will overflow the container. One issue omitted in the book – because it is well outside its scope – is the other side of the ledger: What are the social benefits of free capital flow in its various forms, the analogue of gains from trade? CPK, whose specialties as an economist included international trade, international finance and economic development, would have been sensitive to the need for some pragmatic balancing of risks and benefits. One can only hope that the continued, up-to-date availability of this book will help to spread his open-minded habit of thought.

As he carries the sixth edition up to date, Aliber emphasizes the likelihood that the roughly concurrent credit bubbles in a number of different countries are interrelated events, possibly responses to a common disturbance. It seems implausible that the appearance of housing-based credit bubbles in the United States, the UK, Ireland, and Spain merely reflects independent rolls of the dice. Aliber shows how these events are transmitted internationally through current account imbalances in a world in which capital moves easily across borders. CPK, as a specialist in international economics, would probably have cottoned to this account.

A more complicated question, also surfaced by Aliber, is whether there are successive 'waves' of credit bubbles that are causally related. If this is so, its has important implications for the design of future regulation, both domestically and internationally. We are now well beyond natural history.

It seems to me that the Aliber version preserves this basic Kindleberger orientation but imposes a little more order on CPK's occasionally wayward path through his specimen cabinets. More manias, panics, and crashes may plague us, but readers of this book will at least have been inoculated.

Robert M. Solow

1
Financial Crisis: a Hardy Perennial

The years since the early 1970s are unprecedented in terms of the volatility in the prices of commodities, currencies, real estate, and stocks. There have been four waves of financial crises; a large number of banks in three, four, or more countries collapsed at about the same time. Each wave was followed by a recession, and the economic slowdown that began in 2008 was the most severe and the most global since the Great Depression of the 1930s.

The first crisis wave was in the early 1980s when Mexico, Brazil, Argentina, and ten other developing countries defaulted on their $800 billion of US dollar-denominated loans. The second wave occurred in the early 1990s and engulfed Japan and three of the Nordic countries – Finland, Norway, and Sweden. The Asian Financial Crisis that began in mid-1997 was the third wave; Thailand, Malaysia, and Indonesia were involved initially and subsequently South Korea, Russia, Brazil, and Argentina tumbled. In retrospect the financial crisis that impacted Mexico during its presidential transition at the end of 1994 was the forerunner for the crisis in Southeast Asia thirty months later. The fourth wave began in 2007 and was triggered by declines in the prices of real estate in the United States, Britain, Spain, Ireland, and Iceland – and then by declines in the prices of the bonds of the Greek, Portuguese, and Spanish governments.

Each wave of crises followed a wave of credit bubbles, when the indebtedness of similarly placed groups of borrowers increased at a rate two or three times higher than the interest rate for three, four, or more years. Usually these borrowers used the money to buy real estate – homes and commercial properties. However, the first wave of credit bubbles involved rapid growth in the loans from the major international banks to the governments and to government-owned firms in Mexico and other developing countries that continued for nearly ten years. Japan was the key country in the second wave of bubbles, real estate prices and stock prices increased by a factor of five to six in the 1980s. At about the same time the prices of both types of assets surged in Finland, Norway, and Sweden. The third wave of bubbles initially was centered

on Thailand and some of its neighbors in Southeast Asia. The fourth wave of bubbles primarily was in the real estate markets in the United States, Britain, Spain, Ireland, and Iceland.

Each of these waves of credit bubbles involved cross-border flows of money, which induced large increases in the values of the currencies and increases in the prices of real estate or stocks in the countries receiving the money.

Bubbles always implode, since by definition they involve non-sustainable increases in the indebtedness of a group of borrowers or non-sustainable increases in the prices of stocks. Debt can increase much more rapidly than income for two or three or a few more years, but debt cannot grow more rapidly than income for an extended period. When debt increases at 20 to 30 percent a year, the borrowers have an impeccable record for paying the scheduled interest in a timely way. Eventually the rate of growth of their indebtedness slows, and the 'day of reckoning occurs' when there isn't enough cash from new loans to pay the interest on outstanding loans. Then the prices of real estate and of stocks decline. Moreover when the rate of growth of indebtedness slows, the currencies depreciate, and often very sharply.

When real estate prices decline the borrowers are the first group to incur losses; after they default, the losses cascade to the lenders. The implosion of the real estate and stock bubbles in Japan led to the massive failure of banks and a prolonged period of below-trend growth. The implosion of the asset price bubble in Thailand in mid-1997 triggered declines in currency values and asset prices throughout the region; recessions followed. However, there were no significant failures of US financial firms when the US stock prices declined by 40 percent between 2001 and 2003, and the ensuing recession was brief and shallow.

The range of movement in the values of national currencies since the early 1970s has been much larger than ever before. In 1971 the United States abandoned the US gold parity of $35 an ounce that had been established in 1934. The effort to retain a modified version of the Bretton Woods system of pegged currencies that was formalized in the Smithsonian Agreement of 1972 failed and a floating exchange rate arrangement was adopted by default early in 1973. At the beginning of the 1970s, the dominant market view was that the German mark and the Japanese yen might appreciate by 10 to 12 percent because their inflation rates had been below the US rate in the previous few years. The German mark and the Japanese yen appreciated more rapidly than anticipated through most of the 1970s, and then both currencies depreciated significantly in the first half of the 1980s, although not to the levels of the early 1970s. The Mexican peso, the Brazilian cruzeiro, the Argentinean peso and the currencies of many of the other developing countries depreciated by 30 to 40 percent or more in the early 1980s. The Finnish markka, the Swedish krona, the British pound, the Italian lira and the Spanish peseta lost more

than a third of their value in the last six months of 1992. The Mexican peso lost more than half of its value during the presidential transition at the end of 1994. Most of the Asian currencies – the Thai baht, the Malaysian ringgit, the Indonesian rupiah and the South Korean won – depreciated sharply during the Asian Financial Crisis in the summer and autumn of 1997. The Argentinean peso lost more than two-thirds of its value in the first few months of 2001. The Icelandic krona lost half of its value in 2008. The euro, the new currency that eleven members of the European Union adopted at the beginning of 1999, soon depreciated by 30 percent, and then appreciated by 50 percent beginning in 2002.

The changes in the values of these individual currencies were much larger than those that would have been forecast based on the differences between the US and the foreign inflation rates. The 'overshooting' and 'undershooting' of national currencies were much larger than in any previous period.

The increases in commodity prices in the 1970s were spectacular. The US dollar price of gold increased from $40 an ounce at the beginning of the 1970s to nearly $1000 ten years later; the price was $450 at the end of the 1980s and $283 at the end of the 1990s. The price exceeded $1200 in the summer of 2010. The price of oil was $2.50 a barrel at the beginning of the 1970s and $40 at the end of that decade; in the mid-1980s the oil price was $12 and then at the end of the 1980s the price increased to $40 after the Iraqi invasion of Kuwait. The oil price almost reached $150 in the early summer of 2008, and then declined below $50 and then increased to $80.

The number of bank failures during the 1980s and the 1990s was much, much larger than in earlier decades. Several of these failures were isolated events: both Franklin National Bank in New York City and Herstatt AG in Cologne, Germany, had made large bets on the changes in currency values in the early 1970s that they subsequently lost. Crédit Lyonnais, once the largest bank in France and a government-owned firm, rapidly increased its loans in the effort to become a first-tier international bank and its bad loans eventually cost the French taxpayers the equivalent of more than $30 billion. However, most failures of banks and other financial firms were systemic and reflected dramatic changes in the financial environment. Three thousand US savings and loan associations and other thrift institutions failed in the 1980s, with losses to the American taxpayers of more than $100 billion.

When the bubbles in Japanese real estate and stocks imploded, the losses incurred by the Japanese banks were several times larger than their capital and virtually all the Japanese banks implicitly became wards of the government. Similarly when the Mexican peso and the currencies of the other developing countries depreciated sharply in the early 1980s, most of the banks in these countries went under because of the combination of the large loan losses by their domestic borrowers, in part due to the massive currency revaluation

losses they had incurred. Virtually all of the banks in Finland, Norway, and Sweden went bankrupt when the bubbles in their real estate and stock markets imploded in the first half of the 1990s. Most of the Mexican banks failed at the end of 1994 when the peso depreciated sharply. Similarly most of the banks in Thailand and Malaysia and South Korea and several of the other Asian countries – except for Hong Kong and Singapore – tanked after the mid-1997 Asian Financial Crisis. The sharp declines in prices of residential real estate in the United States, Britain, Ireland, and several other countries that began toward the end of 2006 led to massive government investments – 'bailouts' – of the financial institutions. In 2008 many of the top firms in the US investment banking industry were wiped out or forced to seek a stronger merger partner. The British government 'nationalized' Northern Rock, the largest mortgage lender in the country, and became the dominant shareholder in the Royal Bank of Scotland. The Irish government made massive investments in the six largest banks in the country. The three large banks in Iceland were taken over by the government. Countrywide Financial, the largest mortgage lender in the United States, was acquired by Bank of America, which subsequently acquired Merrill Lynch, one of the largest US investment banks – but then Bank of America required a large injection of capital from the US Treasury. The US government made a massive investment in Citibank. The Dutch government provided capital to ING, the insurance–banking conglomerate.

These financial crises and bank failures resulted from the implosion of the asset price bubbles and from the sharp depreciations of currencies; in many cases the currency crises triggered the banking crises. The cost of these bank crises was extremely high in terms of several metrics – the losses incurred by the banks as a ratio of a country's GDP and as a share of government spending, and the slowdowns in the rates of economic growth and the increases in unemployment and in the output gaps.

The massive number of bank failures, the large changes in currency values, and the asset price bubbles were systematically related and resulted from rapid changes in the global economic environment. The 1970s was a decade of accelerating inflation, the largest-ever sustained increase in the US price level in peace-time. The market price of gold surged because some investors relied on the cliché that 'gold is a good inflation hedge'; however the increase in the gold price was many times larger than the contemporary increase in the US and world price levels. Toward the end of the 1970s investors were buying gold because its price was increasing – and the price was increasing because investors were buying gold.

The prevailing view in the late 1970s was that the US and world inflation rates would accelerate. Some analysts predicted that the gold price would reach $2500 an ounce and that the oil price would reach $80 to $90 a barrel by 1990.

The range of movement in bond prices and stock prices in the 1970s was much greater than in the several previous decades. In the 1970s the real rates of return on both US dollar bonds and US stocks were negative. In contrast in the 1990s the real rates of return on bonds and on stocks averaged more than 15 percent a year.

The foreign indebtedness of Mexico, Brazil, Argentina, and other developing countries as a group increased from $125 billion in 1972 to $800 billion in 1982. One cliché at the time was that 'countries don't go bankrupt'. During this period the borrowers had a stellar record for paying the interest on their loans on a timely basis. Then in the autumn of 1979 the Federal Reserve adopted a sharply contractive monetary policy; interest rates on US dollar securities surged. The price of gold peaked in January 1980 and then began to decline as inflationary anticipations were reversed.

The sharp increase in real estate prices and stock prices in Japan in the 1980s was associated with a boom in the economy; *Japan as Number One: Lessons for America*[1] was a bestseller in Tokyo. The Japanese banks increased their deposits and their loans and their capital much more rapidly than banks headquartered in the United States and in Germany and in the other European countries. At the time seven or eight of the ten largest banks in the world were Japanese. Then at the beginning of the 1990s real estate prices and stock prices in Japan imploded. Within a few years many of the leading Japanese banks and financial institutions were broke, kaput, bankrupt, and insolvent, and remained in business only because of an implicit understanding that the Japanese government would protect the depositors from financial losses if the banks were closed. A striking story of a mania and a crash – but without a panic, because depositors believed that government would socialize the loan losses.

Three of the Nordic countries – Norway, Sweden, and Finland – experienced bubbles in their stock markets and real estate markets at about the same time as a result of money inflows associated with financial liberalization. Their bubbles popped at about the same time as the one in Japan.

Mexico had been one of the great economic success stories of the early 1990s as it prepared to enter the North American Free Trade Agreement. The Bank of Mexico had adopted a tough contractive monetary policy that reduced the inflation rate from 140 percent to less than 10 percent in four years; during the same period several hundred government-owned firms were privatized and business regulations were liberalized. Money flowed to Mexico because the real rates of return on government securities were high and because the prospective profit rates on industrial investments were also high. The universal expectation was that Mexico would become the low-cost base for producing automobiles and washing machines and many other manufactured goods for the US and Canadian markets. The large money inflow led to a real appreciation of the peso, Mexico's trade deficit increased to 7 percent of its GDP and its external

debt to 60 percent of its GDP. Then several political incidents associated with the presidential election in 1994 led to a sharp decline in the flow of money and the Mexican government was unable to support the peso. Once again the depreciation of the peso resulted in large loan losses, and most of the Mexican banks – which had been privatized in the previous several years – failed.

In the mid-1990s real estate prices and stock prices surged in Thailand, Malaysia, and Indonesia; these were the 'dragon economies' that seemed likely to emulate the economic successes of the 'Asian tigers' – Taiwan, South Korea, Hong Kong, and Singapore – of the previous generation . Firms based in Japan, Europe, and the United States invested in these countries as low-cost sources of supply, much as US and foreign firms had invested in Mexico as a source of supply for the North American market. European and Japanese banks rapidly increased their loans to firms and banks in these countries. The domestic lenders in Thailand then experienced large losses on their domestic loans in the autumn and winter of 1996 because they had not been sufficiently discriminating in their evaluations of the willingness of Thai borrowers to pay the interest on their indebtedness. Foreign lenders sharply reduced their purchases of Thai securities, and then the Bank of Thailand, much like the Bank of Mexico thirty months earlier, did not have the money to support its currency. The sharp decline in the value of the baht in early July 1997 led to money outflows from the other Asian countries and their currencies (except for the Hong Kong dollar and the Chinese yuan, which remained rigidly pegged to the US dollar) declined by 30 percent or more. The Indonesian rupiah lost 80 percent of its value. Most of the banks in the area – except for those in Hong Kong and Singapore – would have been bankrupt in any reasonable 'mark-to-market' test. The crisis spread to Russia, there was a debacle in the ruble, and the country's banking system collapsed in the summer of 1998. Investors then became more cautious and they sold risky securities and bought safer US government securities, and the changes in the relationship between the interest rates on these two groups of securities led to the collapse of Long-Term Capital Management, then the largest US hedge fund.

The 1990s bubble in NASDAQ stocks

Stocks in the United States are traded on either the over-the-counter market or on one of the organized stock exchanges, primarily the New York Stock Exchange. The typical pattern was that shares of young firms would initially be traded on the over-the-counter market and then most of these firms would incur the costs associated with obtaining a listing on the New York Stock Exchange because they believed that a listing would broaden the market and lead to higher prices for their stocks. Some very successful new firms associated with the information technology revolution of the 1990s – Microsoft, Cisco, Dell, Intel – were exceptions to this pattern; they

chose not to obtain a listing on the New York Stock Exchange because they believed that trading stocks electronically in the over-the-counter market was superior to trading stocks by the open-outcry method used on the New York Stock Exchange.

In 1990 the market value of stocks traded on the NASDAQ was 11 percent of that of the New York Stock Exchange; the comparable figures for 1995 and 2000 were 19 percent and 42 percent. The annual average percentage rate of increase in the market value of NASDAQ stocks was 30 percent during the first half of the decade and 46 percent during the next four years. A few of the newer firms traded on the NASDAQ would eventually become as successful as Microsoft and Intel and so high prices for their stocks might be warranted. The likelihood that all of the firms whose stocks were traded on the NASDAQ would be as successful as Microsoft was extremely small, since it implied that the profit share of US GDP would be two to three times higher than it ever had been previously.

In part the large number of crashes in national financial markets in the last thirty years reflects that there are more independent countries. Despite the lack of perfect comparability across periods, the conclusion is unmistakable that financial failure has been more extensive and pervasive in the last thirty years.

The bubble in US stock prices in the second half of the 1990s was associated with a remarkable US economic boom; the unemployment rate declined sharply, the inflation rate declined, and the rates of economic growth and productivity both accelerated. The US government developed its largest-ever fiscal surplus in 2000 after having had its largest-ever fiscal deficit in 1990. The remarkable performance of the real economy contributed to the surge in US stock prices that in turn led to the increase in investment spending and consumption spending and an increase in the rate of US economic growth.

US stock prices began to decline in the spring of 2000 and fell by 40 percent in the next three years while the prices of NASDAQ stocks declined by 80 percent.

US real estate prices began to increase at an above-average rate in 2002. Real estate prices increase in the long run, in part because of the increase in the general price level and in part because of the increase in nominal GDP. (Much of the increase in real estate prices reflects increases in the price of land.) The Federal Reserve maintained low interest rates in part because of the sluggishness in the economy, and house prices increased three times as rapidly as the general price level. The sharp increase in prices induced a construction boom, and housing starts reached two million units a year – about 500,000 more units than the number required to satisfy the growth in population and the losses to fires, storms, and similar factors. Part of the increase in demand was from investors who sought profits from the continued increases in prices.

The sharp decline in the price of residential real estate and the debacle in the prices of mortgage-related securities since 2007 has led to many books that are

US-centric, and that seek to explain the bubble in terms of the failure of regulation or the greed of the bankers or the failure of the regulators or vagaries of new financial instruments.

One of the themes of this book is that the credit bubbles that often occur in several different countries at the same time have similar initial causes. Thus the surge in the indebtedness of the developing countries in the 1970s occurred because the major international banks believed that commodity prices would continue to increase and that the growth rates in these countries would remain high. The likelihood that the bubbles in the real estate markets in the United States, Britain, Ireland, Iceland, Spain, South Africa, and several other countries that began about 2002 were independent events seems low. Bubbles in real estate always result from bubbles in the growth of credit. There were unique idiosyncratic aspects in these different national markets; the market in subprime mortgages seems uniquely American. The rapid growth in the supply of credit led to a sharp increase in the demand for mortgages, which was greater than the supply of prime loans and so the mortgage brokers ginned up a large increase in the supply of sub-prime mortgages.

Another theme is that the likelihood that the four waves of bubbles over a thirty-year period were unrelated events is low. Each bubble led to a crisis. Several of these crises appear to have laid the basis for the next wave of bubbles. The financial crisis in the developing countries in the early 1980s had a knock-on effect that contributed to the bubble in Japanese real estate and stocks in the second half of the 1980s. The implosion of the bubble in Tokyo in the early 1990s led to an increase in money flows from Japan to Thailand and Malaysia and Indonesia, which led to the appreciation of their currencies and to increases in the prices of real estate and of securities in these countries. When the bubbles in the countries in Southeast Asia imploded, there was surge in the flow of money to the United States as these countries repaid loans; the US dollar appreciated and the US trade deficit increased by $150 billion a year.

The increase in the flow of money to a country almost always led to increases in value of its currency and to increases in the prices of assets in that country as the domestic sellers of the securities used nearly all of their receipts to buy other securities from other domestic residents. These domestic residents in turn similarly used a large part of their receipts to buy other domestic securities from other domestic residents. These transactions in securities occurred at ever-higher prices. It was as if the cash from the sale of securities to foreigners was the proverbial 'hot potato' that was rapidly passed from one group of investors to others at ever-increasing prices.

Manias and credit and books

The production of books on financial crises is counter-cyclical. A spate of books on the topic appeared in the 1930s following the US stock market bubble in the

late 1920s and the subsequent crash and the Great Depression. Relatively few books on crises appeared during the several decades immediately after World War II.

The first edition of this book was published in 1978, after US stock prices had declined by 50 percent in 1973 and 1974 following a fifteen-year bull market in stocks. The stock market debacle and the US recession led to the bankruptcies of the Penn Central railroad, several of the large steel companies and a large number of Wall Street brokerage firms. New York City was on the verge of default on its outstanding bonds and was saved from insolvency by the State of New York. Not quite a crash, unless you were a senior official or a stockholder in one of the firms that failed or the Mayor of New York City.

The fifth edition was published after the implosion of the dot.com bubble in US stocks in the late 1990s. The innovation since the fifth edition were the bubble in residential real estate in the United States, Britain, Ireland, Spain, Iceland, and several other countries, and the sharp increase in the indebtedness of the governments of Greece, Portugal, Ireland, and Spain.

Each of these waves of bubbles has been global, in that four, five, or more countries have been involved. Moreover the bubbles seem larger, judged by the increase in household wealth.

Books on the 2008 financial crisis

The collapse of US investment banks and commercial banks in 2007 and 2008 and 2009 has led to a large number of books on the financial crisis from three different groups of authors. Many are by journalists, including Gillian Tett, who wrote *Fool's Gold: How Unrestrained Greed Corrupted a Dream, Shattered Global Markets, and Unleashed a Catastrophe*. Andrew Ross Sorkin brought out *Too Big to Fail: The Inside Story of How Wall Street and Washington Fought to Save the Financial System – and Themselves*. Roger Lowenstein authored *The End of Wall Street*, Justin Fox wrote *The Myth of the Rational Market: A History of Risk, Reward, and Delusion on Wall Street*, and Scott Patterson produced *The Quants; How a New Breed of Math Whizzes Conquered Wall Street and Nearly Destroyed It*. The theme of market irrationality is also explored in John Cassidy's *How Markets Fail; The Logic of Economic Calamities*. Michael Lewis's *The Big Short: Inside the Doomsday Machine* focused on a few individuals who early on realized that there was a bubble in the housing market, and that exceptional profits could be achieved from short-selling mortgage-related securities. Then there was Suzanne McGee's *Chasing Goldman Sachs: How the Masters of the Universe Melted Wall Street Down...And Why They Will Take Us to the Brink Again* and *The Meltdown Years: The Unfolding of the Global Economic Crisis* by Wolfgang Munchau, Charles R. Morris's *The Trillion Dollar Meltdown: Easy Money, High Rollers, and the Great Credit Crash*, James Grant's *Mr Market Miscalculates: The Bubble Years and Beyond*, Charles Gasparino's *The Sellout: How Three Decades of Wall Street Greed and Government Mismanagement Destroyed the Global Financial System*, Barry Rithholtz's *Bailout Nation: How GREED and EASY MONEY Corrupted Wall Street And Shook the World Economy*, and

Meltdown: How Greed and Corruption Shattered Our Financial System and How We Can Recover by Katrina vanden Heuvel and the Editors of the *Nation*.

Some of the books are by academics. One of the first was Richard Posner's *The Failure of Capitalism*. Robert J. Shiller produced *The Subprime Solution* and George A. Akerlof together with Shiller wrote *Animal Spirits*; some of the chapters in this book focus on the crisis. Simon Johnson and James Kwak authored *13 Bankers: The Wall Street Takeover and the Next Financial Meltdown*. Raghuram G. Rajan wrote *Fault Lines: How Hidden Fractures Still Threaten the World Economy*, Joseph Stiglitz produced *Freefall: America, Free Markets, and the Sinking of the World Economy* and Nassim Nicholas Taleb brought out *The Theory of Black Swan Events*, a critique of the prevailing consensus in academic finance about market efficiency. Thomas Sowell's contribution was *The Housing Boom and Bust* while a group of fifteen distinguished economists brought out *The Squam Lake Report; Fixing the Financial System*, with more than thirty recommendations for changes in regulations. Amar Bhide wrote *A Call for Judgment*, which integrates modern finance and classical economics, while Nouriel Roubini and Stephen Mihm brought out *Crisis Economics: A Crash Course in the Future of Finance*.

A third group of books are by 'insiders', individuals who had formerly been with a financial firm. Henry Paulson, the US Secretary of the Treasury from 2007 to 2009 and previously the head of Goldman Sachs, authored *On the Brink: Inside the Race to the Stop the Collapse of the Global Financial System*. Lawrence McDonald wrote *A Colossal Failure of Common Sense; the Inside Story of the Collapse of Lehman Brothers*, his former employer. William D. Cohan, a former banker turned journalist, brought out *House of Cards: A Tale of Hubris and Wretched Excess on Wall Street*, while Alex Pollock, another former banker, published *Boom and Bust: Financial Cycles and Human Prosperity*. George Cooper, a financial analyst, wrote *The Origin of Financial Crises*.

Most of these books are US-centric and ignore or minimize the bubbles in the property markets in other countries. However, there have been three books on the crisis in Iceland – *Meltdown Iceland: Lessons on the World Financial Crisis from a Small Bankrupt Island*, by Roger Boyes, a British journalist, *Why Iceland?* by Asgeir Jonsson, a former senior banker with Kaupthing, one of the fallen firms, and *Frozen Assets: How I Lived Iceland's Boom and Bust*, by Eftir Armann Thorvaldsson, a former UK head of the same bank. Moreover the government of Iceland that came to power after the collapse established an Icelandic Special Investigation Commission which produced *Causes of the Collapse of the Icelandic Banks – Responsibilty, Mistakes, and Negligence*; the authors were Pali Hreinsson, Tryggvie Gunnarsson, and Sigridur Benediktsdottir.

The US Congress established a bi-partisan ten-member Financial Crisis Inquiry Commission, which has held extensive hearings and interviewed witnesses from the leading financial firms. Both former Chairman Alan Greenspan and former Secretary of the Treasury Robert Rubin testified before the commission. The FCIC's report was published in December 2010 (and is available at www.fcic.gov/report).

The titles and subtitles of these books express common themes – greed, the malfunctioning of markets, the corruption of Wall Street, and the capture of Washington and the regulators by the bankers. And more greed.

The shortcoming of most of these books is that they give no explanation for why the crisis occurred when it did, nor do they have an explanation for why some countries were involved but not others. Were the problems in Iceland a spinoff from the United States, or were the problems of the largest and the smallest affected countries a result of a common factor? Was there a sharp increase in greed of the bankers

soon after the beginning of the new millennium, or was the greed always there and released by some other event?

Moreover, the focus of these books is usually micro, and centered on the failure of rationality and the inability of the lenders to foresee the consequences of the increase in indebtedness of the borrowers. Asset bubbles – most asset bubbles – are a monetary phenomenon and result from the rapid growth of in the supply of credit. The view that excessively rapid increases in the prices of goods and services result from the rapid growth in the money supply is accepted – the axiom is that 'inflation always is a monetary phenomenon'. The counterpart is that 'real estate bubbles always are a credit phenomenon'.

The sixth edition of this book appears thirty years after the first and also after thirty of the most tumultuous years in global financial markets, a period without a good historical precedent in terms of monetary turbulence.

The big ten financial bubbles

1. The Dutch Tulip Bulb Bubble 1636
2. The South Sea Bubble 1720
3. The Mississippi Bubble 1720
4. The late 1920s stock price bubble 1927–29
5. The surge in bank loans to Mexico and other developing countries in the 1970s
6. The bubble in real estate and stocks in Japan 1985–89
7. The 1985–89 bubble in real estate and stocks in Finland, Norway, and Sweden
8. The bubble in real estate and stocks in Thailand, Malaysia, Indonesia, and several other Asian countries 1992–97 and the surge in foreign investment in Mexico 1990–99
9. The bubble in over-the-counter stocks in the United States 1995–2000
10. The bubble in real estate in the United States, Britain, Spain, Ireland, and Iceland between 2002 and 2007 – and the debt of the government of Greece

The earliest bubble noted in the box involved tulip bulbs in the Netherlands in the seventeenth century. Two of the bubbles – one in Britain and one in France – occurred at the end of the Napoleonic Wars. There were manias and financial crises in the nineteenth century that were mostly associated with the failures of banks, often after an extended investment in infrastructure such as canals and railroads. Currency crises and banking crises were frequent between 1920 and 1940. The percentage increases in stock prices in the past thirty years have been larger than in earlier periods, and six of the ten bubbles in the box have occurred in this period. Bubbles in real estate and in stocks have often occurred together; some countries have experienced a bubble in real estate but not in stocks, while the United States had a stock price bubble

in the second half of the 1990s that had no counterpart in the real estate market.

Manias are dramatic but they have been infrequent; only two have occurred in US stocks in two hundred years. Manias generally have occurred during the expansion phase of the business cycle, in part because the euphoria associated with the mania leads to increases in spending. During the mania the increases in the prices of real estate or stocks or in one or several commodities contribute to increases in consumption and investment spending that in turn lead to quickening of economic growth. Seers in the economy forecast perpetual economic growth and some venturesome ones even proclaim an end to recessions and declare that traditional business cycles have become obsolete. The more rapid growth induces investors and lenders to become more optimistic, and asset prices increase more rapidly.

Manias – especially macro manias – are associated with economic euphoria; business firms become increasingly upbeat and investment spending surges because credit is plentiful. In the second half of the 1980s Japanese industrial firms could borrow as much as they wanted from their friendly bankers in Tokyo and in Osaka; money seemed 'free' (money always seems free in manias) and the Japanese went on both a consumption spree and an investment spree. The Japanese purchased ten thousand items of French art. A racetrack entrepreneur from Osaka paid $90 million for Van Gogh's *Portrait of Dr Guichet*, at that time the highest price ever paid for a painting. The Mitsui Real Estate Company paid $625 million for the Exxon Building in New York even though the initial asking price had been $510 million; Mitsui wanted to get in the *Guinness Book of World Records* for paying the highest-ever price for an office building. In the second half of the 1990s in the United States newly established firms in the information technology industry and in biotech had access to virtually unlimited funds from the venture capitalists who believed they would profit greatly when the shares in these firms were first sold to the public.

During these euphoric periods an increasing number of investors seek short-term capital gains from the increases in the prices of real estate and of stocks. Investors make down payments for the purchase of condominium apartments in the pre-construction phase in the anticipation that they will be able to sell these apartments at handsome profits when the buildings have been completed, or even before.

Then an event – perhaps a change in government policy, an unexplained failure of a firm previously thought to have been successful – occurs that leads to a pause in the increase in asset prices. Soon, some of the investors who had financed most of their purchases with borrowed money become distress sellers because the interest payments on the money borrowed to finance their purchases are larger than the investment income on the assets. The prices of these assets decline below their purchase prices and now the buyers are 'under

water' – the amounts owed on the money borrowed to finance the purchases of these assets are larger than their current market value. Their distress sales lead to sharp declines in the prices of the assets and a crash and panic are likely to follow.

The economic situation in a country after several years of bubble-like behavior resembles that of a young person on a bicycle; the rider needs to maintain the forward momentum or the bike becomes unstable. During the mania, asset prices will decline immediately after they stop increasing – there is no plateau, no 'middle ground'. The decline in the prices of some assets leads to the concern that asset prices will decline further and that the financial system will experience 'distress'. The rush to sell these assets becomes self-fulfilling and so precipitous that it resembles a panic. The prices of commodities – houses, buildings, land, stocks, bonds – crash to levels that are just 30 to 40 percent of their prices at the peak. Bankruptcies surge, economic activity slows, and unemployment increases.

The features of these manias are never identical and yet there is a similar pattern. The increase in prices of commodities or real estate or stocks is associated with euphoria; household wealth increases and so does spending. There is a sense of 'We never had it so good'. Then the asset prices peak and then begin to decline. The implosion of a bubble leads to declines in the prices of commodities, stocks and real estate. Some financial crises were preceded by a rapid increase in the indebtedness of one or several groups of borrowers rather than by a rapid increase in the price of an asset or a security.

The thesis of this book is that the cycle of manias and panics results from the pro-cyclical changes in the supply of credit; the credit supply increases rapidly in good times, and then when economic growth slackens, the rate of growth of credit has often declined sharply. A mania involves increases in the prices of real estate or stocks or a currency or a commodity in the present and near-future that are not consistent with the prices of the same real estate or stocks in the distant future. The forecasts that the price of oil would increase to $80 a barrel after the earlier increase from $2.50 a barrel at the beginning of the 1970s to $36 at the end of that decade was manic. During the economic expansions investors become increasingly optimistic and more eager to pursue profit opportunities that will pay off in the distant future while the lenders become less risk-averse. Rational exuberance morphs into irrational exuberance, economic euphoria develops and investment spending and consumption spending increase. There is a pervasive sense that it is 'time to get on the train before it leaves the station' and the exceptionally profitable opportunities disappear. Asset prices increase further. An increasingly large share of the purchases of these assets is undertaken in anticipation of short-term capital gains and an exceptionally large share of these purchases is financed with credit.

The financial crises that are analyzed in this book are major both in size and in effect and most are international because they involve several different countries either at the same time or in a causal, sequential way.

The term 'bubble' is a generic term for the increases in asset prices in the mania phase of the cycle that cannot be explained by the changes in the economic fundamentals. Recently, real estate bubbles and stock price bubbles have occurred at more or less the same time in Japan and in some of the Asian countries. The sharp increases in the prices of gold and silver in the late 1970s were a bubble, but the increases in the price of crude petroleum in the same years were not; the distinction is that many of the buyers of gold and silver in that tumultuous and inflationary decade anticipated that the prices of both precious metals would continue to increase and that profits could be made from buying and holding these commodities for relatively short periods. In contrast many of the buyers of petroleum were concerned that the disruptions in oil supplies due to actions of the cartel and the war in the Persian Gulf would lead to shortages and increases in prices.

Ponzi finance, chain letters, pyramid schemes, manias, and bubbles

Ponzi finance, chain letters, bubbles, pyramid schemes, finance, and manias are somewhat overlapping terms for non-sustainable patterns of financial behavior, in that asset prices today are not consistent with asset prices at distant future dates. The Ponzi schemes generally involve promises to pay an interest rate of 30 or 40 or 50 percent a month; the entrepreneurs that develop these schemes always claim they have discovered a new secret formula so they can earn these high rates of return. They make the promised interest payments for the first few months with the money received from their new customers attracted by the promised high rates of return. But by the fourth or fifth month the money received from these new customers is less than the monies promised the first sets of customers and the entrepreneurs go to Brazil or jail or both.

A chain letter is a particular form of pyramid arrangement; the procedure is that individuals receive a letter asking them to send $1 (or $10 or $100) to the name at the top of the pyramid and to send the same letter to five friends or acquaintances within five days; the promise is that within thirty days you will receive $64 for each $1 'investment'.

Pyramid arrangements often involve sharing of commission incomes from the sale of securities or cosmetics or food supplements by those who actually make the sales to those who have recruited them to become sales personnel.

The bubble involves the purchase of an asset, usually real estate or a security, not because of the rate of return on the investment but in anticipation that the asset or security can be sold to someone else at an even higher price; the term 'the greater fool' has been used to suggest the last buyer was always counting on finding someone else to whom the stock or the condo apartment or the baseball cards could be sold.

The term 'mania' describes the frenzied pattern of purchases, often an increase in prices accompanied by an increase in trading volumes; individuals are eager to buy before the prices increase further. The term 'bubble' suggests that when the prices stop increasing, they are likely – indeed almost certain – to decline.

Chain letters and pyramid schemes rarely have macroeconomic consequences, but rather involve isolated segments of the economy and involve the redistribution of income from the latecomers to those who came in early. Asset price bubbles have often been associated with economic euphoria and increases in both business and household spending because the futures seem so much brighter, at least until the bubble pops.

Virtually every mania is associated with a robust economic expansion, but only a few economic expansions are associated with a mania. Still, the association between manias and economic expansions is sufficiently frequent and sufficiently uniform to merit renewed study.

Some economists have contested the view that the use of the term bubble is appropriate because it suggests irrational behavior that is highly unlikely or implausible; instead they seek to explain the rapid increase in real estate prices or stock prices in terms that are consistent with changes in the economic fundamentals. Thus, for them, the surge in the prices of NASDAQ stocks in the 1990s occurred because investors sought to buy shares in firms that would repeat the spectacular successes of Microsoft, Intel, Cisco, Dell, and Amgen.

The policy implications

The appearance of a mania or a bubble raises the policy issue of whether governments should seek to moderate the surge in asset prices to reduce the likelihood or the severity of the ensuing financial crisis. Virtually every large country has established a central bank as a domestic 'lender of last resort' to reduce the likelihood that a shortage of liquidity would cascade into a solvency crisis. The practice leads to the question of the role for an international 'lender of last resort' that would assist countries in stabilizing the value of their currencies and reduce the likelihood that a sharp depreciation of the currencies because of a shortage of liquidity would trigger large numbers of bankruptcies.

During a crisis, many firms that had recently appeared robust tumble into bankruptcy because the failure of some firms often leads to a decline in asset prices and a slowdown in the economy. When asset prices decline sharply, government intervention may be desirable in order to provide the public good of stability. During financial crises the decline in asset prices may be so large and abrupt that the price changes become self-justifying. When asset prices tumble

sharply, the surge in the demand for liquidity may drive many individuals and firms into bankruptcy, and the sale of assets in these distressed circumstances may induce further declines in their prices. At such times a lender of last resort can provide financial stability or attenuate financial instability. The dilemma is that if investors knew in advance that governmental support would be forthcoming under generous dispensations when asset prices fall sharply, markets might break down somewhat more frequently because investors will be less cautious in their purchases of assets and of securities.

The role of the lender of last resort in coping with a crash or panic is fraught with ambiguity and dilemma. Thomas Joplin commented on the behavior of the Bank of England in the crisis of 1825: 'There are times when rules and precedents cannot be broken; others, when they cannot be adhered to with safety.' Breaking the rule establishes a precedent and a new rule that can be adhered to or broken as occasion demands. In these circumstances intervention is an art rather than a science. The general rules that the state should always intervene or that the state should never intervene are both wrong. This same issue of intervention reappeared with the question of whether the US government should have rescued Chrysler in 1979, New York City in 1975 and the Continental Illinois Bank in 1984. (In fact Continental Illinois failed, although the depositors in the bank were made whole.). Similarly, should the Bank of England have rescued Baring Brothers in 1995 after the rogue trader Nick Leeson in its Singapore branch office had depleted the firm's capital through hidden transactions in option contracts? The question appears whenever a group of borrowers or banks or other financial institutions incurs such massive losses that they are likely to be forced to close, at least under their current owners. The United States acted as the lender of last resort during the Mexican financial crisis at the end of 1994. The International Monetary Fund acted as the lender of last resort during the Russian financial crisis of 1998, primarily after prodding by the US and German governments. Neither the United States nor the International Monetary Fund was willing to act as a lender of last resort during the Argentinean financial crisis at the beginning of 2001. This list highlights that coping with financial crises remains a major problem.

The conclusion of *The World in Depression, 1929–1939* was that the 1930s depression was wide, deep and prolonged because there was no international lender of last resort.[2] Britain was unable to act in that capacity because it was exhausted by World War I, obsessed with pegging the British pound to gold at its pre-1914 parity and groggy from the aborted economic recovery of the 1920s. The United States was unwilling to act as an international lender of last resort; at the time few Americans had thought through what the United States might have done in that role. This book extends the analysis of the responsibilities of an international lender of last resort (Chapter 12).

The monetary aspects of manias and panics are important and are examined at length in several chapters. The monetarist view – at least one monetarist view – is that the mania would not occur if the rate of growth of the money supply were stabilized or constant. Many of the manias are associated with the surge in the growth of credit, but some are not; a constant money supply growth rate might reduce the frequency of manias but is unlikely to consign them to the dustbins of history. The rate of increase in US stock prices in the second half of the 1920s was exceptionally high relative to the rate of growth of the money supply, and similarly the rate of increase in the prices of NASDAQ stocks in the second half of the 1990s was exceedingly high relative to the growth of the US money supply. Some monetarists distinguish between 'real' financial crises that are caused by the shrinkage of the monetary base or high-powered money and 'pseudo' crises that do not. The financial crises in which the monetary base changes early or late in the process should be distinguished from those in which the money supply did not increase significantly.

The earliest manias discussed in the first edition of this book were the South Sea and Mississippi bubbles of 1719–20. The earliest manias analyzed in this edition are the *Kipper- und Wipperzeit*, a monetary crisis (1619–22) that occurred at the outbreak of the Thirty Years War, and the much-discussed 'tulipmania' of 1636–37. The view that the trade in tulip bulbs in the Dutch Republic constituted a bubble followed from widespread recognition, even at the time, that exotic specimens of tulips are difficult to breed, but once bred propagate easily – and hence eventually their prices would decline sharply.[3]

The early-historical treatment focusses on European experiences. The most recent crisis covered in this edition centers on the real estate markets in the United States, Britain, Ireland, Spain, and Iceland. The concentration on the financial crises in Britain in the nineteenth century reflects both the central importance of London in international financial arrangements and the abundant writings by contemporary analysts. In contrast, Amsterdam was the dominant financial power for much of the eighteenth century, but events there are rather glossed over because of the difficulties in accessing the Dutch literature.

The waves of credit bubbles and crises since the mid-1970s suggest that these market events were much more global then in the past. Most of the countries that have experienced credit bubbles also have received an inflow of money. Because few currencies are pegged, the inflows have led to an appreciation of their currencies and – since the money has to go somewhere – to increases in asset prices. Each of these countries has experienced an economic boom – 'the times could not be better'. Perhaps because the currencies have appreciated, upward pressures on prices of goods and services have been smaller than they would have been had the currencies been pegged. Nevertheless the central

banks in many countries raised interest rates – which had the impact of attracting more money from abroad.

One unique feature of these monetary developments is that the links among the global markets means that money is more likely to move abroad for a smaller differential in anticipated returns. The innovation is that there is a larger pool of liquid funds denominated in the US dollar that investors can tap when they want to buy assets and securities in countries whose financial and economic prospects suddenly look much brighter.

The story chapter by chapter

A stylized model of speculation, credit expansion, financial distress, and then crisis that ends in a panic and crash is presented in Chapter 2. The model follows the early classical ideas of 'over-trading' followed by 'revulsion' and 'discredit' – musty terms used by earlier generations of economists including Adam Smith, John Stuart Mill, Knut Wicksell and Irving Fisher. These concepts were developed further by Hyman Minsky, who argued that the financial system in a market economy is unstable, fragile, and prone to crisis. The Minsky model has great explanatory power for earlier crises in the United States and in Western Europe, for the asset price bubbles in Japan in the second half of the 1980s, and for the bubbles in real estate in the United States, Britain, Ireland, Spain, and Iceland between 2002 and 2007.

The mania phase of the economic expansion is the subject of Chapter 3. The central issue is whether markets are always rational, or whether speculation can be destabilizing – do investors in real estate and in stocks develop estimates of the anticipated prices on the basis of recent increases in the prices of these assets or are their anticipations of prices based on their estimates of their earning power. The nature of the outside, exogenous shock that triggers the mania is examined in different historical settings including the onset and the end of a war, a series of good harvests and a series of bad harvests, the opening of new markets and of new sources of supply and the development of different innovations – the railroad, electricity and e-mail. A particular recent form of displacement that shocks the system has been financial liberalization or deregulation in Japan, the Nordic countries, some of the Asian countries, Mexico, Russia, and Iceland. Deregulation has led to monetary expansion, foreign borrowing and speculative investment.[4]

Investors have speculated in commodities, agricultural land, urban building sites, railroads, new banks, discount houses, stocks, bonds (both foreign and domestic), glamour stocks, conglomerates, condominiums, shopping centers and office buildings. Moderate excesses burn themselves out without damage to the economy although individual investors encounter large losses. One question is whether the euphoria of the economic upswing endangers financial

stability only if it involves at least two or more objects of speculation, a bad harvest, say, along with a railroad mania or an orgy of land speculation, or a bubble in real estate and in stocks at the same time.

The monetary dimensions of both manias and panics are analyzed in Chapter 4. The occasions when a boom or a panic has been triggered by a monetary event – a re-coinage, a discovery of precious metals, a change in the ratio of the prices of gold and silver under bimetallism, an unexpected success of some flotation of a stock or bond, a sharp reduction in interest rates as a result of a massive debt conversion, or a rapid expansion of the monetary base – are noted. Innovations in finance, as in productive processes, can shock the system and lead to overinvestment in some types of financial services.[5] The financial sectors expanded rapidly in the United States, Britain, Ireland, and Iceland during their real estate booms because of the rapid increases in mortgages and mortgage-related securities. A sharp increase in interest rates may induce some investors to withdraw money from banks and thrift institutions, which are then squeezed because the prices of their long-term securities decline when they need to downsize.

The problems associated with managing the monetary mechanism to avoid manias and bubbles is stressed in this edition. Most bubbles, and especially those since the 1970s, have resulted from the rapid increase in the credit available to a group of borrowers, often the buyers of real estate. Monetary control by central banks limits the growth of money and credit. Money is a public good but monetary arrangements often have been exploited by private parties. Banking, moreover, is difficult to regulate because new institutions are developed that circumvent the regulations. Many monetarists insist that many, perhaps most, of the cyclical difficulties of the past have resulted from mismanagement of the monetary mechanism. Such mistakes were frequent and serious. The argument advanced in Chapter 4, however, is that even when the supply of money was adjusted to the demands of an economy the monetary mechanism did not stay right for very long. When government produces one quantity of the public good, money, the public may proceed to produce many close substitutes for money, just as lawyers find new loopholes in tax laws almost as fast as older ones are closed. The evolution of money from coins to bank notes, bills of exchange, bank deposits and finance paper illustrates the point. The Currency School may have been right about the need for a fixed supply of money, but it was wrong to believe that the money supply could be fixed forever.

The domestic aspects of the crisis are reviewed in Chapter 5. One question is whether manias can be halted by official warnings – moral suasion and jawboning. The evidence suggests that they cannot, or at least that many crises followed warnings that were intended to head them off. One widely noted remark was that of Alan Greenspan, chairman of the Federal Reserve Board, who stated on 6 December 1996 that he thought that the US stock market was

'irrationally exuberant'. The Dow Jones industrial average was 6600; subsequently the Dow peaked at 11,700. The NASDAQ had been at 1300 at the time of the Greenspan remark and peaked at more than 5000 four years later. A similar warning had been issued in February 1929 by Paul M. Warburg, a private banker who was one of the fathers of the Federal Reserve system, without slowing the stock market's upward climb. The nature of the event that ultimately produces a turning point is discussed: some bankruptcy, defalcation or troubled area revealed or rumored, or a sharp rise in the central bank discount rate to halt the hemorrhage of cash into domestic circulation or abroad. And then there is the interaction of falling prices – a crash – and its impact on the liquidity in the economy.

The impacts of the mania on domestic spending and the resulting euphoria are discussed in Chapter 6. Bubbles lead to extravagant expenditure. Malaysia, Dubai, and a few other places have all built the tallest building in the world – to show that they could do it and afford to pay for it. The Japanese imported French art in the 1980s – because other Japanese were importing French art. Money seems 'free' as if the fundamental laws of economics no longer apply. But observations of extravagant expenditures eventually lead to questions of whether there is an underlying bubble.

The implosion of a bubble always leads to the discovery of frauds and swindles that developed in the froth of the mania; these events are reviewed in Chapter 7. Fraud and corruption are based on mis-information – both falsification and misrepresentation; some fraud also involves the theft of private information before it becomes publicly available. Some of the fraud is personal, some is corporate. Bernie Madoff ran one of the largest Ponzi schemes ever, investors lost more than $20 billion. The owners of some business conglomerates in Iceland had 'captured' control of the banks and then borrowed from the banks to increase their consumption and their investments. The combination of failed thrift institutions and the rapid growth of junk bonds in the 1980s cost American taxpayers more than $100 billion; some of these thrifts had been acquired by individuals who relied on the junk bonds for their financing. Enron, MCIWorldCom, Tyco, Dynegy, and Adelphia were a rogue's gallery of the 1990s. Many of the large US mutual fund families were exposed because they provide favored treatment to hedge funds. Crashes and panics are often precipitated by the revelation of some misfeasance, malfeasance or malversation (the corruption of officials) that occurred during the mania. One inference is that the swindles are a response to the appetite for wealth (or plain greed) stimulated by the boom; the Smiths want to keep up with the Joneses and some Smiths engage in fraudulent behavior in order to do so. As the monetary system gets stretched, institutions lose liquidity and as unsuccessful swindles seem about to be revealed, the temptation to take the money and run becomes irresistible.

The international contagion of manias and crises from the seventeenth to the first half of the twentieth century is the subject of Chapter 8; two, three, four, or more countries experienced similar bubble symptoms at the same time. There are two competing narratives. One involves the metaphor of the sun and the moon, with the mania initiated in one country, often a large one, and then 'beamed' to its smaller neighbors or trading partners. Its alternative posits that these countries are subject to the same shock or innovation. There are many possible linkages among countries, including trade, arbitrage of securities, capital flows, changes in central bank reserves of gold or other international reserve assets, and direct contagion of speculators in euphoria or gloom. Some bubbles are national, others global. Some crises are national, others international. What constitutes the difference? Did, for example, the 1907 panic in New York precipitate the collapse of the Società Bancaria Italiana via pressure on Paris communicated to Turin by withdrawals of bank deposits? There is a fundamental ambiguity – tight money in one financial center can serve either to attract funds or to repel them, depending on the expectations that a rise in interest rates generates. With inelastic expectations – no fear of crisis or of currency depreciation – an increase in the discount rate attracts funds from abroad and helps provide the cash needed to enhance liquidity; with elastic expectations of changes – of falling prices, bankruptcies, or currency depreciation – raising the discount rate may suggest the need to take more funds out rather than bring new funds in. The dilemma is familiar in economic life generally. A rise in the price of a commodity may lead consumers to postpone purchases in anticipation of a decline, or to speed up purchases before prices rise further. And even where expectations are inelastic, and the increase in the discount rate at the central bank sets in motion the right reactions, lags in responses may be so long that the crisis begins before the Marines arrive.

One complex but not unusual trigger that leads to financial crisis is a sudden halt to foreign lending, perhaps because of a domestic boom; thus the boom in Germany and Austria in 1873 led to a decline in money outflows and contributed to the difficulties of Jay Cooke in the United States. Similar developments occurred with the Baring crisis in 1890, when troubles in Argentina led to a sudden decline in money flows to South Africa, Australia, the United States and other Latin American countries. The stock market boom in New York in the late 1920s led Americans to buy fewer of the new bond issues of Germany and various Latin American countries, which in turn caused these countries to slide into depression. A halt to foreign trading is likely to precipitate depression abroad, which may in turn feed back to the country that launched the process.[6]

The discussion in Chapter 9 highlights the four waves of credit bubbles since the mid-1970s, and the relationships among the successive waves. The likelihood that these four waves are independent and unrelated seems low.

The first of these waves involved the surge in bank loans to governments and government-owned firms in the larger developing countries in the 1970s, and the second bubble was in Tokyo in the second half of the 1980s; was there a connection between the developing country debt crisis and the bubble in Japan? The third wave was in Thailand, Malaysia, Indonesia, Russia, and other developing countries in the mid-1990s, while the fourth wave occurred in the real estate markets in the United States, Britain, Ireland, Iceland, and Spain ten years later. The crisis in the debt of the governments of Greece and other Mediterranean countries may be a follow-on to the fourth wave. Asset bubbles would not occur without the rapid growth of credit, either in the aggregate or to particular groups of borrowers. The theme of this chapter is that there was a systematic relationship among these waves, or at least between the first and the second, and the second and the third. The credit bubble that centered on Mexico and the other developing countries was *sui generis*. When the bubble in bank loans to the developing countries imploded in 1982, the currencies of these countries depreciated sharply – but the Bank of Japan resisted the pressure toward the appreciation of the yen in the second half of the 1980s, which led to a rapid increase in the money supply in Japan. When the bubble in Japan imploded at the beginning of the 1990s, the yen appreciated, and Japanese firms rapidly increased their investments in productive facilities in Southeast Asia. The currencies of these countries appreciated, and property prices increased. When the bubble in stock prices and real estate prices in Bangkok and the other Asian capitals imploded in 1997 and 1998, there was a surge in the flow of money to New York as Asian borrowers repaid their foreign loans; the Asian currencies depreciated sharply.

Jail time and financial penalties

Enron was the poster-child of the 1990s boom; the company had transformed itself from the owner of regulated natural gas pipelines into a financial firm that traded natural gas, petroleum, electricity, and broadband as well as owning water systems and an electrical power generating system. The top executives of Enron felt the need to show continued growth in profits to keep the stock price high, and in the late 1990s they began to use off-balance sheet financing vehicles to obtain the capital to grow the firm; they also put exceptionally high prices on some of their long trading positions so they could report that their trading profits were increasing. The collapse of Enron led to the failure of Arthur Andersen, which previously had been the most highly regarded of the global accounting firms.

MCIWorldCom was one of the most rapidly growing telecommunications firms. Again the need to show continued increases in profits led the managers to claim that several billion dollars of expenses should be regarded as investments. Jack Grubman had been one of the sages in Salomon Smith Barney (a unit of the Citibank Group); he continually promoted MCIWorldCom stock. Henry Blodgett was a security analyst

for Merrill Lynch who was privately writing scathing e-mails about the economic prospects of some of the firms that he was otherwise promoting to investors; Merrill Lynch paid $100 million to move the story off the front pages. Ten investment banking firms paid $1.4 billion to forestall trials. The chairman and chief executive officer of the New York Stock Exchange resigned soon after it became known that he had a compensation package of more than $150 million; the NYSE served both as a tent for trading stocks and as a regulator and the managers of some of the firms that were being regulated served as directors of the exchange and participated in determining the compensation package. Then a number of large US mutual funds were revealed to have allowed firms to trade on stale news.

A greater number of individuals have already gone to prison following these events than in the aftermath of any previous crisis, and a number are still awaiting trial. Six Enron senior managers already have been jailed. One Arthur Andersen partner who worked on the Enron account went to prison. Two of the senior financial officials of MCIWorldCom have gone to jail. As a sign of the reach of this crisis, the television and style celebrity (and former stockbroker) Martha Stewart was charged over a share deal and found guilty of obstruction of justice and imprisoned for five months.

Domestic crisis management is reviewed in Chapters 10 and 11. The first of these two chapters considers the range of domestic responses to a crisis; at one extreme the government may take a hands-off position, at the other there is a range of miscellaneous measures. Those who believe that the market is rational often prefer the hands-off approach; according to one formulation, it is healthy for the economy to go through the purgative fires of deflation and bankruptcy to get rid of the mistakes and excesses of the boom. Among the miscellaneous devices are holidays, bank holidays, the issue of scrip, guarantees of liabilities, issuance of government debt, deposit insurance and the formation of special institutions like the Reconstruction Finance Corporation in the United States in 1932 and the Istituto per la Ricostruzione Industriale (IRI) in Italy in 1933. The Italian literature calls the process the 'salvage' of banks and companies; the British in 1974–75 referred to saving the fringe banks as a 'lifeboat' operation.

Issues related to a domestic lender of last resort are the focus of Chapter 11 – primarily whether there should be such a lender, who this lender should be and how it should operate. A key topic is 'moral hazard' – if investors are confident that they will be 'bailed out' by a lender of last resort, their self-reliance may be weakened. But on the other hand, the priority may be to stop the panic, to 'save the system today' despite the adverse effects on the incentives of investors. If there is a lender of last resort, however, whom should it save? Insiders? Outsiders and insiders? Only the solvent, if illiquid? The distinction between solvent and insolvent firms depends on the scope and duration of the panic. These are political questions, and they are raised in particular when it becomes necessary to legislate to increase the capital of the Federal Deposit Insurance

Corporation (FDIC) or the Federal Savings and Loan Insurance Corporation (FSLIC) when one or the other runs out of funds to bail out failed banks. The issue was particularly acute in the 1990s in Japan, where the collapse of the Nikkei stock bubble in 1990 uncovered all sorts of bad real estate loans by banks, credit unions and other financial houses, confronting the government with the headache of deciding how much of a burden to put on the taxpayer. If the taxpayers didn't take the hit, then the owners of bonds and stocks would incur very large losses because the prices of these securities would decline further. Particularly troubling in this context was the catatonic state of government in Japan in the 1990s – slow to decide how to meet the crisis and even slower to act.

The role of an international lender of last resort in providing global monetary stability is the focus of Chapter 12. The problem is that there is no responsible government or agency of government with the *de jure* responsibility for providing this public good. US government support for Mexico, first in 1982 and again in 1994 was justified on the grounds that countries of the North American Free Trade Agreement (NAFTA) should stick together and that assistance to Mexico would dampen or neutralize the contagion effect and prevent a collapse of lending to Brazil and Argentina and other developing countries. The sharp depreciation of the Thai baht in the early summer of 1997 triggered crises in nearby Asian countries, including Indonesia, Malaysia, and South Korea as well as in Singapore, Hong Kong and Taiwan.

The thrust of Chapter 13 – a new chapter for the sixth edition – is the severity of the credit crisis that began in September 2008 – after the US government decided not to provide the financial assistance that would have allowed Lehman Brothers to remain a viable concern. In February the US Treasury and the Federal Reserve made an arrangement for JPMorgan Chase by agreeing that Morgan could put up to $29 billion of 'toxic securities' with the Fed. Shareholders of Bear Stearns received $10 a share – many of the shareholders were employees of Bear and this price was viewed as a retention bonus. During the first week of September Fannie Mae and Freddie Mac, the US government-sponsored lenders that accounted for more than 50 percent of the credit risk on home mortgages, were taken over by the US Treasury – both institutions were bankrupt and the owners of their common shares and their preferred shares lost virtually all of their money. It appeared as if the US government had in effect adopted a 'too big to fail' policy – these institutions would continue in business, although there would be a dramatic change in ownership. At the outset of the crisis, the markets viewed the decision not to 'save Lehman Brothers' as a major change in policy, but a day after Lehman closed its doors, there was a run on AIG, then the largest insurance company in the world, and the 'too big to fail' policy was resurrected. Nevertheless, investors still panicked, evidenced by the sharp increase in the spreads on riskier assets.

The final chapter seeks to answer two questions; the first is why there has been so much economic turmoil in the international financial economy in the last thirty years, and the second is whether an international lender of last resort would have made a difference. The International Monetary Fund was established in the 1940s to act as an international lender of last resort and to fill an institutional vacuum; the view was that financial crises in the 1920s and the 1930s would have been less severe had there been an international lender of last resort. The large number of crises in the last thirty years leads to the question of whether the presence of the IMF as a supplier of national currencies to countries with financial crises encouraged profligate national financial policies. Financial arrangements need a lender of last resort to prevent the escalation of the panics that are associated with crashes in asset prices. But a commitment to the view that a lender is needed should be distinguished from the view that individual borrowers will be 'bailed out' if they become over-extended. For example, uncertainty about whether New York City would be helped, and by whom, may have been the appropriate policy in the long run, so long as help was finally provided, and so long as there was doubt right to the end as to whether it would be forthcoming. This is a neat trick: always come to the rescue, in order to prevent needless deflation, but always leave it uncertain whether rescue will arrive in time or at all, so as to instill caution in other speculators, banks, cities or countries. The economic implosion triggered by the failure of Lehman Brothers could have been avoided if the firm had been acquired by another private firm much as Bear Stearns or if the government had taken over ownership of the firm. In Voltaire's *Candide*, the head of a general was cut off 'to encourage the others'. A sleight of hand may be necessary to 'encourage' the others (without, of course, cutting off actual heads) to participate in lender-of-last-resort activities because the alternative is likely to have very expensive consequences for the economic system.

Is the fourth wave of asset price bubbles since the mid-1970s the last in this series, or instead is it likely that the crisis that followed the collapse of real estate prices and the measures adopted to dampen the hardships associated with the failure of many banks in various countries have laid the foundation for a fifth wave. An epilogue summarizes some of the proposals for reform of the US financial system and asks whether the bubble and crisis of 2002–08 would have been significantly different if the legislation adopted in 2010 had been adopted in 2000.

2
The Anatomy of a Typical Crisis

History vs economics

Historians view each event as unique. In contrast economists search for the patterns in the data, and the systematic relationships between an event and its antecedents. History is particular; economics is general. The business cycle is a standard feature of market economies; increases in investment spending lead to increases in household income and in GDP growth. Macroeconomics focuses on the explanations for the cyclical variations in the growth of GDP relative to its long-run trend.

An economic model of a general financial crisis is presented in this chapter, while the various phases of the speculative manias that lead to crises are illustrated in the following chapters. This model of a crisis covers the boom and the subsequent bust and centers on the episodic nature of the mania and the subsequent events. This model differs from those that focus on the variations and the periodicity of economic expansions and contractions, including the Kitchin inventory cycle of thirty-nine months, the Juglar cycle of investment in plant and equipment that has a periodicity of seven or eight years and the Kuznets cycle of twenty years that highlights the rise and fall in housing construction.[1] In the first two-thirds of the nineteenth century, crises occurred regularly at ten-year intervals (1816, 1826, 1837, 1847, 1857, 1866), thereafter less regularly (1873, 1907, 1921, 1929).

The model

A model developed by Hyman Minsky helps explain the financial crises in the United States, Britain and other market economies. Minsky highlighted that the changes in the supply of credit were pro-cyclical and increased when the economy was booming and decreased during slowdowns. During the expansions investors became more optimistic and they revised upward their

estimates of the profitability of a wide range of investments and became more eager to borrow. At the same time, both the lenders' assessments of the risks of individual investments and their risk averseness declined and they became more willing to make loans, including some for investments that previously had seemed too risky.

When the economy slowed, investors became less optimistic and more cautious. The lenders also became more cautious as their loan losses increased, especially if the losses led to declines in their capital.

Minsky believed that the increases in the supply of credit in good economic times and the subsequent decline in the supply led to fragility in financial arrangements and increased the likelihood of a crisis. His model is in the tradition of the classical economists, including John Stuart Mill, Alfred Marshall, Knut Wicksell and Irving Fisher, who focused on the variability in the supply of credit. Minsky followed Fisher and attached great importance to the behavior of heavily indebted borrowers, particularly those that increased their indebtedness to buy real estate or stocks or commodities in search for short-term capital gains. Their motive was the profits from the increases in the prices of these assets, which they anticipated would greatly exceed the interest payments on the borrowed money. When the economy slowed many of these borrowers would become distress sellers because the prices of these assets would be falling.

Minsky suggested that the events that lead to a crisis start with a 'displacement' or innovation, some exogenous shock to the macroeconomic system.[2] If the shock was sufficiently large and pervasive, the economic outlook and the anticipated profit opportunities would improve in at least one important sector of the economy. Business firms and individuals would borrow to take advantage of the increase in the anticipated profits in this sector. Economic growth would quicken and in turn there might be a feedback to even greater optimism. It's 'Japan as Number One' or the 'East Asian Miracle' or 'The New American Economy' or 'the Geyser Tiger' – much more profound optimism about the economy. The words differ across the countries but the tune is the same.

The shock varies from one speculative boom to another. The shock in the United States in the 1920s was the rapid expansion of automobile production and development of highways together with the electrification of much of the country and the large increase in households with telephones. The shocks in Japan in the 1980s were rapid increases in the supplies of money and of credit and financial liberalization that enabled the banks to increase their real estate loans at a rapid rate. The shock in the Nordic countries in the 1980s was financial liberalization, which permitted the domestic banks to borrow in the offshore market. One of the shocks the preceded the Asian financial crisis was the discovery of 'emerging market equities as a new asset class' which led to

sharp increases in the purchases of these securities by mutual funds and pension funds headquartered in the United States, Britain, and other industrial countries. The shock in the United States in the 1990s was the revolution in information technology and the sharp declines in the costs of communication. The shock in the US housing market in the 2002 was securitization which involved the packaging of mortgages with similar attributes into bundles that provided the basis for issuing collateralized mortgage obligations (CMOs) and collateralized debt obligations (CDOs); the amount of money available for purchases of homes increased sharply. The shock in Iceland after 2000 was the privatization of the banks. At times the shock has been outbreak of war or the end of a war, a bumper harvest or a crop failure, the widespread adoption of an invention with pervasive effects – canals, railroads. Unanticipated changes of monetary policy like the one in the autumn of 1979 have also constituted major shocks.

If the shock is sufficiently large and pervasive, the anticipated profit opportunities improve in at least one important sector of the economy: the profit share of GDP increases. In the early 1980s, US corporate profits were 3 percent of GDP; toward the end of the 1990s this ratio had increased to 10 percent. The significant increase in stock prices associated with the dot.com boom at the end of the 1990s resulted in part from the much more rapid increase in corporate profits than in GDP.

The boom in the Minsky model is fueled by the expansion of credit. In the pre-banking seventeenth and eighteenth centuries personal credit or vendor financing fueled speculative booms. Once banks had been developed they expanded the supply of credit; in the first several decades of the nineteenth century banks increased the supplies of credit by issuing more notes and subsequently they increased the supply by adding to the deposit balances of borrowers. In addition new banks were formed and their efforts to increase market share led to rapid growth of credit because the established banks usually were reluctant to accept a decline in market share. In the 1970s the European banks began to poach on the turf of the US banks in making loans to the governments in Latin America.

One central policy issue centers on the control of credit from banks and from other suppliers of credit. Often the authorities in a country have applied strict controls that limited the ability of banks to make certain types of loans. The banks then set up subsidiaries that could make these loans, or their holding companies made these loans. Even if the credits from the financial institutions were controlled, increases in the supply from non-bank sources could lead to a boom.

Assume an increase in the demand for goods and services. Eventually the increase in demand presses against the productive capacity, prices increase and the more rapid increase in profits attracts both more investment and

more firms. Positive feedback develops as the increase in investment leads to increases in the growth rate that in turn induces additional investment.

Minsky noted that 'euphoria' might develop at this stage. Investors buy goods and securities to profit from the capital gains associated with the anticipated increases in their prices. The authorities recognize that something exceptional is happening and while they are mindful of earlier manias, 'this time it's different', and they have extensive explanations for the difference. The Chairman of the US Federal Reserve, Alan Greenspan, discovered a surge in US productivity in 1997, about a year after he first became concerned that US stock prices might be too high; the increase in productivity meant that profits would increase at a more rapid rate, and the higher level of stock prices relative to corporate earnings might not seem unreasonable.

Minsky's three-part taxonomy of finance

Minsky distinguished among hedge finance, speculative finance, and Ponzi finance – on the basis of the relation between the operating income of various firms and their debt service payments. A firm is in the hedge finance group if its anticipated operating income is more than sufficient to pay both the interest and scheduled reduction in its indebtedness. A firm is in the speculative finance group if its anticipated operating income is sufficient to pay the interest on its indebtedness; however the firm must use cash from new loans to repay part or all of the amounts due on maturing loans. A firm is in the Ponzi group if its anticipated operating income is smaller than the amount needed to pay all of the interest on its indebtedness on the scheduled due dates, so the firm must either increase its indebtedness or sell some assets to get the cash for these payments.

Minsky's hypothesis is that when the economy slows, some of the firms that had been involved in hedge finance are shunted to the speculative finance group and that some of the firms that had been involved in this group would move to the Ponzi finance group.

Minsky's use of the term 'Ponzi finance' memorializes Carlo Ponzi, who operated a small deposit-taking firm in a Boston suburb in the early 1920s. Ponzi promised his depositors to pay interest at the rate of 30 percent a month and his financial transactions went smoothly for three months. In the fourth month however the inflow of cash from new depositors was smaller than the interest payments promised to the older depositors and soon Ponzi was on his way to prison.

The term Ponzi finance is now a generic term for a non-sustainable pattern of finance. The borrowers can only meet their commitments to pay the high interest rates on their outstanding loans if they can obtain cash from new loans. Since in many arrangements the interest rates are very high, often 30 to 40 percent a year, the continuation of the arrangement requires a continuous inflow of new money and often at an accelerating rate. Initially many of the existing depositors are so pleased with their high returns that they allow their interest income to compound; the cliché is that they are 'earning interest on the interest'.

As a result the inflow of new money can be below the promised interest rate for a few months. The arrangement can operate only as long as the cash withdrawals are smaller than the inflow of new money.

The continuation of the process leads to what Adam Smith and his contemporaries called 'overtrading'. This term is not precise and includes speculation about increases in the prices of assets or commodities, an overestimate of prospective returns, and 'excessive leverage'.[3] Speculation involves buying assets for resale at higher prices rather than for their investment income. The euphoria leads to an increase in the optimism about economic growth and about the increases in corporate profits. In the late 1990s Wall Street security analysts projected that US corporate profits would increase by 15 percent a year for five years. (If their forecasts had been correct, then the ratio of US corporate profits to US GDP would have been 40 percent higher than ever before at the end of the fifth year.) Loan losses incurred by the lenders decline because asset prices are increasing and they become more optimistic and reduce the minimum down payments and the minimum margin requirements. Even though bank loans are increasing, the leverage – the ratio of debt to capital or to equity – of many of their borrowers may decline because the rapid increase in the prices of real estate or securities leads to increases in their net worth.

A follow-the-leader process develops as firms and households see that the speculators are making a lot of money. 'There is nothing as disturbing to one's well-being and judgment as to see a friend get rich.'[4] Unless it is to see a non-friend get rich. Similarly banks may increase their loans because they are reluctant to lose market share to other lenders. More and more firms and households begin to participate in the scramble for high profits. Making money never seemed easier. Speculation for capital gains leads from normal, rational behavior to what has been described as a 'mania' or a 'bubble'.

The word 'mania' emphasizes irrationality; 'bubble' foreshadows that some values will eventually decline and probably sharply. The term bubble means any significant increase in the price of an asset or a security or a commodity that cannot be explained by the 'fundamentals'. Small price variations around the values based on the fundamentals are 'noise'. In this book, a bubble is an increase in indebtedness at a rate of 20 or 30 percent a year for an extended period that cannot be sustained. Someone with 'perfect foresight' should have foreseen that the price increases were not sustainable and that an implosion was inevitable.

In the twentieth century most of the manias and bubbles have centered on real estate and stocks. There was a real estate mania in southeast Florida in the mid-1920s and an unprecedented bubble in US stocks in the last three years of the 1920s; US real estate prices also increased significantly in late 1920s although not as rapidly as stock prices. In Japan in the 1980s the speculative purchases of real estate induced a boom in the stock market. Similarly the bubbles in the Asian countries in the early 1990s involved both real estate and stocks, and generally increases in real estate prices pulled up stock prices.

The US bubble in the late 1990s centered on stocks, although the increases in household wealth in Silicon Valley and several other regions led to surges in real estate prices. The oil price shocks of the 1970s led to sharp increases in real estate activity in Texas, Oklahoma, Louisiana, and other oil-producing areas. Similarly the rapid increases in the prices of cereals in the inflationary 1970s led to surges in land prices in Iowa, Nebraska and Kansas and other Midwest farm states.

International propagation

Minsky focused on the pro-cyclical changes in the supply of credit in a single country. Historically euphoria has often spread from one country to others through one of several channels. The bubble in Japan in the 1980s had significant impacts on South Korea and Taiwan, which were parts of the Japanese supply chain; if Japan had been doing well economically, its former colonies also did well. Moreover the surge in asset prices in Japan had a dramatic impact on real estate prices in Hawaii, which is to Tokyo as Miami is to New York; Japanese travel to Hawaii for rest and recreation in the sun and the state experienced a real estate boom as the Japanese bought second homes and golf courses and hotels.

Arbitrage is one conduit – investors ensure that the increases in the price of a commodity in one national market lead to comparable increases in the prices of the same commodity in other national markets. Thus changes in the price of gold in Zurich, Beirut, and Hong Kong are closely tied to changes in the price of gold in London. Similarly increases in the prices of securities in one national market lead to nearly identical changes in the prices of the same securities in other countries. The prices of securities which are more distant substitutes for each other also are affected – the cliché that 'the rising tide lifts all ships' applies.

In addition increases in GDP in a large country induce increases in imports and hence increases in counterpart exports in other countries and in their GDPs. Capital flows constitute a third link; the increase in foreign purchases of securities available in a country lead to increases in both their prices and in the value of the country's currency.

Moreover there are psychological connections, an increase in investor euphoria or pessimism in one country affects investors in others. The declines in stock prices on 19 October 1987 were practically instantaneous in all national financial centers (except Tokyo), far faster than can be accounted for by arbitrage, income changes, capital flows, or money movements.

In the textbook model an increase in the flow of gold and hence in the money supply in one country would be matched by a corresponding decline in the gold supply in a foreign country, and the increase in the money supply and in credit in the first country would be offset by the contraction of

credit and money in the other. In the real world, however, the increase in the credit expansion in the first country may not be followed by a contraction of credit elsewhere, because investors in the second country may respond to rising prices and profits abroad by demanding more credit so they can buy the assets and securities whose prices they anticipate will increase. The potential contraction from the shrinkage in the monetary base in the second country may be overwhelmed by the increase in the speculative demand for credit.

As the boom continues, interest rates, the speed of payments and the commodity price level increase. The purchases of securities or real estate by 'outsiders' means that the 'insiders' – those who own these assets – sell them and realize profits – if the outsiders are buyers, then the insiders must be sellers. At every moment the purchases of real estate or stocks by the outsiders are necessarily balanced by sales of the insiders – usually at ever-higher prices. In 1928 the market value of the transactions on the New York Stock Exchange increased at an annual rate of 36 percent, and in the first eight months of 1929 at an annual rate of 53 percent. Similarly in 1998 the market value of the stocks traded on the NASDAQ increased by 41 percent and in the subsequent fifteen months the market values doubled. Investors rush to get on the train even as it accelerates. As long as the outsiders are more eager to buy than the insiders are to sell the prices of the assets or securities increase.

As the buyers become less eager and the sellers become more eager, an uneasy period of 'financial distress' follows; the corporate finance term reflects that a firm is unable to fulfill its debt servicing commitments. For the economy as a whole, the equivalent is the awareness by many firms and individual investors that it is time to become more liquid – to increase their money holdings. They sell real estate and stocks – their prices may fall sharply. Some highly leveraged investors may go bankrupt because the decline in prices is so large that the market value of these assets declines below their indebtedness – they have 'upside down mortgages'. Some investors continue to hold the assets because they believe that the price decline is temporary, a hiccup.

The prices of the securities may increase again, at least for a while. There were four 'bear market rallies' in Tokyo in the 1990s; each involved increases in stock prices of more than 20 percent despite the downward price trend. But some investors believed that stock prices had declined too far, and they wanted to be among the first to buy the stocks while they were still 'cheap'.

As the decline in prices continues, more and more investors realize that prices are unlikely to increase and that they should sell before prices decline further; in some cases this realization occurs gradually and in others suddenly. The race into money may turn into a stampede.

The specific signal that precipitates the crisis may be the failure of a bank or of a firm, the revelation of a swindle or defalcation by an investor who sought to escape distress by dishonest means, or a sharp fall in the price of a security

or a commodity. The rush is on – prices decline and bankruptcies increase. Liquidation sometimes is orderly but may escalate into a panic as investors realize that only a relatively few can sell while prices remain not far below their peak values. The term 'revulsion' was used to describe this behavior in the nineteenth century. The banks become much more cautious in making new loans based on the collateral of commodities and securities and real estate because their values are more uncertain. In the early nineteenth century this condition was known as 'discredit'.

'Overtrading', 'revulsion', and 'discredit' have a musty, old-fashioned flavor; they convey a graphic picture of the decline in investor optimism. Revulsion and discredit may lead to panic (or as the Germans put it, *Torschlusspanik*, 'door-shut-panic') as investors crowd to get through the door before it slams shut. The panic feeds on itself until prices have become so low that investors are tempted to buy the less liquid assets, or until trade in the assets is stopped because limits have been placed on maximum daily price declines, or because the exchanges have been closed, or because a lender of last resort succeeds in convincing investors that money will be available to meet the demand for cash. Confidence may be restored even without a large increase in the volume of money once investors conclude that credit is now more readily available.

Whether a lender of last resort should provide liquidity to forestall a panic and the decline in prices of real estate and stocks has been debated for more than two hundred years. Those who oppose the provision of liquidity from a lender of last resort argue that the knowledge that such credits will be available encourages speculation. Those who want a lender of last resort worry more about coping with the current crisis and reducing the likelihood that a liquidity crisis will cascade into a solvency crisis and trigger a severe recession – like the one that began in 2008. At the international level, there is neither a world government nor any world bank adequately equipped to serve as a lender of last resort. The International Monetary Fund has not met the expectations of its founders as a lender of last resort.

The validity of the model

Three types of criticism have been directed at the Minsky model. One is that each crisis is unique so that a general model is outdated. A second is that this type of model is no longer relevant because of changes in the economic environment. A third is that asset price bubbles are highly improbable because 'all the information is in the price' – the mantra of the efficient market view of finance.

Each criticism merits a response. The first criticism is that each crisis is the product of a unique set of circumstances, or that there are such wide differences among economic crises so they should be arranged into various species,

each with its own particular features. Financial crises were frequent in the first two-thirds of the nineteenth century and in the last third of the twentieth century. In this view, each unique crisis is a product of a specific series of historical accidents – which was said about 1848 and about 1929,[5] and may be inferred from the historical accounts of separate crises noted throughout this book. Each crisis also has its unique individual features – the nature of the shock, the object of speculation, the form of credit expansion, the ingenuity of the swindlers and the nature of the incident that touches off revulsion. But if one may borrow a French phrase, the more something changes, the more it remains the same. Details proliferate; structure abides. Most of these crises followed from the implosion of a credit bubble.

More compelling is the suggestion that the genus 'crises' should be divided into commercial, industrial, monetary, banking, fiscal, and financial (in the sense of financial markets) sub-species or into local, regional, national, and international groups. Taxonomies along such lines abound. This view is not accepted because the primary concern is with international financial crises that involve a number of critical elements – speculation, the increase in credit, an increase in the prices of securities or of real estate or commodities followed by a sharp fall as investors rush to sell. The test is whether the Minsky model provides insights about the broad features of crises.

The second criticism is that the Minsky model of the instability of the supply of credit is no longer relevant because of structural changes in the institutional features of the economy, including the rise of large corporations and of big labor unions and big governments, modern banking and speedier communications. The financial debacles in Mexico, Brazil, Argentina, and more than ten other developing countries in the early 1980s are consistent with the Minsky model; the external indebtedness of these countries increased much more rapidly than the interest rates on their loans. The bubble in real estate prices and stock prices in Japan in the second half of the 1980s and the subsequent implosion of asset prices is consistent with the Minsky model since the annual increases in the prices of stocks and of real estate were three or four times higher than the interest rates on the money borrowed to buy these assets. The booms and the subsequent busts in the second half of the 1990s in Thailand and Hong Kong and Indonesia and then in Russia feature the same pattern of cash flows.

The third criticism is that bubbles are impossible because market prices always reflect the economic fundamentals, and that sharp declines in asset prices usually reflect 'policy switches' by government or central banks. Those who take this position suggest that the alleged bubble appears to be the result of herd behavior, positive feedback or bandwagon effects – credulous suckers following smart insiders. These critics suggest that the model is 'mis-specified', that is, that something was going on and not taken into account by the theory,

and that more research is needed.[6] Some of the research ignored by those with this belief is offered in this book.

A more cogent attack on the Minsky model was by Alvin Hansen who claimed that the model was relevant prior to the middle of the nineteenth century but ceased to be so because of changes in the institutional environment.

> Theories based on uncertainty of the market, on speculation in commodities, on 'overtrading,' on the excesses of bank credit, on the psychology of traders and merchants, did indeed reasonably fit the early 'mercantile' or commercial phase of modern capitalism. But as the nineteenth century wore on, captains of industry ... became the main outlets for funds seeking a profitable return through savings and investments.[7]

Hansen – who was a foremost expositor of the Keynesian model of the business cycle and especially of persistent high levels of unemployment – wanted to downplay the significance of alternative explanations for declines in economic activity other than a high level of saving. Hansen's emphasis on the importance of the relation between savings and investment does not require the rejection of the view that changes in the supply of credit can have important impacts on the prices of securities and economic activity.

The model's relevance today

The Minsky model can be readily applied to the currency market and to periods of overvaluation and undervaluation of the German mark, the Swiss franc, the Japanese yen, the euro, and other national currencies that are associated with 'overshooting' and 'undershooting'. Changes in the values of national currencies have been large relative to long-run equilibrium values despite extensive intervention by central banks to dampen these deviations. Speculation in currencies has resulted in substantial profits for the large international banks although a few banks and industrial firms have incurred large losses.[8]

Consider the growth in the external debt of Mexico, Brazil, Argentina, and the other developing countries in the 1970s; bank loans to these countries increased by 30 percent a year and the external debt of these countries increased by 20 percent a year. The bank loans generally had a maturity of eight years and interest rates were floating and set with a specified markup over London Interbank Offer Rate (LIBOR). The interest rates averaged 8 percent although they trended upward during the decade. Since the cash that the borrowers received from new loans was substantially larger than the interest payments on their outstanding loans, they incurred no burden or hardship in making their debt service payments on a timely basis.

The inflow of foreign funds led to a real appreciation of the currencies of the money-receiving countries which was necessary so that the increase in their trade and current account deficits would match the increase in their capital account surpluses. It was inevitable and hence predictable that at some future date the receipts of money from new loans would decline below the interest payments on the outstanding loans, and at that time the value of their currencies would decline; the counterpart of the decline in the money inflows was that these countries would need trade surpluses to get some of the money necessary to pay the interest to their foreign creditors. Most of these borrowers effectively defaulted on their loans when the lenders stopped making new loans. These defaults cost the lenders $250 billion in the form of the reduction in the face value of the loans and what in effect was a reduction in the interest rates to below market levels. The lenders had failed to ask, 'Where will the borrowers get the cash to pay us the interest if we stop supplying them with the cash in the form of new loans?'.

During the 1980s real estate prices and stock prices in Japan surged; the country experienced an economic boom. Real estate investors earned 30 percent a year – and much more after adjusting for leverage. Business firms recognized that the profits on real estate investments were substantially higher than the profits from making steel or automobiles or TV sets and so they bought real estate. Real estate prices were increasing many times more rapidly than rents. At some stage, the net rental income declined below the interest payments on the money borrowed to buy the real estate and the borrowers had a 'negative carry'. The borrowers might obtain the funds to make the interest payments by increasing their loans against some of the properties that they already owned. At the beginning of 1990, the incoming governor of the Bank of Japan instructed the banks to limit the growth in real estate loans. Once the bank loans for real estate began to increase at a much less rapid rate, some of the firms and investors that needed money from new loans to pay the interest on the outstanding loans were no longer able to obtain new loans. They sold real estate and the bubble imploded.

The saga of Iceland's 'perfect' bubble

Iceland is the smallest country – 300,000 people – with its own currency. Between 2002 and 2007, prices of residential real estate doubled, which was not exceptional, while the prices of stocks increased by a factor of nine, which was exceptional. In 2002, the value of the assets owned by the three Icelandic banks was 150 percent of the country's GDP, while in 2007 the value of bank assets was eight times Iceland's GDP. Iceland's banks were the most rapidly growing in the world. And then the bubble popped in September 2008; the Icelandic krona lost half of its value, stock prices

declined by 90 percent, and economy went into a recession – severe by Iceland's standards but modest by those in other countries.

Because Iceland is so small, the financial processes associated with a bubble are more readily visible than in larger countries. The perfect bubble began in 1998 when a market-oriented government began to privatize the banks, a process that was completed in 2002. The buyers of the shares were the entrepreneurs who owned the firms engaged in importing.

In 2002, the Icelandic kronor began to appreciate in response to an increase in the foreign demand for Icelandic securities; a flow that was motivated by the higher anticipated returns on securities denominated in the kronor. The flow of money to Iceland was part of a global flow of money to the United States, Britain, Ireland, Spain, South Africa, and Australia. Iceland's trade balance morphed from a small surplus to a deficit of more than 20 percent of its GDP. The supply of kronor securities was limited, in part because the Icelandic government had had modest fiscal deficits and the various pension funds had bought many of these securities. The increase in the foreign demand for kronor securities led to higher prices.

The Icelandic banks owned relatively modest amounts of stocks, less than 20 percent of their assets, and as stock prices increased, both the profits of the banks and their capital increased. The banks increased their loans at a pace that was almost as rapid as the increase in their capital. Some of those who borrowed from the banks used the money to buy stocks. Stock prices were increasing by 70 percent a year, and bank capital was increasing at the same rate. Banks were buying more Icelandic stocks, and they owned an increasingly large share of these stocks. The external indebtedness of Iceland was increasing by nearly 30 percent a year. The flow of money to Iceland was three to four times larger than the interest payments that Icelandic residents were making to their foreign creditors. Similarly the domestic indebtedness of households and firms was increasing at a rapid rate, and these borrowers could readily obtain the money to pay the interest on their outstanding loans with the money from new loans. At some stage, the entrepreneurs who owned the Icelandic banks took the initiative and began to borrow from the international banks. Most of this money was lent to the Icelandic entrepreneurs who used the money to buy firms and banks in Britain and other countries in Northern Europe. The branches of the Icelandic banks in London and elsewhere in Europe rapidly increased their deposits. Some of the money that these branches received from the sale of deposits was used by the head offices to pay the interest on their indebtedness.

During this stage, there was chatter in the financial circles that Reykjavik would become a new international financial center, like Switzerland, Hong Kong, or Luxembourg.

The earnings of the Icelandic banks primarily came from the increases in the prices of Icelandic stocks that they owned. The prospect that earnings would continue to grow supported the increasingly high level of stock prices. Iceland experienced a consumption boom in response to the increase in household wealth.

When Lehman Brothers failed in mid-September 2008, the Icelandic banks were unable to borrow in the foreign markets, so they no longer had access to the money to pay the interest on their indebtedness. Stock prices stopped increasing, and some of the borrowers were no longer able to get the cash to pay the interest on their indebtedness and they defaulted. The loan losses of the Icelandic banks soared. The banks failed, and they were taken over by the Icelandic government.

The US international financial position after 2000 in some ways parallels that of Mexico, Brazil, and Argentina in the 1970s. These countries had unsustainably large current account deficits and obtained the money to pay the interest to their foreign creditors from new loans. The implication is that the US external payments position is not sustainable.

This book is a study in financial history, not economic forecasting. But investors seem not to have learned from experience.

3
Speculative Manias

Rationality of markets

The word 'mania' suggests a loss of a connection with rationality, perhaps mass hysteria. Economic history is replete with canal manias, railroad manias, joint stock company manias, real estate manias and stock price manias – surges in investment in a particular activity. Economic theory assumes that men and women are rational – and hence manias would not occur. There is a disconnect between the observations that manias occur episodically, and the rationality assumption. This chapter focuses on the investor demand for a particular type of asset or security and the next chapter is centered on changes in the supply of credit.

The 'rational expectations' assumption is that investors react to changes in economic variables as if they are always fully aware of the long-term implications of each of these changes, either because they are clairvoyant or because they have Superman-like vision. Thus the cliché that 'all the information is in the price' reflects the view that prices in each market change immediately and fully to every news item so that no 'money is left on the table'.

Contrast the rational expectations assumption with the adaptive expectations assumption that the values of certain variables in the future are extensions of their recent values – the assumption that is implicit in the view that there are 'trends' in the changes in asset prices. The cliché that 'the trend is your friend' reflects that prices will continue to increase if they have been increasing. In contrast the thrust of the rational expectations view is that the prices that are anticipated next week and next month determine the prices that prevail today, in effect a backward-looking view from the future to the present. Thus the price of gold in the spot market today is derived from the anticipated prices of gold two or three or more years from now as revealed in a futures market discounted to the present by an appropriate interest rate, usually the one on risk-free US Treasury securities. The Canadian dollar price of the US

dollar today is the anticipated price of the US dollar for a distant future date discounted to the present by the difference between US and Canadian interest rates. If a government reduces tax rates to stimulate consumption spending or investment spending, the rational expectations view implies that investors immediately will realize that the larger fiscal deficit today implies higher tax rates tomorrow and so they will increase the amounts they save in anticipation of the projected increase in their tax bills.

What does it mean to say that investors are rational?[1] One assumption is that most investors behave rationally most of the time. A second is that all investors behave rationally most of the time. A third is that each and every participant in the markets has the same intelligence, the same information, the same objectives and uses the same economic model. A fourth is that all investors behave rationally all the time.

Each of these assumptions has different implications for the way that investors behave. Obtaining agreement on the assumption that most investors behave rationally most of the time is easier than obtaining agreement on the assumption that each investor behaves rationally all of the time. Frequently the argument is between two polar positions, one that holds that no investor is ever rational while the other asserts that all investors are always rational. Harry G. Johnson offered this description of the difference between an older group of economists and a younger group interested in international monetary reform:

> The difference can be encapsulated in the proposition that whereas the older generation of economists is inclined to say 'the floating rate system does not work the way I expected, therefore the theory is wrong, the world is irrational and we can only regain rationality by returning to some fixed rate system to be achieved by cooperation among national governments' while the younger group is inclined to say 'the floating rate system is a system that should be expected to operate rationally, like most markets; if it does not seem to work rationally by my standards, my understanding of how it ought to work is probably defective; and I must work harder at the theory of rational maximizing behavior and its empirical consequences if I am to achieve understanding'. This latter approach is the one that is being disseminated, and intellectually enforced, through the [younger] network.[2]

Rationality is an a priori assumption about the way the world should work rather than a description of the way the world has actually worked. The assumption that investors are rational in the long run is a useful hypothesis because it illuminates understanding of changes in prices in different markets; in the terminology of Karl Popper, it is a 'pregnant' hypothesis. Hence it is useful to

assume that investors are rational in the long run and to analyze economic issues on the basis of this assumption. One interpretation of the rationality assumption is that prices in a particular market today must be consistent with the prices one and two months from now and one and two years from now adjusted for the 'costs of storage'; otherwise there would be a profitable and relatively riskless arbitrage opportunity.

Ragnar Nurkse summarized his survey of changes in the values of the French franc and the German mark in the 1920s with the statement that speculation in the currency market had been destabilizing. Milton Friedman asserted that destabilizing speculation cannot occur – or at least is unlikely to persist – because investors that bought as prices were increasing and sold as prices were declining 'would be buying high and selling low'; because of these losses, they would either go out of business or change their strategy. The Friedman view is that since the destabilizing speculators would fail to survive, destabilizing speculation cannot occur.[3] One response is that from time to time some investors follow strategies that would lead to losses.

There have been many historic episodes of destabilizing speculation, although at times the language has been imprecise and at other times hyperbolic. Consider some of the phrases in the literature: *manias ... insane land speculation ... blind passion ... financial orgies ... frenzies ... feverish speculation ... epidemic desire to become rich quick ... wishful thinking ... intoxicated investors ... turning a blind eye ... people without ears to hear or eyes to see ... investors living in a fool's paradise ... easy credibility ... overconfidence ... overspeculation ... overtrading ... a raging appetite ... a craze ... a mad rush to expand.*

Fernand Braudel used the terms 'craze' and 'passion' when he discussed everyday life in Europe from the fifteenth to the eighteenth centuries, largely in connection with consumption but also extended to spices, styles of dress, craving for knowledge, and purchases of land.[4]

The principals in the London banking firm of Overend, Gurney, which crashed on Black Friday in May 1866, were said to be 'sapient nincompoops'.[5] 'These losses,' said Bagehot, 'were made in a manner so reckless and so foolish that one would think a child who had lent money in the City of London would have lent it better.'[6]

Clapham's description of the Baring firm in 1890 is understated in a characteristic British fashion: 'They had not considered these enterprises or the expected investors in them coolly or wisely enough [but had] gone far beyond the limits of prudence.'[7]

Consider Adam Smith's comment on the South Sea Bubble: 'They had an immense capital dividend among an immense number of proprietors. It was naturally to be expected, therefore, that folly, negligence, and profusion should prevail in the whole management of their affairs. The knavery and

extravagance of their stock-jobbing operations are sufficiently known [as are] the negligence, profusion and malversation of the servants of the company.'[8]

And finally a description by the usually restrained Alfred Marshall:

> The evils of reckless trading are always apt to spread beyond the persons immediately concerned ... when rumors attach to a bank's credit, they make a wild stampede to exchange any of its notes which they may hold; their trust has been ignorant, their distrust was ignorance and fierce. Such a rush often caused a bank to fail which might have paid them gradually. The failure of one caused distrust to rage around others and to bring down banks that were really solid; as a fire spreads from one wooden house to another until even fireproof buildings succumb to the blaze of a great conflagration.[9]

Rationality of the individual, irrationality of the market

Manias are associated on occasion with 'irrationality' or mob psychology. The relationship between rational individuals and an irrational group of individuals can be complex. A number of distinctions can be made. One assumption is that of mob psychology, a 'group think' when virtually each of the participants in the market changes his or her views at the same time and moves as a 'herd'. Alternatively various individuals change their views about prospective developments in markets at different times as part of a continuing process; most start rationally and then more of them lose contact with reality, gradually at first and then more quickly. A third possible case is that rationality differs among different groups of traders, investors, and speculators, and that an increasing number of individuals in these groups succumb to the hysteria as asset prices increase. A fourth case is that all the market participants succumb to the 'fallacy of composition', the view that from time to time the behavior of the group of individuals differs from the sum of the behaviors of each of the individuals in the group. The fifth is that there is a failure of a market with rational expectations as to the *quality* of a reaction to a given stimulus to estimate the appropriate *quantity*, especially when there are lags between the stimulus and the reactions. Finally irrationality may exist because investors and individuals use the wrong model, or fail to consider a particular and crucial bit of information, or suppress information that does not conform to the model that they have implicitly adopted. The irrationality of the gullible and greedy in succumbing to swindlers is discussed in a later chapter.[10]

Mob psychology or hysteria is well established as an occasional deviation from rational behavior. Some economic models highlight the demonstration effect, which leads the Smiths to spend more than their incomes – at least for a while – as they seek to keep up with the Joneses. Another is the Duesenberry effect: both the Smiths and the Joneses increase their consumption expenditures

when their incomes increase and both are reluctant to reduce their spending when their incomes decline. Politics has its 'bandwagon effects' when individuals back the most probable winners (or 'rats desert the sinking ship' when they turn from losers – though if the ship is really sinking, the rational rats leave). The French historian Gustave LeBon discussed this subject in *The Crowd*.[11] Charles MacKay in his discussion of the South Sea Bubble[12] mentioned a banker who purchased £500 worth of South Sea stock in the third subscription list of August 1720, saying, 'When the rest of the world are mad, we must imitate them in some measure.'[13] In 2008 Chuck Prince, then the Chair of Citigroup, said 'You have to keep dancing as long as the music is playing'.

Minsky highlighted a mild form of this type of irrationality in his discussion of 'euphoria' in markets. In an earlier day, such waves of excessive optimism perhaps followed by extreme pessimism might have been explained by 'sunspots'[14] or the paths of Venus or Mars through the heavens. In Minsky's formulation these waves of optimism start with a 'displacement' or shock, which leads to an increase in optimism of investors and business firms and of bank lenders. More confident expectations of a steady stream of prosperity and of an increase in profits induce investors to buy riskier stocks. Banks make riskier loans in this more optimistic environment. The optimism increases and may become self-fulfilling until it evolves into a mania.

The 1970s surge in the price of gold

On 1 January 1970 the market price of gold was less than $40 an ounce, on 31 December 1979 the price was $970. Between 1934 and 1970, the market price of gold had been linked to the US gold parity of $35 an ounce. In the early 1970s the formal link between gold and the US dollar was broken and gold seemingly became 'just another commodity' like petroleum or pork bellies or eggs, freely traded on one of the commodity exchanges. (Obviously gold had a very different history from these other commodities; very few books have been written about the monetary history of pork bellies or of eggs.) The decade of the 1970s was one of accelerating inflation although not in a linear way, the price of gold increased to $200 an ounce in 1973 and then declined to $110 and then surged.

One of the clichés is that 'gold is a good inflation hedge'; for four hundred years the real price of gold or its purchasing power in terms of a market basket of commodities had been more or less 'constant' over the long run although not in the short run. In the 1970s, in contrast, the annual percentage increase in the market price of gold was many times greater than the annual percentage increase in the consumer price level. The prices of petroleum and copper and wheat and other primary commodities were increasing in this inflationary episode, but the price of gold increased much more rapidly.

At some stage in the late 1970s, investors were extrapolating from the increase in the market price of gold from Monday to Tuesday to project the market price on Friday; they purchased gold on Wednesday in anticipation that they could sell at a

higher price on Friday. The 'greater fool theory' was at work, some of these buyers may have realized that the increase in price was a bubble and anticipated that they would be able to sell at a profit before the bubble imploded.

At the end of the 1990s the market price of gold was a bit less than $300 an ounce, and once again the cliché that gold is a good inflation hedge seemed valid; the price of gold had increased by a factor of fifteen since 1900 and the price of a market basket of US goods had increased by about the same amount.

The market price of gold surged after 2008, apparently in expectation that national monies would become virtually worthless as banks failed; some investors wanted a secure store of value. Or maybe some investors believed that central banks would follow policies that would be inflationary as they increased the supply of credit in response to the failure of the banks.

An earlier alternative explanation for this un-sober upswing were provided by Irving Fisher and by Knut Wicksell who emphasized that the real rate of interest was too low.[15] Consumer prices increase in economic expansions and while interest rates increase, they increase less rapidly than the inflation rate so the real rate of interest declines. Lenders have 'money illusion' and ignore the decline in the real rate of interest. In contrast borrowers do not have money illusion; they recognize that the real rate of interest has declined. Rational investors buy more stocks or real estate in this environment of increases in anticipated profits and lower real interest rates. (The Fisher and Wicksell explanations were effective descriptions of the changes in nominal and real interest rates in the 1970s.)

This model relies on the ad hoc assumption that two groups of market participants systematically differ in their susceptibility to money illusion.

Too low an interest rate is a special case of what is perhaps a wider phenomenon – the pricing of financial innovations. Initially these innovations may be under-priced as 'loss leaders' so they will be more readily accepted, but the low price also may lead to excess demand. Alternatively, undue risks may be taken by recent entrants in an industry as they reduce their selling prices to increase their market shares. One notable example is that of Jay Cooke, the last prominent banker of the early 1870s to back a railroad, the Northern Pacific.[16] Other examples include Rogers Caldwell in the municipal bond market of the late 1920s,[17] Bernard K. Marcus of the Bank of the United States in mortgages in the same era,[18] and Michele Sindona of the Franklin National Bank in the early 1970s.[19]

Speculation often develops in two stages. In the first, sober stage households, firms and investors respond to a shock in a limited and rational way; in the second, the anticipations of capital gains play an increasingly dominant role. 'The first taste is for high interest but that taste soon becomes secondary. There is a second appetite for large gains to be made by selling the principal.'[20] In

the 1830s in the United States investors initially bought land to cultivate more high-priced cotton; thereafter they purchased land for the anticipated capital gains they would achieve when they sold the land. In the 1850s farmers and planters both cultivated and speculated in land. In ordinary times they bought more land than they cultivated as a hedge against the declining value of the acres they planted; in booms this more or less sound basis was discarded, and farms were heavily mortgaged to buy *more* land, which in turn was mortgaged so they could buy still more land to profit from anticipated increases in land prices.[21] The 1830s railway boom in Britain also had two stages: the first prior to 1835 when the projects were not bubbles, and a second after 1835 when they were. In the first phase, shares were sold by promoters to local chambers of commerce, Quaker capitalists, and hard-headed Lancashire businessmen, both merchants and industrialists – that is, to men of substance who anticipated benefits from the construction of the railroads. These groups were able to meet both the initial 5 to 10 percent payment and the subsequent calls for payment as the construction progressed. In the second phase, professional company promoters – many of them rogues interested only in quick profits – tempted a different class of investors, including ladies and clergymen.[22] The same stages were evident in building sites in Vienna in the early 1870s; initially these sites were bought for construction but then later were snapped up like speculative poker chips for profitable resale.[23] Ilse Mintz noted a two-stage process in the sale of foreign bonds in New York in the 1920s; these bonds were sound prior to 1924 and the Dawes loan (which touched off the boom) and inferior thereafter.[24] The loans to Mexico and Brazil in the early 1970s were based on the realistic assessments of the credit standing of the borrowers; thereafter the banks wanted to increase their loans and they had less concern with the quality of the projects that were being financed.

In the Florida real estate bubbles between 2003 and 2006 speculators traded titles to condominium apartments in buildings that were still under construction. One wag developed a 'Condo-Flip' website.

Essentially there was a reversal between the objective and the process, and in the end the objective became the process. The lenders became so enthusiastic about the process that they failed to appreciate the end-game and provide an answer to the question, 'Where will the borrowers get the cash to pay the interest if they no longer had access to cash from new loans?' Initially the junk bond market may have been rational, but then the supply of junk bonds surged and the creditworthiness of the borrowers declined more or less continuously. A lot of attention in the US housing debacle that began in 2007 has been given to the defaults on 'subprime mortgages', whose share of the market for new mortgages had increased from 8 percent in 2004 to 20 percent in 2005 and 2006. This sharp increase reflected that there weren't enough prime mortgage loans to satisfy the demands of those who wanted to buy mortgages.

The market in just-built and unfinished houses in southern California, sold from one person to another at ever-higher prices with the help of an active market in second mortgages, peaked in 1981 and then collapsed, with price declines of 40 percent.[25] There was a condominium 'craze' in Boston in 1985 and 1986; 60 percent of the buyers intended to sell the units. The condo market turned soft in 1988,[26] in a pattern similar to the 'flat craze' in Chicago in 1881.[27] A similar boom and dip occurred in the apartment market in Chicago in 2003.

The analysis in terms of two stages suggests two groups of speculators, the insiders and the outsiders. The insiders destabilize by driving the price up and then sell at or near the top to the outsiders. The losses of the outsiders necessarily are equal to the gains of the insiders. Johnson pointed out that for every destabilizing speculator there must be a stabilizing one.[28] But the professional insiders initially destabilize by exaggerating the upswings; these insiders follow the mantra that the 'trend is my friend'. At one stage, these investors were known as 'tape watchers'; more recently they have been called 'momentum investors'. The outsider amateurs who buy high and sell low are the victims of euphoria that affects them late in the day. After they lose, they go back to their normal occupations to save for another splurge in five or ten years.

The evidence from the gold panic of 1869 is consistent with the hypothesis that Gould and Fisk first drove up the gold price and then sold at the top in a manner that is consistent with destabilizing speculation, although Larry Wimmer does not agree that speculation had been destabilizing.[29] The information available to the two groups of speculators differed. In the early stage, Gould tried to persuade the US government of the desirability of forcibly depreciating the US dollar by driving up the 'agio' or premium on gold to increase grain prices, while the outsider speculators operated on the expectation derived from past performance that the US government would seek to drive the agio down so that greenbacks would again be convertible into gold at the pre-Civil War parity. On 16 September the outsiders abandoned this expectation and adopted Gould's; they bought gold and the price went up. On 22 September Gould learned from his associate, President Grant's brother-in-law, that the outsiders originally had been right and that his plan was not going to be adopted; Gould then sold. Belatedly the outsiders saw they were wrong. The result was the Black Friday of 23 September 1869 when stock prices collapsed.

Another case that involves two sets of speculators, insiders and outsiders, is the 'bucket shop'. This term has practically disappeared from the language since the Securities and Exchange Commission declared the practice illegal, but the men and women who run the boiler shops are the children of those who ran bucket shops. Bucket shops are described in novels; a classic picture is given in Christina Stead's excellent *House of All Nations*.[30] The insiders in a

bucket shop take orders from the public to buy and sell securities but do not execute these orders because they assume that the outsider's bet will prove to be wrong. And the bucket shop has the advantage of a hedge. If the outsiders should turn out to be right by 'buying low and selling high', the bucket-shop operators decamp. In *House of All Nations*, Jules Bertillon in 1934 fled to Latvia.

Bucket shops evolved into boiler shops that hustled untutored investors with promises of quick sure-fire gains. The owners of the boiler shops had brought forth their own firms; initially they owned nearly all or all of the shares in these firms. Robert Brennan of First Jersey Securities owned and operated or was associated with a series of boiler shops; the names kept changing but the scam was always the same. Their buddies hustled increases in the prices of stocks of very small, little-known firms; once the stock prices were increasing, they used tele-marketing to sell the stocks to dentists and undertakers in small towns all over America. They managed to increase the prices of the stock day by day until most of the shares in the firms had been sold to the gullible investors who were congratulating themselves on how much money they had made. When one of these investors tried to convert the paper profits into cash, there suddenly were no buyers.

For a further example of an outside destabilizing speculator who bought high and sold low, there is the story of the great Master of the Mint, Isaac Newton, the world-class scientist. In the spring of 1720 he stated: 'I can calculate the motions of the heavenly bodies, but not the madness of people.' On 20 April, accordingly, he sold his shares in the South Sea Company for £7000, a 100 percent profit. Later an infection from the mania gripping the world that spring and summer caused him to buy a larger number of shares near the market top and he lost £20,000. In the irrational habit of so many who experience financial disaster, he put it out of his mind and for the rest of his life he could never bear to hear the name South Sea.[31]

Yet euphoric speculation with insiders and outsiders may also lead to manias and panics when the behavior of every participant seems rational. Consider the fallacy of composition when the whole differs from the sum of its parts. The action of each individual is rational – or would be if many other individuals did not behave in the same way. If an investor is quick enough to get in and out ahead of the others, he may do well, as insiders generally do. Carswell quotes a rational participant on the South Sea Bubble:

> The additional rise above the true capital will only be imaginary; one added to one, by any stretch of vulgar arithmetic will never make three and a half, consequently all fictitious value must be a loss to some person or other first or last. The only way to prevent it to oneself must be to sell out betimes, and so let the Devil take the hindmost.[32]

'Devil take the hindmost', *'sauve qui peut'*, *'die Letzen beissen die Runde'* ('dogs bite the laggards'), and the like are recipes for a panic. The analogy is someone yelling 'fire' in a crowded theater. The chain letter is another analogy; because the chain cannot expand infinitely, only a few investors can sell before the prices start declining. It is rational for an individual to participate in the early stages of the chain and to believe that all others will think they are rational too.

Closely akin to the fallacy of composition is the standard 'cobweb' demonstration in elementary economics in which demand and supply are linked with a lag rather than simultaneously, as in an auction that clears the market at each moment. 'Displacement' consists of events that change the situation, extend the horizon, and alter expectations. In such cases, otherwise rational expectations of some investors fail to recognize the strength of similar responses by others. When there appears to be a shortage of physicists or mathematicians or schoolteachers many young people enter graduate school to prepare for one of these professions; by the time they have finished their studies, there may be an 'excess supply' of individuals trained for careers in this field. After the belated surge in supply, job opportunities suddenly become scarce. But the excess supply becomes known only after the long gestation period. Responses to shortages of coffee, sugar, cotton, or some other commodity may be similarly excessive. The price increases sharply in response to the initial surge in demand and then declines even more rapidly as the new supply becomes available.

The history of manias and panics is full of examples of destabilizing 'cobweb' responses to exogenous shocks. When Brazil opened as a market for British goods in 1808, more Manchester goods were sent to the market in a few weeks than had been consumed there in the previous twenty years, including ice skates and warming pans that, as Clapham noted, proved to be the accepted illustration of commercial madness among nineteenth-century economists.[33] In the 1820s, independence for the Spanish colonies triggered a boom in lending to new Latin American governments, investing in mining shares and exporting to the area; the surge in investment proved excessive. 'The demand is sudden, and as suddenly stops. But too many have acted as if it were likely to continue.'[34]

In the 1830s the cobweb fluctuation had a two-year periodicity. 'Each merchant would be ignorant of the amounts that other merchants would be bringing forward by the time his own merchandise was on the market.'[35] The same was true in the United States in the 1850s following the discovery of gold in California:

The extraordinary and undue expectations entertained not only in the United States but in this country [Britain] as to the capability of California – after the 1849 gold discovery – unquestionably aided in multiplying and extending the disaster consequent on the American crisis. When it was again

and again stated, both in London and in Boston, in regard to all shipments to San Francisco, that six, or at most eight, moderately-sized or assorted cargos per month were all that were required or could be consumed; instead of that eastern shippers dispatch twelve to fifteen first-class ships a month, fully laden.[36]

A rather far-fetched line of reasoning led from the phylloxera that ruined many vineyards and set back wine production in France to the 1880s boom in brewery shares in Britain, as one after another, private breweries sold shares to investors for the first time in the public-companies mania. Among them, Arthur Guinness and Co. was bought for £1.7 million and sold for £3.2 million.[37] 'The success of the issue was like the firing of a starting pistol; by November 1890, 86 other brewery companies had issued new shares to the public for the first time.'[38]

There was a boom in Britain at the end of World War I when businessmen thought victory would ensure the elimination of German competition in coal, steel, shipping, and cotton textiles. Prices of industrial assets, ships, stocks, and even houses increased. Companies were merged; many of the mergers were financed with large amounts of credit. Then sober realization set in from the summer of 1920 to the coal strike of the second quarter of 1921.[39]

Three more cases are on the borderline of rationality. The first deals with target workers, so to speak – individuals who get used to a certain level of income and find it difficult to adjust their spending downward when their incomes decline. In consumption theory, this is the Duesenberry effect that was noted earlier. In labor supply, it constitutes the 'backward-bending supply curve', which suggests that higher wages or salaries produce not more work but less and that the way to increase effort is to lower the wage per unit of time. In economic history books, this principle is known as 'John Bull can stand many things but he cannot stand 2 percent'. John Stuart Mill put it thus:

> Such vicissitudes, beginning with irrational speculation and ending with a commercial crisis, have not hitherto become less frequent or less violent with the growth of capital and the extension of industry ... Rather they may be said to have become more so: in consequence, it is often said, of increased competition; but, as I prefer to say, of a low rate of profit and interest, which makes the capitalists dissatisfied with the ordinary course of safe mercantile gains.[40]

In France at the end of the Restoration and the beginning of the July Monarchy – that is, between 1826 and 1832 – speculation was rife despite the 'distrust that the French always feel toward ill-gotten money'. Landowners earned 2.25 to 3.75 percent on their assets; industrialists tried to do better

than the long-run interest rate on their fixed investments by 2 to 4 percentage points and earn 7 to 9 percent. Merchants and speculators in raw materials sought returns in the range of 20 to 25 percent on their investments.[41] Charles Wilson noted that earlier the Dutch were converted from merchants into bankers (accused of idleness and greed); they developed habits of speculation because of the decline in the rate of interest in Amsterdam to 2.5 and 3 percent.[42] Large-scale conversions of public debt in 1822 and 1824 and again in 1888 led to a decline in the rate of interest and induced British investors to buy more foreign securities.[43] Andréadès observed that 'When interest goes down, the English commercial world, unable to reduce its mode of life, deserts its usual business in favour of the more profitable, but on that very account more risky undertakings... speculation leads to disaster and ultimately must be borne by the central bank.'[44]

The boom in Third World bank-syndicated loans in the 1970s followed a sharp decline in interest rates on US dollar securities in the spring of 1970 as the Federal Reserve adopted a more expansive policy. Banks were highly liquid and looked for attractive borrowers which they found in Third World governments and government-owned firms, mostly in Latin America. The 1960s had been a decade of accelerating internationalization for the major US banks and they had rapidly increased the number of their foreign branches. Because of the sharp increase in commodity prices, nominal and real incomes in Mexico, Brazil, and most other developing countries were expanding at above-trend rates. Commodity prices declined sharply in the early 1980s in response to the surge in US interest rates, and then nominal and real incomes declined in these indebted countries. Should the banks have foreseen that the decline in commodity prices was inevitable?

The second borderline case involves hanging on in the hope of some improvement or failing to take a specific type of action when circumstances change. On the first score, note the failures of the New York Warehouse and Security Company, of Kenyon, Cox & Co., and of Jay Cooke and Co. on 8, 13, and 18 September 1873, because of loans made to railroads (respectively, the Missouri, Kansas and Texas, the Canada Southern, and the Northern Pacific) with which they were associated. These railroads were unable to sell bonds to obtain the funds they needed to complete construction because Berlin and Vienna had stopped lending to the United States.[45] Similarly, when US long-term lending to Germany stopped in 1928, as US investors turned to stocks and stopped buying bonds, New York banks and investment houses continued to make short-term loans to German borrowers. When riding a tiger or holding a bear by the tail, it seems rational to hang on – at least for a while.

For an error of omission, note the plight of Hamburg banks that had made large loans to Swedish banks that were financing smuggled goods into Russia during the Crimean War; the Hamburg banks failed to cancel these loans when

peace came. The Swedes used the money to speculate in shipbuilding, factories, and mining, which helped embroil Hamburg in the world crisis of 1857.[46]

The third borderline case is to have a rational model in mind, but the wrong one. The most famous example in another field is the French 'Maginot Line psychology', though this may be thought of less as a case of irrational expectations than one of an undistributed lag. ' "When a man's vision is fixed on one thing," thought Ponzi, "he might as well be blind".'[47] Or Bagehot on Malthus: 'Scarcely any man who has evolved a striking and original conception ever gets rid of it.'[48] In the 1760s, Hamburg merchants were not hurt by the fall in commodity prices until the end of the Seven Years' War. Thus in 1799 while the Napoleonic Wars were continuing they were unprepared for the decline in prices that came with the penetration of the blockade of Napoleon's 1798 Continental system.[49] Or take the French bankers and industrialists who formed the copper ring in 1888, patterned after the cartel movement in iron and steel, steel rails, coal, and sugar in the early part of the decade, attracted by the successes of the diamond syndicate in South Africa and of the Rothschilds' mercury monopoly in Spain. (Many economists and analysts extrapolated from the apparent success of the Organization of Petroleum Exporting Countries in increasing the price of petroleum in the 1970s to assume that successful price-fixing cartels would reduce the output of practically every other raw material and foodstuff and lead to much higher prices for these products.) By 1890 the French syndicate owned 60,000 tons of high-priced copper plus contracts to buy more; the older mines were reworked and firms began to process scrap while the copper price was declining rapidly. The collapse of the copper price from £80 to £38 a ton in 1889 almost took with it the Comptoir d'Escompte, which was saved by an advance of 140 million francs from the Bank of France, reluctantly guaranteed by the Paris banks.[50]

Financial innovation in the form of deregulation or liberalization has often been a shock. In the early 1970s Ronald McKinnon led an intellectual attack on 'financial repression', that is, the segmentation of financial markets in developing countries that led to preferential treatment of government borrowers that were involved in foreign trade, and large firms.[51] The message appealed particularly to Latin American countries already influenced by the Chicago doctrine of liberalism. Some countries deregulated their financial systems, which was followed by a rapid growth of new banks and rapid increases in credit and in inflation; subsequently some of the new banks collapsed.[52] McKinnon felt that the lesson was that the several steps in the process of deregulation should be staged carefully.[53]

The same questions surfaced again in Poland and in the former Soviet Union in the early 1990s in fierce debates over whether the shift from command to market economies should be carried through rapidly or slowly. The success of a transition from a command economy seems to depend on the extent to which

individuals in the socialist economy remember the institutional background of its early capitalism. The memory of the market economy was far greater in Poland than in Russia; long years of socialism and corruption had eradicated the memory of market institutions in Russia. Such memory is more important to transitional success than the speed of decontrols and of the privatization of state monopolies.

Charlie Ponzi was alive and well and living in Tirana

The transition from the command economies to the market economies in what had been Eastern Europe in the early 1990s meant that the financial structures were no longer regulated. Entrepreneurs – some of them former members of the army in Albania – started firms that promised high rates of return, often 30 percent a month. The public in these countries had accumulated lots of currency and lots of deposits in the state-owned banks; the interest rates on these deposits were extremely low. The public was attracted to the high rates of returns promised by these newly established financial institutions. Competition among the several different 'banks' kept the promised interest rates high.

Some Albanians sold their homes to get the cash to buy these bank deposits and then rented the same properties from the buyers; the 'interest income' on their deposits was much higher than the rent they had to pay for the same homes. Often the buyers of the apartments were the same entrepreneurs who owned and managed the deposit banks. Albanians in its diaspora sent money from New York and Chicago and Frankfurt to their relatives in Tirana to be deposited in these new institutions. Some Albanians stopped working because the interest income on their deposits was so much higher than their wages and salaries.

Alas it was too good to be true and it wasn't.

One purely irrational case involves a society that pins its hopes on some outstanding event of limited relevance to its current economic circumstances and another is when a society ignores evidence that it would prefer not to think about. Many Austrian enterprises had invested extensively in anticipation of the increase in business activity that would follow from the opening of World Exhibition in Vienna on 1 May 1873; their liquid liabilities greatly exceeded their liquid assets and they had acute financial distress. The objective of these world's fairs is to increase business activity, so there is significant investment in facilities designed to accommodate those who will attend the fairs. The credit at banks was stretched to the limit; a move from commodities, land, shares, and debt back into money was under way and the chain of accommodation bills was extended as far as it would go. Nonetheless the banks and the firms hung on, waiting for the exhibition to open, because they thought or at least hoped that the increase in sales would save the situation. When the exhibition opened and the increase in sales was disappointing, the market collapsed in early May.[54]

As an illustration of repression of contradictory evidence – the cognitive dissonance case – consider J.W. Beyen's analysis of the German failure to restrict short-term borrowing from abroad at the end of the 1920s. He suggested that the dangers were not faced, even by Schacht, the German finance minister, and added: 'It would not have been the first nor the last time that consciousness was being "repressed".'[55] These examples suggest that despite the general usefulness of the assumption of rationality, markets have on occasions – infrequent occasions – acted in ways that were irrational even when each participant in the market believed he or she was acting rationally.

Displacements

A displacement is an outside event or shock that changes horizons, expectations, anticipated profit opportunities, behavior – 'some sudden advice many times unexpected'.[56] A surge in the oil price is a displacement. An unanticipated devaluation is another – although most devaluations have been anticipated. A change in financial regulations, especially the liberalizations on restrictions to loans to particular groups of borrowers is a displacement. The shock must be sufficiently large to have an impact on the economic outlook. Each day's events produce some changes in outlook, but few are significant enough to qualify as a shock.

War is a major shock. Some crises occur immediately at the beginning or end of a war, or soon enough after the end to permit a few expectations to be falsified. For beginnings, the most notable is the crisis of August 1914. The displacements at the end of wars include the crises of 1713, 1763, 1783, 1816, 1857, 1864, 1873, and 1920. Moreover there have been an impressive series of crises seven to ten years after the end of a war, long enough for expectations formed at the end of the original crisis to be falsified; these included 1720, 1772, 1792, 1825, 1873 in the United States, and 1929.

Far-reaching political changes may also change expectations. The Glorious Revolution of 1688 gave rise to a boom in company promotion. By 1695 there were 140 joint stock companies with a total capital of £4.5 million; more than 80 percent had been formed in the previous seven years. By 1717 total capitalization had reached £21 million.[57] In July 1720 the Bubble Act forbade formation of new joint-stock companies without explicit approval of parliament, a limitation that lasted until 1856. Although this regulation has normally been interpreted as a reaction against the South Sea Company speculation, Carswell asserts that it was undertaken to support the South Sea Company, as king and parliament sought to repress the development of rival companies that might attract cash that was intensely needed by the South Sea promoters.[58]

The events of the French Revolution, Terror, Directorate, Consulate, and Empire, along with incidents of the Napoleonic Wars themselves, set in motion

large-scale specie movements in 1792–93 and 1797 and opening and closing markets in Europe and elsewhere for British and colonial goods. Further political events of the kind in France were the Restoration (1815), the July Monarchy (1830), the February 1848 revolution, and the Second Empire (1852). The Sepoy Mutiny in India in May 1857, followed by a Hindustan military revolution, contributed to the distress of London financial markets.[59] These events were a precedent for the Invergordon disorder of September 1931, when a contingent of British sailors came close to striking over reductions in pay decreed by the new national government. Continental Europeans interpreted this response as a mutiny by a great British institution, the navy, and this interpretation contributed to the British decision to stop pegging the pound to gold.[60]

War, revolution, restoration, change of regime, and mutiny come largely from outside the system; they are beyond the usual set of risky events that can be modeled as a probability distribution – hence they qualify as a 'black swan'. Monetary and financial displacements are more difficult to describe as exogenous. But maladroit recoinage, tampering with gold-silver ratios under bimetallism, conversions undertaken to economize on government revenue that unexpectedly divert investor attention to other avenues, new lending that proves successful beyond all anticipation – these also are displacements.

The *Kipper- und- Wipperzeit* of 1619–23 (noted earlier) got its name from the action of money changers who took the debased coins that were coming from the increase in the number of princely mints and rigged their scales as they sought to exchange bad money for good with naive peasants, shopkeepers, and craftsmen. Rapidly rising debasement spread from state to state until the coins used in daily transactions became worthless.[61]

Two later German recoinages provide a study in contrast. In 1763, Frederick II of Prussia bought silver in Amsterdam on credit to provide for a new coinage to replace that which had been debased during the Seven Years' War. He withdrew the old debased money from circulation before the new money was issued, which precipitated a deflationary crisis and the collapse of a chain of discounted bills.[62] More than 100 years later, after the Franco-Prussian indemnity, the German authorities issued new money but this time before the old money was withdrawn to save on their interest payments. In three years the circulation of coins rose threefold from 254 million thalers (762 million marks). The result was inflation.[63]

The crisis of 1893 in the United States, which arose from the threat to gold convertibility from the Sherman Silver Act of 1890, has already been noted. So have the British debt conversions of 1822, 1824, 1888, and 1932, although the last was associated with a boom in housing that did not lead to a crisis. In France, conversion of the 5 percent *rente* was discussed after 1823 as the money supply expanded and the rate of interest would have fallen had investors not been reluctant to buy *rentes* at a premium. Each of three bankers had a

different idea of the purpose of the conversion: Rothschild wanted to sell more *rentes*; Greffuhle (and Ouvrard) hoped to attract investors into canals while Laffitte wanted to ensure the development of industry. In the event, political obstacles prevented passage of the necessary legislation, and the market finally gave up its objection to maintaining the *rente* at a premium. This sharp decline in interest rates touched off speculation.[64] Canals were built by the government with private money[65] and the faint glow of a railroad boom could be seen in France along the Loire, the Rhone, and the Seine. But the main object of speculation was building in and around the major cities – Toulouse, Lyons, Marseilles, Le Havre.[66] Honoré de Balzac's novel *César Birotteau*, written in 1830, was inspired by this experience; he recounted the doleful story of a perfumer who was enticed into buying building lots in the vicinity of the Madeleine on borrowed money for 'one quarter of the value they were sure to have in three years'.[67]

The success of loans in recycling reparations or indemnities after the Napoleonic and Franco-Prussian wars and World War I has been mentioned. Any surprising success of a security issue, with a large multiple over-subscription and a quick premium for subscribers, attracts borrowers, lenders, and especially investment bankers. The Baring loan of 1819 – 'the first important foreign loan contracted by a British bank'[68] – led quickly to a series of issues for France, Prussia, Austria, and, later, after independence, the countries that had been Spanish colonies. The Thiers *rente* made French banking houses salivate in the hope of foreign loans, a hunger that received a further fillip from the 1888 conversion loan for czarist Russia that bailed out German investors and sent French investors down a trail that was to end, after revolution in 1917, with a whimper rather than a bang. The Dawes loan in 1924 opened the eyes of American investors to the romance of buying foreign securities. The Thiers *rente* was oversubscribed fourteen times, and the Dawes loan eleven times; far more important than the size of the multiple, however, was its relation to expectations. Rosenberg described the three French loans of 1854 and 1855 as sensational, since they were oversubscribed almost two to one (468 million francs on an offering of 250 million), four to one (2175 million francs for an issue of 500 million), and five to one (3653 million against 750 million). In Austria and Germany, however, when the speculative boom of the 1850s was under way, the Credit Anstalt opening stock sale was oversubscribed 43 times, largely by people who had stood in line all night; and when the Brunswick Bank sought 2 million thalers in May 1853, it was offered 112 times that amount in three hours.[69]

Among major recent displacements, as noted earlier, have been deregulation of banks and financial institutions; such innovations as derivatives (which existed earlier but only on a much modest scale); mutual and hedge funds, offering new opportunities to acquire wealth, with, however, the risk of

loss; REITs (Real Estate Investment Trusts); bank flotation of loans and mortgages as marketable securities; and initial public offerings (IPOs) of private companies.

The deregulation of financial institutions was a major contributory factor to the asset price bubble in Japan in the 1980s and especially during the second half of that decade. Each Japanese bank was keenly interested in its position on the hit parade in terms of assets and deposits; each wanted to move to a higher position on the hit parade ladder – which meant that each had to 'grow its loans' more rapidly than the banks that had higher positions.

The technological revolution in the 1920s – the sharp increase in automobile production, the electrification of much of America, the rapid expansion of the telephone system, the increase in the number of movie theaters, and the beginning of radio – was a major shock. Investment surged. Similarly, in the 1990s, especially in the second half of the decade, there was a major technological information revolution. The venture capital firms, especially those in the San Francisco Bay area, were eager suppliers of finance to many of the engineers who had ideas. Then, at a later stage, these firms received 'mezzanine financing'. The next stage was that the firms had an initial public offering (IPO) arranged by one of the major investment banks such as Merrill Lynch, Morgan Stanley or Credit Suisse First Boston. The investment banks would arrange 'road shows' for these firms as they were about to go public; entrepreneurs would visit the mutual funds and the pension funds and the managers of other pools of cash. Based on the demand, the investment banks would price the shares at $19 or $23 or $31 and perhaps 20 percent of the firms' outstanding shares would be sold. Often the price of the shares at the end of the first day's trading would be three or four times the IPO price.

The 'pop' in the share price on the first day's trading was an advertisement that stock prices only increase. During the late 1990s an extremely high proportion of new stock issues experienced these large price pops on the first day of trading. The price pops encouraged lots of new stock offerings.

'Dow at 36,000', 'Dow at 40,000', 'Dow at 100,000'*

Three books with nearly identical titles were published in 1999. Their themes were almost identical – if interest rates remained low and corporate earnings continued to increase, then eventually the Dow Jones index of stock prices would reach much higher levels than ever before. The logic was irrefutable, more or less an extension of the Archimedes principle that he could move the world if he had a large enough lever. In the long run the level of stock prices reflects three factors: the rate of growth of GDP, the profit share of GDP and the relation of stock prices to corporate earnings or the price-earnings ratio. The profit share of US GDP has been remarkably constant in the long run at about 8 percent and the price-earnings ratio has averaged about 17 per cent.

Investors continually choose between buying bonds and buying stocks. The interest rate on bonds has averaged about 5 percent; the earnings yield on bonds, the reciprocal of the interest rate, is thus 20.

Those who forecast the Dow at 36,000 believed that the price-earnings ratio should be much higher because stocks were no more risky than bonds.

* James K. Glassman and Kevin A. Hassett, *Dow 36,000: the New Strategy for Profiting from the Coming Rise in the Stock Market* (Random House, 1999); David Elias, *Dow 40,000: Strategies for Profiting from the Greatest Bull Market in History* (McGraw-Hill, 1999); Charles W. Kadlec, *Dow 100,000: Fact or Fiction* (Prentice Hall, 1999).

Objects of speculation

In the last several decades of the twentieth century investors speculated primarily in real estate or stocks; in earlier periods the objects of speculation were more diverse. A stylized table of cycles is presented in the Appendix. The list shows the tendency for these objects to move from a few favored items at the beginning of the period of observation to a wide variety of commodities and other assets and securities at the end. The list is partial but suggestive.

How likely is it that a displacement will lead to a shock that induces individuals to invest for capital gains and especially short-term gains? Assume that destabilizing speculation can occur in a world of normally rational individuals. Then assume a shock; these individuals misjudge its impact. Then assume a shock; these individuals misjudge the impacts of the shocks. There are many shocks: only a relatively small proportion of shocks lead to a speculative mania.

One question is whether two or more objects of speculation such as real estate and stock are likely to be involved before 'overtrading' reaches sufficient dimensions to result in a crisis. Consider several occasions when two or more objects have been involved in speculation.

The 1720 South Sea and Mississippi bubbles were related, and fueled by monetary expansion in Britain and France that supported a high head of speculative steam. Speculation that started in the securities of the South Sea Company and the Sword Blade Bank in England and in those of the Mississippi Company and John Law's banques in France spread rapidly to other ventures and to commodities and land; many of these other ventures were swindles. The South Sea Company was brought down by its attempt to suppress rival speculations, bringing proceedings under the Bubble Act of June 1720 against York Buildings, Lustrings, and Welsh Copper. The effort boomeranged.[70] The spread of speculation from one object to another, to generalize the rise of prices, occurred because the speculators that sold South Sea stock when prices were approaching their peak purchased banks and insurance stocks and country houses.[71] Then the price of land began to move with the price of the South Sea

stock quotations.[72] In France land prices rose in the fall of 1719 and speculators started to take their profits from the Mississippi Bubble.[73]

The 1763 boom was based exclusively on government war expenditure and its finance through chains of discount bills. The DeNeufville Brothers, whose failure set off the panic, sold 'commodities, ships, and securities like so many Dutch firms',[74] with hundreds of thousands of florins in acceptance liabilities against which they rarely kept more than a few thousand guilders in cash reserves.

Some contribution to the downturn in business may have been brought on by an unparalleled drought in England in 1762, with a shortage of hay and scarcities of meat, butter, and cheese.[75] The crisis of 1772 was precipitated by speculation in Amsterdam and London in the stock of the East India Company and by the collapse of the Ayr Bank (Douglas, Heron & Co.). Numerous complex details were involved, including the political reverses of the East India Company and restriction on its credit by the Bank of England; the practice of the thrusting new Ayr Bank (which was left bad loans by the established banks) in borrowing from London when its acceptances came due; and the flight in July 1772 of Alexander Fordyce, who had lost his firm's money selling East India Company stock prematurely. When the stock fell in the autumn, Clifford & Co., the Dutch bank that had headed a syndicate trying to push the price up, failed. These phenomena seem superficial, however. Extensive investment in Britain in houses, turnpikes, canals, and other public works had put a strain on resources.[76] One source relates the fall in coffee prices beginning in 1770 to the financial crisis of 1772–73,[77] but this is not mentioned by Wilson, the standard source, or by Ashton, Clapham, or Buist.[78]

In 1793 there were several causes – country banks, canals, the Reign of Terror – that stimulated a flow of money to Britain, as well as bad harvests. In 1799 there was one cause, the tightening and loosening of the blockade. Contrariwise, the crisis of 1809–10 had 'two separate causes: a reaction from the speculation in South America; and a loosening and then tightening of the continental blockade'.[79] There was a postwar boom in exports to Europe and the United States in 1815–16 that was larger then the amount that could be sold, plus a fall in the price of wheat. Canals and South American government bonds and mines combined in 1825; British exports, cotton, land sales in the United States, and the beginning of the railroad mania contributed to the crisis in the mid-1830s. The crisis of 1847 was caused by the railway mania, the potato blight, a wheat crop failure one year and a bumper crop the next, followed by revolution in Europe.

Thus at least two objects of speculation were involved in most of the significant crises. Just as the national markets were connected, so the speculation was connected by the underlying credit conditions. But when a crisis like that of 1847 arises from objects as disparate as railroads and wheat, there is some basis

for suggesting that the crisis is accidental in origin unless the monetary weakness that feeds it is systematic.

In Japan and in the Asian countries, bubbles in real estate and stocks have generally occurred together. In some countries, especially small ones, the market value of real estate companies is a high proportion of the market value of all stocks as a group. When real estate prices increase, the value of the assets owned by real estate companies increases, and the market value of the real estate companies also increases. Those investors who have sold the real estate stocks have cash to invest and much of their cash is likely to be invested in stocks of firms not involved in the real estate business. Moreover when the real estate prices increase, then the construction business is likely to boom and the market value of the construction companies is likely to increase. The bank loan losses are likely to be below trend when real estate prices are increasing. And the symbiotic relationship is symmetric; when real estate prices decline, stock prices also decline.

National difference in speculative temperament

One suggestion is that investors in some countries are more likely to speculate than those in others. Despite Ruth Benedict's distinction between cultures with Apollonian (balancing) and those with Dionysian (orgiastic) temperaments,[80] the proposition is dubious. And despite this implausibility, the opinion among historians is that the Brabanters had a strong gambling temperament in the sixteenth century, and that those tens of thousands who migrated to the United Provinces after the sack of Antwerp in November 1576 and its devastating siege in 1585 took it with them.[81] In the Dutch Republic, the gambling instinct of bankers, investors, and even common folk existed in great tension with Calvin and Lutheran frugality and abstemiousness.[82] There may be substance to the view that banking institutions facilitate more speculation in some countries than in others. Juglar, for example, claims the French crises in the eighteenth century were less abrupt and less violent than those of Britain because (after the John Law affair) credit in France was less used and less abused.[83] A different view ascribes French experience to a more severe bankruptcy law:

> Whether by the education forces of law and established institutions, or by tradition, a high standard of business honesty prevails in France. The act of sons in toiling for years to pay the debts of their fathers, and of notaries in paying for the defalcations of one of their number, for the sake of the profession, although without personal association with him, indicates a standard of compliance with business obligations which cannot be without influence upon the material prosperity of a people. It may be surprising that the nation whose soldiers are so noted for dash in war should furnish

financiers and business men who are the embodiment of conservatism in their methods, but such is clearly the case.

This author continues, 'England is the country in which a spirit of adventure and speculation has done most to promote crises and depressions.'[84]

One historian has suggested that mining and sheep grazing contributed to a love of gambling, and that Australians, starting with the gold discoveries of 1851–52, developed a particular love of gambling, expressed both through horse racing and speculation in land.[85]

A common view is that the United States is 'the classic home of commercial and financial panics', presumably because of wildcat banking.[86] This was observed in the 1830s by Michel Chevalier who contrasted French moderation with American speculation (but who believed, however, that the latter was a stimulus to the production of canals, railroads, roads, factories, and villages).[87] Letter 25 of his letters from America to France is devoted to a discussion of speculation: 'All the world speculates and it speculates on everything. From Maine to the Red River (in Arkansas) the United States has become an immense Rue Quincampoix [the Wall Street of the Mississippi Bubble].'[88] Partly the origins lie in permissive institutions. But it is easy to find abundant and contradictory views on the demand side for other countries. 'The French nation is prudent and economical, the English nation is enterprising and speculative.'[89] 'France has not shown proofs of prudence equal to those of Scotland; its nerves are extremely susceptible, impressionable in matters of credit.'[90] 'The character of this nation [Britain] is in carrying everything to excess…virtue, vice.'[91] After 1866, a new arrogance was said to have taken hold of the Germans, but they surpassed the French only in 'stock-market swindling and speculation horrors'.[92] Morgenstern found ten panics in France, two more than in the United States, which is 'not surprising, given the unstable character of French politics'.[93] (To be sure, this addresses displacements rather than love of speculation.) Contrast, however, the opinion of a French financier who claims that 'the French love money not for the possibilities of action which it opens, but for the income it assures'.[94] Or consider two views, at the level of a Harvard–Yale debate, from a fictional Frenchman and an Englishman in 1931:

> WILLIAM BERTILLION: England's such a Christmas tree for sharepushers. Noble lords will sit on the board of any company for a couple of quid a sitting. And the public. Loco or idiotic. God, I've never heard of such people, except perhaps some peasants in Bessarabia, or the niggers in the Cameroons, who believe in what they believe in. Magic. Put up any sort of business that sounds utterly impossible and they gulp it down.[95]

STEWART: England's the world's banker. Never failed yet, never failed yet. She keeps her word, that's why...None of this – none of this speculation you get in the American stock market. Every Tom, Dick and Harry trying to make a pile – like in France.[96]

It's a stand-off. The speculative temperament may differ among countries. The amount of speculation in a country may change as the national mood swings between elation and depression.

4
Fueling the Flames: the Expansion of Credit

Axiom number one: You can't have inflation without the rapid growth of money.

Axiom number two: You can't have a real estate bubble without the rapid growth of credit.

Speculative manias gather speed through expansion of credit. Most increases in the supply of credit do not lead to a mania – but nearly every mania has been associated with rapid growth in the supply of credit to a particular group of borrowers. In the last hundred years, the increases in supplies of credit have been provided in part by the banks and in part by the development of new financial instruments. The mania in tulip bulbs in the seventeenth century occurred because the sellers of the bulbs provided credit to the buyers.[1] John Law had his Banque Générale, later the Banque Royale, as his source of credit while the South Sea Company relied on the Sword Blade Bank. In 1763 credit expansion in Holland was financed by the *Wisselruiti*, or chains of accommodation bills (think post-dated checks) from one merchant to another. The British canal mania of 1793 was fed by spending facilitated by loans from many newly established country banks to the entrepreneurs who were developing the canals.

In many cases the expansion of credit resulted from the development of substitutes for more traditional monies. In the United States in the first part of the nineteenth century, the innovation was that bills of exchange (again think post-dated checks) replaced silver in the triangular trade among the United States, China, and Britain. The United States had a bilateral trade deficit with China and China had a bilateral trade deficit with Britain. Previously the United States bought silver from Mexico which was then shipped to China to finance the US trade deficit; then the silver was shipped to Britain to finance China's trade deficit. The institutional innovation was that American merchants sent

bills of exchange that specified payment in the British pound for China's imports; the Chinese then shipped these bills to Britain to finance their trade deficit. The transactions costs involved in making cross-border payments using bills of exchange were much smaller than those that involved the shipment of silver. The result of this innovation was that the silver stayed in the United States so that the US money supply did not decline.[2]

The global boom of the 1850s followed from the combination of new gold discoveries, the formation of new banks in Britain, France, Germany, and the United States, the establishment of clearing-houses by the banks in New York and in Philadelphia and the expansion of the London clearing-house. The growth in bank clearing-houses led to the increased use of credit in the transactions among the banks which were its members; payments imbalances among them were settled by transfer of clearing-house certificates – a new form of 'money'. The expansion of credit in Britain in 1866 resulted from the increase in loans by the newly formed joint-stock discount houses. The boom in Central Europe in the 1870s was based on the gold reparations payments from France to Prussia and on the creation of *Maklerbanken* (brokers' banks) in Germany, which spread into Austria, and *Baubanken* (construction banks) in Austria, which expanded into Germany.

One of the many different institutional avenues for the expansion of credit that occurred in France in 1882 was based on a system of fortnightly clearing of stock exchange transactions that provided credit to speculators through a system of delayed payments called *reportage*. The buyers of stocks had up to fourteen days before they had to pay for their purchases; in effect they got interest-free loans until the date of payment (although the value of the loans may have been reflected in the prices paid for the stocks).[3] Similarly the expansion of credit in the call-money market in New York helped finance the stock market boom in the late 1920s. The catalyst for the expansion of credit in the United States in 1893 was the addition of silver coins to the US money supply; in 1907, the increase in the supply of credit resulted from the increase in the loans by the trust companies. In the years before and after World War I, the development of the gold exchange standard led to an expansion of the international credit base so that a larger volume of international trade could be financed with the existing stock of monetary gold. The rapid increase in installment credit in the United States in the 1920s contributed to the surge in automobile ownership (the dramatic increases in the number of automobiles and other consumer durables led to rapid growth in demand for staggered payments arrangements).

After World War II, the development of negotiable certificates of deposit (CDs) contributed to the expansion of credit. In the 1870s the Austrian banks had developed a new financial instrument (the *Cassenscheine*) that paid interest; the expansion in the demand for these instruments led to an increase in

the supply of credit and hence to an increase in spending. In the 1950s and the 1960s the large US banks adopted the practice of liability management which meant that the growth of their loans no longer depended on the increases in their deposits; instead they funded their loans by borrowing in the inter-bank money market. Liability management enabled the banks to be much more aggressive in managing the growth of credit.

One unique form of the expansion of 'bank credit' occurred in Kuwait between 1977 and 1982 when shares and real estate were bought and sold on the Kuwaiti Souk al-Manakh (stock exchange) with post-dated checks. Eventually the value of these checks in circulation increased to billions of dinars – nearly US$100 billion. The values of post-dated checks written by the buyers of the shares and real estate were much, much larger than their bank deposits. The sellers of the shares and real estate increased their spending as their wealth increased; they hoped that there would be money in the bank accounts of the payors when the due dates of the checks that they had received for the sale of shares and real estate arrived. In July 1982 the checks bounced when some sellers of stocks tried to collect on checks on the due dates.[4]

These examples suggest that the expansion of credit is a systematic development that has continued for several hundred years as the participants in financial markets seek to reduce the costs both of transactions and of holding liquidity and money balances. The form each event takes may seem accidental – the substitution of bills of exchange for silver in payments to China, or the development of deposits in the Eurocurrency market because ceilings prevented US banks from increasing the interest rates that they could pay on deposits in New York and other US cities. The development of new credit instruments occurs episodically in response to changes in institutional arrangements. Monetary expansion is systematic and endogenous rather than random and exogenous.

During economic booms the money supply defined as means of payment has increased and the existing stock of money has been used more efficiently to finance both increases in economic activity and the purchases of real estate and securities. The efforts of central bankers to limit and control the growth of money have been offset in part by the development of new and very close substitutes for money. Such efforts have a long history, including the resumption of specie payments and the return to convertibility of national currencies into gold after the end of wars. The demonetization of the lesser metallic monies – initially copper was displaced by silver and subsequently silver's monetary role was eclipsed by gold – was an effort to obtain greater control over the money supply. Central banks wanted a monopoly on the issue of currency notes, which restricted and then eliminated the rights of private, country, and joint-stock (corporate) banks to issue currency notes. Legislation and custom limited the amounts of deposit money that could be

issued against primary bank reserves, starting shortly after the Bank Act of 1844 and continuing through the application by the Federal Reserve System of reserve requirements on both demand and time deposits (as embodied in the Federal Reserve Act of 1913) and then on certificates of deposit and subsequently on the borrowings by US banks from their branches in offshore financial centers including London. The process is Sisyphean, a *perpetuum mobile*; the history of money is a story of continuing innovations so that the existing supply of money can be used more efficiently and the development of close substitutes for money that circumvent the regulatory requirements applied to the creation of money.

The Eurocurrency deposit market surged in the 1960s as an end-run around the regulations imposed on US banks by the Federal Reserve and the Federal Deposit Insurance Corporation; the US dollar deposits produced by the branches of US banks in London and other offshore centers were not subject to interest rate ceilings, reserve requirements, and deposit insurance premiums. The US stock brokerage firms developed money market funds that were a close substitute for bank deposits in the 1970s and paid interest on these deposits (the deposits were not guaranteed by any agency of the US government – at least not until the financial crisis in the autumn of 2008).

Currency school vs banking school

One feature of the history of monetary theory is a continuing debate between two different views – the Currency School and the Banking School – about how to manage the growth of the money supply. The proponents of the Currency School advocated a firm limit on the expansion of money to avoid inflation. The adherents of the Banking School believed that increases in the supply of money would not lead to inflation as long as they corresponded with increases in business transactions. The same distinction in ideology and economic analysis separated the hard money school from the populists in the 1890s – the former were concerned with inflation while the latter believed prices would not change as long as economic activity increased. The debate between these two views of managing the growth in the money supply has continued for at least three hundred years.

The Currency School wanted a simple rule that would fix the growth rate of the money supply at 2 or 4 or 5 percent, much like today's monetarists.[5] Viner's discussion of the nineteenth-century controversy is succinct:

> The currency school tended also to minimize or to deny the importance of bank credit in other forms than notes as a factor affecting prices, or as in the case of Torrens, to claim that the fluctuations in the deposits were governed closely by the fluctuations in the note issues. They had a hankering

also for a simple automatic rule, and could find none suitable for governing the general credit operations of the Bank. They also had laissez faire objections to extending legislative control of the banking system any further than seemed absolutely necessary.[6]

Neither the Currency School nor the Banking School paid much attention to the expansion of non-bank credit. The Bank of Amsterdam, founded in 1609, issued notes against deposits of precious metals; in effect these notes were warehouse receipts and the amounts of these notes were tied one-for-one to the deposits of the metal. Initially the Bank of Amsterdam did not expand credit; subsequently in the eighteenth century the bank increased its loans in the effort to rescue the Dutch East India Company during the fourth Anglo-Dutch War. The Bank of Amsterdam was also a *Wisselbank* where bills of exchange (*Wissel* in Dutch, *Wechsel* in German) were paid. Merchants kept deposits at the Bank of Amsterdam to meet bills presented for collection. Deposits of precious metals enabled the Bank of Amsterdam to earn seignorage – a type of profits – when it produced coins and it paid a low interest rate on deposits. In 1614 the Municipality of Amsterdam established a Bank of Lending (*Huys van Leening*), that enabled merchants to establish their own credit efficiently but it was not an active lender.[7] This credit created by the merchants led to an excessive expansion of the *Wisselruiti*; when the chain of bills of exchange broke in 1763 because one of the merchants did not have the money to pay on a maturing bill, the DeNeufville bank failed.

The Swedish Riksbank, established in 1668, had two departments, a Bank of Exchange patterned after the Bank of Amsterdam (*Wisselbank*) and a Bank of Lending (*Liinebank*).[8] These two departments foreshadowed the Bank Act of 1844 in Britain, which was a compromise between the two schools; an Issue Department would provide bank notes against deposits of coin or bullion above a specified fiduciary issue that represented the Bank of England's holdings of British government debt and a Banking Department that would make loans and discounts up to a multiple of its reserves of bank notes that had been produced by the Issue Department. The establishment of the Issue Department was a victory for the Currency School, which had criticized the loans made by Bank of England after the suspension of the gold standard in 1797. (The defense of the Bank of England was that these loans did not lead to an increase in the inflation rate because they financed trade.) The establishment of the Banking Department was a victory for those who believed that an expansion of credit would help finance the initial upswing in the early stages of an economic recovery.

The view of the Currency School that the expansion of credit would eventually lead to inflation was correct. The Banking School's view that an increase in the supply of credit was needed at the start of an economic expansion was also

correct. The Currency School's view that the discounts be limited to accept-ances that were related to actual commercial transactions became known as the 'real bills doctrine'. The larger the number of business opportunities, the greater the scope for discounting, and the larger the increase in the money supply and eventually the higher the inflation rate. The central policy ques-tions, once an expansion of credit has started, are whether it is practicable to decree a stopping place and whether this limit could be determined by an automatic rule.

The core issue is that it is easier to define money than it is to measure the effective money supply. Walter Bagehot wrote 'Men of business in England do not...like the currency question. They are perplexed to define accurately what money is: *how* to count they know, but *what* to count they do not know.'[9]

The stylized fact is that every time the monetary authorities stabilize or control some quantity of money, M, either in absolute volume or at a predeter-mined rate of growth, more of the money and the near-money substitutes will be produced in periods of euphoria. If the definition of money is fixed in terms of designated liquid securities the euphoria may lead to the 'monetization' of credit in ways that are beyond the definition; the velocity of money (velocity is defined either as total spending or national income divided by the money sup-ply) will increase even if the amount of money defined in the traditional way remains unchanged. The debate was over whether money should be defined as M_1, currency plus demand deposits; M_2, equal to M_1 plus time deposits; M_3, consisting of M_2 plus highly liquid government securities; or some other designation.

The process seemed endless; fix any M_i and in economic booms the market will create new forms of money and near-money substitutes to get around the limit. The Radcliffe Commission in Britain in 1959 claimed that in a devel-oped economy there is 'an indefinitely wide range of financial institutions' and 'many highly liquid assets which are close substitutes for money, as good to hold, and only inferior when the actual moment for a payment arrives'. The Radcliffe Commission did not use the concept of velocity of money because it 'could not find any reason for supposing, or any experience in monetary history indicating, that there is any limit to the velocity of circulation'.[10] The commission recommended that a complex of controls of a wide range of finan-cial institutions be developed as a substitute for the traditional control of the money supply: 'Such a prospect would be unwelcome except as a last resort, not mainly because of its administrative burdens, but because the further growth of new financial institutions would allow the situation continually to slip out from under the grip of the authorities.'[11]

Economists have debated which assets should be included in 'money' for more than two centuries. One view is that the most appropriate definition is the one that provides the strongest correlation with changes in economic

activity. Measuring economic activity is relatively unambiguous. The identification of the monetary variables that have the highest correlation with the economic activity variable might change over time and differ across countries. 'In common parlance, *bank currency means circulating bank notes* – "paper money." Yet some writers include checks and *promissory notes*, if not also loans and deposits' (italics in original) in their measure of money.[12]

The debate was neatly summarized by John Stuart Mill:

> The purchasing power of an individual at any moment is not measured by the money actually in his pocket, whether we mean by money the metals, or include bank notes. It consists, first, of the money in his possession; secondly, of the money at his banker's, and all the other money due him and payable on demand; thirdly of whatever credit he happens to possess.[13]

One approach is to include all assets in the measurement of the money supply. But in theory the question is the one raised by Mill – how much credit can households, firms, and governments command at a given time? The amounts are certain to vary because access to credit depends on satisfying certain conditions, which is likely to be easier in euphoric periods than at other times.[14] But in theory the analyst wants to know, with Mill, what credit a household, firm, or government would be able to command at a given time, and the amount is almost certain to vary over a wide range because the access to credit depends on satisfying certain conditions and both households and firms are better able to satisfy these conditions in euphoric periods. Banks and other lenders have often extended credit lines to firms and household borrowers, but the amount of credit available under the lines at each moment may require that the borrowers satisfy specific requirements.

Consider the rapid growth of US dollar deposits in London and other offshore banking centers in the 1960s, the 1970s, and the 1980s, which was a response to the increases in interest rates on these deposits relative to interest rates on bank deposits in New York and Chicago and Los Angeles that were subject to regulatory ceilings. The banks that sold these offshore deposits used the funds to make US dollar loans to American and non-American firms that they might otherwise have made from one of their US offices. The firms that borrowed US dollar funds from the offshore banks in London were as likely to spend these funds in the United States as if they had borrowed the US dollar funds in New York or some other US city. Should the US dollar deposits produced in London and other offshore banking centers be included in the US money supply?

The home-equity credit line is a recent financial innovation; banks and other lenders offer to lend homeowners an amount that may be equal to the value of the equity in their homes or in some cases a modestly larger amount. (At

an earlier period the loans that used the equity in the home as collateral were known as second mortgages; a home-equity credit line represents potential borrowing until the homeowner draws on the line, while the second mortgage was an actual loan.) The availability of home equity credit lines means that homeowners economize on their holdings of money and near-monies; households engage in liability management of the type that banks developed thirty and forty years earlier, and the increase in these lines leads to higher levels of spending with the same money supply.

The purchasing power of the household cannot readily be extrapolated to that for a country since an increase in the amount of credit extended to one individual may or may not subtract from the amounts of credit available to others, depending on both banking institutions and on the scope of euphoria. As one novelist wrote about credit:

> Beautiful credit! The foundation of modern society. Who shall say this is not the age of mutual trust, of unlimited reliance on human promises? That is a peculiar condition of modern society which enables a whole country to instantly recognize point and meaning to the familiar newspaper anecdote, which puts into the speculator in lands and mines this remark: 'I wasn't worth a cent two years ago, and now I owe two million dollars.'[15]

The basis for this generalization is the historical development of close substitutes for money that led to increases in the amount of credit and total spending. Consider only bills of exchange, call money, and the gold-exchange standard from a list that also includes bank notes, bank deposits, clearinghouse certificates, the liabilities of specialized banks (for example, *banques d'affaires, Maklerbanken,* or *Baubanken*), the liabilities of trust companies, negotiable CDs, the Eurocurrency deposits, installment credit, credit cards, and NOW accounts.

Quality of debt[16]

The credit-rating agencies were established to rank the quality of the debt of individual borrowers – firms, governments, and even households. Minsky's taxonomy of corporate debt used a three-part distinction based on the relationship between cash flows to the borrowers from their operating activities and their projected debt-servicing payments. 'Hedge finance' occurs when the cash from the firm's operating activities is larger than the cash needed for its scheduled debt-servicing payments. 'Speculative finance' occurs when the cash from the firm's operating activities is large enough to enable the firm to pay the interest on its debt on a timely basis; however, the firm would need to borrow to get the cash to pay some or all of the principal due on maturing

loans. 'Ponzi finance' occurred when the cash from the firm's operating activities is not large enough to pay all of the interest due on debt on a timely basis. The firms involved in Ponzi finance either will need to borrow or they will need to sell assets to get the cash to pay the interest.[17] (This distinction between Ponzi finance and speculative finance is comparable to that used in the public finance literature between a 'primary fiscal balance' which involves the relationship between the government's tax and other receipts and its total payments exclusive of those for interest. A government with a primary fiscal deficit does not have enough money from its fiscal revenues for all of its current scheduled interest payments.)

Minsky emphasized the 'quality' of debt to gauge the fragility of the credit structure; the terms 'speculative' and 'Ponzi' highlight fragility. The implication of the term 'Ponzi finance' is that the firm may not be able to make a debt service payment on a timely basis in the absence of a 'miracle'.[18] An edifice of debt contracted to finance risky ventures is inherently unstable. Buyers of homes in California and Nevada and Arizona who used subprime mortgages were betting the increase in the property prices would be sufficient so they could borrow against the increase in the equity of their homes to get the money to pay the interest. So were their lenders.

The model in the previous chapter emphasizes that in periods of economic euphoria the quantity of debt increases because the lenders and investors become less risk-averse and more willing – or less unwilling – to make loans that had previously seemed too risky. During economic slowdowns, many firms experience less rapid increases in their revenues than they had anticipated with the result that some that had been in the hedge finance group are shunted into the speculative finance group while some firms that had been in the speculative finance group move into the Ponzi finance group.

Drexel Burnham Lambert, Michael Milken, and 'junk bonds'

One of the great financial innovations in the 1980s was the development of 'junk bonds' – the bonds of firms that had not been ranked by one of the major credit-rating agencies. The interest rates on these bonds were generally three to four percentage points higher than interest rates on the bonds that had been ranked in one of the 'investment grades'. Many of the junk bonds had been 'fallen angels' – issued by firms when their economic circumstances were more favorable so they received a credit rating. A series of mishaps would lead to a reduction in the credit rating and eventually to the lowest investment grade; one more mishap and the credit-rating agencies would move the firm to the non-investment grade or speculative ranking.

Many financial institutions are prohibited by the regulatory authorities from holding bonds that are below investment grade, and once this threshold was crossed, these banks and insurance companies were required to sell these bonds; the interest rates on these bonds then increased sharply.

The sales pitch was that the buyers of junk bonds – say of a diversified portfolio of these bonds – had a 'free lunch' because the additional interest income would be

more than enough to cover the losses when one or several of these bonds tanked because the borrowers went bankrupt.

The innovation in the 1970s and 1980s was that Drexel Burnham Lambert, then a third-tier investment bank, began to issue junk bonds, known in more polite circles as high yield bonds; the mastermind of this innovation was Michael Milken. The firms that issued these bonds had to pay high interest rates to attract buyers. Many firms issued junk bonds to get the cash to finance leveraged buyouts; often the senior executives of a firm would seek to buy all of its publicly traded shares. Or Firm A might issue high yield bonds to get the cash to acquire Firm B before Firm B got the cash to buy Firm A.

So much for the facts that are not in dispute. What is in dispute is whether some underwriting transactions by Milken were illegal or unethical. The polite critics note that many of the firms that were buyers of junk bonds were savings and loan associations and other thrift institutions and insurance companies; the managers and the owners of some of these firms had used Drexel as the underwriter to raise the money so they might buy ownership and control of various firms. The captive thrift institutions relied on the deposit guarantee of the US government; they offered high interest rates and used the money from the sale of deposits to buy the junk bonds that Drexel had underwritten.

About half of the firms that had issued the junk bonds through Drexel Burnham Lambert went bankrupt and as a consequence the thrift institutions incurred large losses[19]; many of the institutions that had provided the ready market for the high yield bonds went bankrupt with losses to the American taxpayers of many tens of billions. But it was all legal.

In a Cassandra-like book, Henry Kaufman decried the increase of all kinds of debt – consumer, government, mortgage, and corporate, including junk bonds; Kaufman argued that the quality of debt declined as the quantity of debt increased.[20] Felix Rohatyn, a distinguished investment banker and the head of the US office of Lazard Frères, called the United States 'a junk-bond casino'.

Still, the owners of junk bonds were earning much higher interest rates than the owners of traditional bonds – at least for a while.

In the economic slowdown of the late 1980s and early 1990s, many of the firms that had issued junk bonds went bankrupt. A new set of studies showed that the owners of junk bonds on average lost one-third of their money and that the additional three to four percentage points of interest income per year of these bonds was insufficient to compensate for the losses on the defaulted bonds.

The large number of failures among the issuers of junk bonds was consistent with Minsky's taxonomy: many of these bonds would have been in his speculative group when they were initially issued. When the US economy moved into a recession, their cash receipts declined, and the bonds would have shifted to the Ponzi group. An economic miracle would have been necessary to avoid a default.

A very expensive free lunch.

Bills of exchange

Bills of exchange were claims for future payment written by a seller of goods and were a form of vendor financing.[21] The seller provided a 90 or 120-day credit to the buyers to facilitate the sales. These bills of exchange were frequently

discounted with the banks that provided the holder of the bills with cash in the form of bank notes or coin and then in the nineteenth century, bank deposits. The bills of exchange often were used directly in payment. Once the seller of the goods had received a bill of exchange from the buyer, the seller transferred the bill to someone else in payment. Each recipient of a bill would add its name to the bill, much like endorsing a check; there were five or ten endorsers on some bills. 'The bill was now money.' Ashton said that even if some of the parties in the chain of endorsers were of doubtful credit, the bill would still circulate as if it were a bank note.[22] In the first half of the nineteenth century, some bills for as little as £10 circulated with fifty or sixty names.

Payment practices differed. Bank notes were disliked in Lancashire, and at the beginning of the nineteenth century coins and bills of exchange were the primary items used for payments.[23] Because of the increase in the use of bills of exchange in payments, Bank of England note circulation declined by £9 million during the economic expansion from 1852 to 1857, even though the demand for money increased. The deposits of five banks in London increased from £18 million to £40 million. The average volume of bills of exchange in circulation, however, expanded from £66 million to £200 million, according to the contemporary estimates of Newmarch.[24]

Initially the bills of exchange were issued in connection with specific transactions and the amount of the bill matched the exact value of the sale. Subsequently the link between the sale of goods and the issue of a bill of exchange was relaxed. In 1763 in Sweden, Carlos and Claes Grill bills on Lindegren in London could not be identified with particular shipments, which often were made in rapid succession, but were drawn when the firm needed to pay creditors.[25] Thus the credit of a firm or individual was gradually separated from particular transactions and the bill had become 'accommodation paper' or a promissory note – an IOU.

Some economists were firmly opposed to 'accommodation paper' because it was believed to be of lower quality than self-liquidating, commercial bills since there was less assurance that the firms that issued the bills would have the cash to pay the holders when the bills matured.[26] In a period of falling prices, however, the merits of the higher quality commercial bills were exaggerated, since the buyers of the goods might not have the cash to settle their obligations because they might not be able to sell the goods at a profit.[27] The ratio of the debt to the debtor's income or wealth is a more meaningful measure of the quality of credit.

The bill of credit, as Franklin said,

> is found very convenient in Business; because a great Sum is more easily counted in them, lighter in Carriage, concealed in less Room, and therefore safer in Travelling or Laying Up, and on many other Accounts they are

very much valued. The Banks are the General Cashiers of all Gentlemen, Merchants and Great Traders... This gives Bills a Credit; so that in *England* they are never less valuable than Money, and in *Venice* and *Amsterdam* they are generally more so.[28]

The statement that in Britain bills were 'never less valuable than Money' is optimistic, but the efficiency of bills when they were as good as money is clear. During the first half of the nineteenth century there was a continuous debate as to whether bills of exchange were 'money', 'means of payment', or 'purchasing power'. The members of the Currency School agreed that only the supply of bank notes needed to be controlled, and that there was no need to control or limit the amounts of bills of exchange and of bank deposits.[29]

Problems were likely to arise when the ratio of the debt represented by the outstanding value of bills of exchange issued by a borrower became large relative to the borrower's wealth, which often happened in periods of euphoria. Drawing of bills of exchange in chains was infectious. Described by Adam Smith as a normal business practice, it could easily be overdone.[30] A draws on B, B on C, C on D, and so on, which increases the supply of credit. The vice of the accommodation bill, according to Hawtrey, was its 'use for construction of fixed capital when the necessary supply of bona-fide long-run savings cannot be obtained from the investment market'. Hawtrey claimed the system was particularly abused in the London crisis of 1866 and in the New York crisis of 1907.[31] Each endorser on the bill was liable for the full payment. The spectacular failure of the DeNeufvilles in Amsterdam in 1763 has been noted, which produced panics in Hamburg and in Berlin, and to a lesser extent in London and Amsterdam because a particularly impressive chain of bills was unraveled. If one bank in the chain of banks that had endorsed the bill failed, the chain collapsed and might bring down good names, those with a relatively low ratio of outstanding IOUs to capital as well as those with much higher ratios. Accommodation bills enabled traders with limited capital to borrow large amounts of money, and these short-term loans stretched into longer-term loans because they often were rolled over when they matured. During the period when the gold standard was suspended at the beginning of the nineteenth century, there was no need to be concerned about the impacts of the expansion of credit on the value of the pound. Sir Francis Baring knew of clerks worth less than £100 who were allowed discounts of £5000 to £10,000. The 'phrenzy of speculation' during this period strongly influenced the Currency School.[32] In 1857 John Ball, a London accountant, reported knowing firms with capital under £10,000 and obligations of £900,000 and claimed it was a fair illustration.[33] In Hamburg during the same boom, Schäffle reported a man with capital of £100 and £400,000 worth of acceptances outstanding.[34]

Long-Term Capital Management borrowed more than $125 billion; its capital was $5 billion. Its leverage ratio of 25 to 1 was much higher than the ratios of most other hedge funds, which generally were less than 10 to 1. In the eighteenth century, however, many firms, according to Wirth, speculated for ten to twenty times their capital during the boom of 1763.[35] Lehman Brothers had capital of three percent in the several years prior to its collapse – and it tended to transfer some of its assets to affiliates at the end of each month to reduce its leverage ratios.

Finance or accommodation bills could lead to excessive credit expansion. From time to time fictitious names were introduced into the chain so they would appear more creditworthy. Moreover, such bills were written for odd amounts to suggest an underlying commercial transaction. Claims were sometimes made (for example, by German banks drawing on Dutch and American banks after the halt in American lending) that the banks knew it was finance paper disguised as commercial bills.[36]

Securitization

Securitization contributed to the bubble in US real estate between 2002 and 2006; the investment banks created new Asset Backed Securities (ABSs) which were claims to the interest and principal of securities with similar attributes that had been placed in trusts.

Mortgages, credit card debt, and student loans were securitized. Mortgage Backed Securities (MBSs) were one type of ABS; they were much more liquid than the individual mortgages that were in the trust. Moreover they provided a way for firms to diversify the credit risks attached to individual mortgages. Because of these advantages, the supply of credit for mortgages was much larger.

Call money

The expansion in the use of call money was important prior to the crash of 1882 in France, which was classic mania financed by call money, or money lent to stockbrokers by banks 'on call', that is, for one day (in French, *reports*).[37] The stockbrokers used the money to finance their holdings of an inventory of stocks and they anticipated that they would be able to renew the one-day loans day after day.

The Union Générale was a bank started by Eugène Bontoux, an engineer who had worked with Rothschild and then left to initiate rival operations in Austria, Serbia, and southeastern Europe. An earlier Union Générale, founded in 1875, was not a success. Bontoux started his Union Générale in Paris in 1878 as France was entering a boom based on the expansion of the railroads and the

construction of the Suez Canal. The boom peaked in December 1881 and the crash followed in the next month. Bouvier's interest was in whether Bontoux, a Catholic, failed because of his mistakes as a lender or whether he was 'done in' by a conspiracy of establishment Jewish and Protestant bankers that resented an intruder. Bouvier concluded with a Scottish verdict of 'not proven'.

Bontoux's Union Générale initially was capitalized at 25 million francs, which was increased in the spring of 1879 to 50 million francs and increased again in January 1881 to 100 million francs; a third increase, planned for January 1882, would have raised the capital to 150 million francs. Only one-fourth of the capital was paid in.[38] With each increase in capital, the investors had to pay a premium above the par value of 500 francs into the reserves of the bank in response to the increase in the market price of the stock. These premiums were 20, 175, and 250 francs, respectively. Shares were registered in the names of buyers even though they still owed three-quarters of the par value for the purchases of shares; nevertheless, roughly half of the 200,000 original shares floated in the trading market.

Trading in securities was conducted through fortnightly settlements in both Paris and Lyon. A purchaser would pay 10 percent down, borrow 90 percent from an *agent de change* or broker who in turn borrowed the money in the call-money market. Money was invested in *reports* by banks, by special *caisses* (funds created especially by banks and other investors for this outlet) and by individuals. A bank and *caisse* moreover could favor brokers who specialized in trading in a particular stock. Thus banks such as the Union Générale and the Banque de Lyon et de la Loire – not to mention three or four less success-ful if less spectacular banks created during the boom – could support their own stock indirectly. When changes in stock prices were modest, speculators realized small gains and small losses, and the brokers would typically pay out or receive very little money. If stock prices increased, some speculators would withdraw money from the market. Then more capital would need to be invested. Assume a speculator bought a share for 100 francs, paid 10 francs down and borrowed 90 francs. If the speculator sold the share at 110 francs and withdrew his 20 francs, 11 francs of the 20 francs withdrawn from the market would come from the new speculator, and 9 francs had to be new *report* and borrowed from the call market. As stock prices increased the interest rates on call money (*taux des reports*) increased to attract new money; the interest rate on call money increased from 4 to 5 percent at the end of 1880 to 8 to 10 percent in the spring of 1881, and reached a peak of 12 percent in the autumn of 1881.[39]

When the share prices declined the speculators needed more money to comply with margin requirements. If the speculator bought a share at 100 francs, again with 10 francs down and 90 francs in *reports*, and the share price fell to 90 francs, the speculator had to produce 9 more francs to comply with

the 10 percent margin requirement. If the speculator had been fully lever-aged earlier and did not meet the margin call, the broker sold the speculator's stocks. If the price dropped below 90 francs, the broker, bank, or individual that had made the loan lost money. The stock of the Union Générale went from 1250 francs in March 1881 to a peak of 3040 on 14 December as the mania gathered speed. Thereafter a period of distress followed with quota-tions at 2950 francs on 10 January 1882, and 2800 francs on 16 January. On 19 January the price declined to 1300 francs. Brokers were 18 million francs short on that day, as speculators could not produce the cash needed to meet margin requirements, and 33 million francs short when it came to the month-end liquidation on 31 January.[40]

The collapse of Banque de Lyon et de la Loire was more spectacular as the share price went from a peak of 1765 francs on 17 December 1881 to 1550 francs on 28 December, when it was supporting its own stock, to 1040 francs on 4 January 1882, 650 francs on 10 January, and finally 400 francs on 19 January, the day after it closed its doors.[41] The signal for the collapse was the approval given to Bontoux to establish a Banque de Crédit Maritime at Trieste, which was announced on 4 January 1882. The coup boomeranged.[42] Investors with losses on their holdings of shares in Banque de Lyon et de la Loire sold their Union Générale shares. The combination of highly leveraged speculation and a credit mechanism that rested on bank and personal credit cycled through the call loan market and spread collapse among banks, caisses, brokers, individuals, and businesses in a few days. Economic activity was adversely affected in advance of the collapse of the stock market, not because of a change in the money supply but because at the peak of the fever the busi-ness world of Lyons turned to speculation in Union Générale: 'silk merchants, cloth manufacturers, industrialists, tradesmen, dry-goods merchants, grocers, butchers, people with fixed incomes, janitors, shoemakers'; 'A lot of capital was diverted from regular business to stock market both in securities and in call money.'[43]

Some of the features of the US stock market crash of 1929 were similar to those in the collapse of the French banks nearly fifty years earlier: a preoc-cupation with speculation and a decline in economic activity as stock prices peaked; more money was needed to support the higher level of stock prices and less money was available for economic activity. Another similarity is that as stock prices peaked, call loans of 'all others' as opposed to New York banks and banks outside New York increased from just under $2 billion at the end of 1926 to nearly $4 billion two years later and to more than $6.6 billion on 4 October 1929. Margin credit was extensive. Brokerage firms might require that the buyers of stocks pay 10 percent down; the remaining 90 percent was bor-rowed. Meanwhile, brokers' loans of New York banks declined from $1.6 billion at their height at the end of 1928 to $1.1 billion on 4 October 1929.[44]

With the crash, 'all others' and banks outside New York took call funds from the market. They were fearful that the Stock Exchange might be closed as it had been in 1873, which would have frozen their liquid day-to-day loans.[45] At this stage, New York banks slightly increased their brokers' loans. Similarly, in 1882, one consortium of Paris banks headed by the Banque de Paris et des Pays-Bas (Paribas) advanced 18 million francs in five loans to the Union Générale directly, while another group headed by the Rothschild bank loaned 80 million francs to the company of brokers to get them through the end of January settlement and allow them time to work out their payments arrangements. In both crises many brokers, clients, and (in 1882) banks and their caisses went bankrupt. The money market banks eased the adjustment, but in Paris in 1882 they did not save the Union Générale.

In the United States buying stocks on margin was curbed in the 1930s by a Federal Reserve regulation that fixed margin requirements at 50 percent. Financial institutions were able to make end runs about this regulation. The regulation applied to organized exchanges for shares, including the New York Stock Exchange, but not to the Chicago Mercantile Exchange, which dealt in Standard and Poor's (S&P) 500 stock index futures and where the margin requirement was 10 percent on the value of positions in the futures contracts. Arbitrageurs linked prices in two markets so in effect they were one market. An investor who bought an S&P 500 futures contract in Chicago with a 10 percent margin was in effect buying stocks in New York with a 10 percent margin; as the price of futures contracts increased in Chicago, arbitrageurs would sell the futures contracts and at the same time buy a representative basket of stocks of the companies that dominated the index.

In the aftermath of the 19 October 1987 collapse, there was some support for the idea of regulating the Chicago and New York markets by a single agency – the Federal Reserve Board or the Securities and Exchange Commission – and some for tightening margin regulations in the futures market, and some for banning futures trading in stocks altogether.

The gold-exchange standard

A third example of the expansion of credit on a fixed money base is represented by the gold exchange standard, which involved central banks' holdings of liquid assets denominated in the British pound and the US dollar. This practice began before World War I, although it is sometimes thought to have been based on the recommendations of the Genoa Conference of 1922 and the Gold Delegation of the League of Nations. This innovation had been strongly supported by Governor Montagu Norman of the Bank of England who sought to increase foreign holdings of British pound securities to provide relief for the Britain's balance of payments and especially its modest holdings of gold.[46] The

boom in world lending of 1913–14 was financed by increases in central bank holdings of securities denominated in the British pound, the French franc and the German mark.

Just as it is less costly to use currency notes and bills of exchange rather than coins in payments, ownership of securities denominated in the currency of a country with a large capital market dominate holdings of monetary gold as long as financial conditions are stable. These assets are easier to use in transactions, free of the need for transport, safeguarding, and assay, and directly useful without conversion into national money. A country can increase its holdings of international reserve assets by selling bonds in London or in New York and then use the money receipts to buy short-term securities denominated in the British pound or the US dollar. Since the British or US monetary authorities are not likely to regard the increase in the foreign holdings of liquid securities denominated in their currencies as a reason to contract their own credit superstructure, these transactions lead to an expansion of credit.

International lending on the gold standard may have the same unstable character as on the gold-exchange standard. Before countries borrowed foreign money that they would use as reserves for domestic monetary expansion, they borrowed gold. During the nineteenth century, American firms and state governments borrowed in London during economic upswings both to finance imports and to increase the gold base of the US banking system. Transactions in gold helped with the transfer of goods and services; gold loans enabled the borrowers to expand credit without inducing credit contractions in the lending countries.[47]

Instability of credit and the Great Depression

The notion that manias and crashes result from the instability of the supply of credit is an old one. Alvin Hansen, writing on business cycles, discussed this view in a general survey of 'early concepts' and in a chapter on mid-nineteenth-century economists – John Stuart Mill and Alfred Marshall – entitled 'Confidence and Credit'.[48] Hansen believed that these views had become obsolete because they neglected the investment and savings decisions of large firms. Perhaps. But theories that attach importance to the instability of credit persisted into the twentieth century. Hawtrey was a classic economist in this vein, and so was A.C. Pigou, whose book *Industrial Fluctuations* (1927) has a chapter dealing with panics.[49] The paradox is that the role of the instability of credit began to be neglected about the time of the Great Depression of the 1930s.

The monetarist view of the Great Depression is set out in a monumental work by Milton Friedman and Anna Schwartz. They maintained that the sharp decline in economic activity in the first half of the 1930s was the result of

policy mistakes by the Federal Reserve; they focused on the decline in the money supply from August 1929 to March 1933. Whenever the question turns on the start of the depression, they note that the money supply did not increase in 1928 and 1929, and that it declined 2.6 percent from August 1929 to October 1930 when it should have been increased to offset the weakness in the economy. Friedman and Schwartz contend that the stock market crash of October 1929 had little or nothing to do with the intensity of the decline in output and that the depression was the consequence of US domestic policies and had only a tangential connection with international capital movements, changes in currency values, and deflation abroad.[50] Their view of the depression has held sway in the United States for an extended period.[51]

Peter Temin challenged this monetarist view from a Keynesian point of view. He asked whether the decline in spending resulted from a decline in the money supply or whether instead the decline in the money supply followed from the decline in spending; he used sophisticated econometrics to choose between these two views. Much of Temin's argument centered on how much actual consumption deviated from forecasts of the levels based on the relationships among consumption, income, wealth, and similar variables that should be specified in forecasting a 'normal' trend in consumption. In addition he examined the relationship between the timing of changes in the money supply and changes in interest rates; thus if the increases in spending preceded the increase in the money supply then interest rates should have increased whereas if the increase in the money supply preceded the increase in spending then interest rates should have declined. Since interest rates declined sharply after the 1929 crash (except for interest rates on high-risk bonds which increased because of the greater concern with default risk) he concluded that the decline in spending preceded the decline in the money supply. Temin also examined the changes in real money balances (the nominal money supply adjusted for changes in the consumer price level) and concluded that real balances increased between 1929 and 1931 by 1 to 18 percent, depending on the choice between M_1 and M_2 and between a wholesale price deflator and a consumer price deflator. Annual averages further dampened the movement; on the basis of monthly figures, and averaging M_1 and M_2 expressed as percentages or relatives of a base year and of the two price indexes, the money supply increased 5 percent between August 1929 and August 1931, if seasonal influences are minimized. Temin concluded that there is no evidence that changes in the money supply between the stock market crash and the British departure from the gold standard in September 1931 caused the depression.[52]

Temin's analysis did not provide an explanation of the depression even though it was a strong challenge to the monetarist view. One analyst claimed that the stock market crash led banks to ration credit to borrowers and thus started the depression without reducing the money supply.[53] Another suggested

that the sharp decline in share prices reduced nominal wealth and household spending; the implicit assumption was that the decline in consumption followed from the decline in the real value of wealth.[54]

These arguments ignore the speed of the decline in industrial production in 1929 that began four or five months before the stock market crash. Industrial production fell from 127 in June to 122 in September to 117 in October, 106 in November, and 99 in December; automobile production declined from 660,000 units in March 1929 to 440,000 in August, 319,000 in October, and 92,500 in December. These declines are much too large to be explained by changes in the money supply.

Instead these declines are best explained by the instability in the supply of credit. As the stock prices increased in the first ten months of 1929, funds were channeled to the call-money market from consumption and production; the volume of call money rose from $6.4 billion at the end of December 1928 to $8.5 billion in early October 1929. Moreover, first the New York banks and then banks headquartered in other US cities became more cautious lenders to the stock market participants and to other borrowers. When the stock market crashed, the credit system suddenly froze. Loans to finance imports declined sharply, in part because of the sharp decline in the prices of imports.

The debate between the monetarists and the Keynesians ignores the instability of credit and the fragility of the banking system and the negative impacts on production and prices when the credit system became paralyzed because declines in the prices of many commodities and goods caused many borrowers to default on their loans – which explains the events in the early stages of the 1929 depression. This view was largely ignored except by Minsky and Henry Simons, the Chicago economist who thought the Great Depression was caused by declines in business confidence that led through an unstable credit system to changes in liquidity and consequent effects on the money supply.[55]

Simons's views were set forth in *Economic Policy for a Free Society*,[56] written after World War II under the strong influence of the 1930s depression. He recommended a 100 percent currency reserve against bank deposits to prevent changes in deposits arising from changes in the willingness of the public to hold currency and a vigorous effort to stamp out the variability of credit elsewhere in the system. He proposed restrictions on open-book credit and installment loans as well as limitation of government debt to non-interest-bearing money at one end of the spectrum and very long-term debt (ideally, perpetual obligations) at the other. Simons advocated a system in which all financial wealth would be held in equity form with no fixed money contracts so that only banks could create effective money substitutes. He was concerned about the speculative temper of the community and the ease with which short-term non-bank borrowing and lending made society vulnerable to changes in business confidence.

Simon's recommendations to limit the character of money and financial assets were very different from Friedman's liberal propensity to let market forces determine the demand and supply for different types of financial securites.[57] Friedman was confident that control of the growth of the money supply would prevent major business cycles and that there was no need to fear that the credit mechanism would not be stable. Even if Simons's positive proposals were desirable (and there is grave doubt on this score) they are utopian and hence not practical. Nevertheless Simons's diagnosis of the tendency of the system toward unstable short-term borrowing and repayment is on target.

In the last several decades still another viewpoint on the relations of money and banking to economic stability has come into prominence from the neo-Austrian school that wants to decontrol money and banking entirely. This group, led by Friedrich Hayek in Britain, Roland Vaubel in Germany and Richard Timberlake, Leland Yeager, Lawrence White, and George Selgin in the United States, seeks to abolish an active monetary policy. Any bank, company, or person would be allowed to issue 'money'; they believe that market forces will determine which firms issue good money. Each firm will compete to ensure that its money is accepted, and so good money will drive out bad. White defends the position on the basis of the experience of the Scottish banks that were unregulated between the failure of the Ayr Bank in 1772 and the Bank Act of 1845 which applied the Bank Act of 1844 to Scotland.[58]

During this period the leading commercial banks accumulated the notes of the lesser ones and were ready to convert them to specie if they thought the supply of currency notes issued by any bank was increasing too rapidly. The large banks served as informal controllers of the money supply. But several historical experiences – the country banks in England from 1745 to 1835, wildcat banking in Michigan in the 1830s, and the latest experience with bank deregulation in Latin America and especially in East Asia – do not support the view that 'good money drives out bad'.

The global inflation of the 1970s resulted from a combination of expansive US monetary policy and expansive monetary policies in Europe and Japan as their large balance of payments surpluses led to rapid growth in their holdings of international Reserve assets. Central bank holdings of international reserve assets again increased rapidly in the mid-1990s and the late 1990s.[59] Three years later a French economist, Pascal Blanqué, wrote of a US credit bubble.[60] In a similar view, Graciela Kaminsky and Carmen Reinhart blame foreign countries for printing money and the United States for running a persistent balance-of-payments deficit.[61]

The central question is whether a central bank can restrain the instability of credit and slow speculation to avoid its dangerous extension. If the monetary authorities fix some proxy for the money supply or for liquidity, or if they focus directly on the rate of interest, can the upswing and decline of the

crisis be moderated or eliminated entirely? There is no a priori way to determine whether a central bank policy of holding the money supply constant, limiting the liquidity of the money market, or raising the discount rate at the first sign of euphoric speculation would prevent the manias that lead to crises. Economists cannot conduct carefully controlled experiments. Nevertheless, the weight of historical evidence favors the view that a somewhat different monetary policy would have moderated booms even if it would not have eliminated them completely.

The Bank of England was severely attacked for its blindness to an approaching crisis in 1839 and its failure to raise interest rates sooner. The general view that it had been dilatory is said, in fact, to have been the proximate cause of the Bank Act of 1844.[62] In the early 1850s the glut of gold led to declines in interest rates in 1852 and 1853. Thereafter interest rates increased although not by enough to forestall the severe crisis of 1857.[63] The number of bills of exchange in circulation kept on increasing as the discount rate rose, and declined as it fell, rather than responding to policy changes in the opposite direction; speculation based on bill creation seems to have been uninhibited by the increases in discount rates.[64] One mid-1850s proposal was that the Bank of England should vary its discount rate as its reserves change so the public will have a clear idea of what to expect – a suggestion Elmer Wood claims shows no grasp of Bank of England transactions.[65] In 1863 and 1864 the Bank of England twice raised the interest rate to 9 percent, which perhaps delayed but did not prevent the 1866 crash. Liquidation was completed in France in 1864; there were two shakeouts in Britain that year, but the main deflation was delayed.[66] In July 1869 the National Bank of Austria-Hungary raised interest rates but the increase came too late to forestall a crash in the fall of 1869 which was a pale forerunner of the Great Crash of 1873 in Vienna.[67] The Bank raised its discount rate again in 1872, although Wirth said the rate was too low.[68] In similar fashion, with the same timing and the same absence of result (unless it precipitated the crisis), the Federal Reserve Bank of New York raised its discount rate from 5 to 6 percent on 9 August 1929.

In 1873 the Bank of England changed its discount rate twenty-four times and avoided the financial crises that seized Austria and Germany in May and the United States in September. In November the rate was raised to 9 percent to prevent the Germans from withdrawing money that had been in London since the French-German reparations.[69] Whether this represented successful fine-tuning against the possibility of a crisis or merely increasing sensitivity to short-term capital movements cannot be determined from the secondary sources.

In the panic of 1907, the preparatory expansion had involved lending of unknown amounts by out-of-town banks in New York, together with large borrowing in London by New York banks – a combination of two of the methods

of expansion discussed earlier, if one equates out-of-town banks lending in New York to the gold-exchange standard. Since the United States did not have a central bank, it was not possible to take discretionary actions to change interest rates. American borrowing led to gold exports from London and to increases in interest rates, and the Bank of England advised the market that further acceptance of American finance bills was a menace to stability and unwelcome.[70] This slowed the boom but failed to prevent the 'rich man's panic' of March 1907 and the full-scale panic of October.

If central bankers were omniscient and omnipotent, they might be able to manage interest rates and reserve requirements to stabilize the credit system; they could then correct the instability in the supply implicit in the infinite expansibility of credit. But 'there are no positive limitations to the expansion of individual credit'.[71]

Central banking developed to impose control on the growth and instability of credit. The evolution of central banking from profit-oriented private banking is a remarkable achievement. By 1825 there was implicit agreement on the division of labor between the private banks and the central bank; private bankers in London and the provinces would finance the boom while the Bank of England would 'finance the crisis' so it would not become self-justifying. In the United States, which was without a central bank after 1837, the major banks in New York were caught between the profit-making role that led them to contribute to the instability of credit and their role as holders of deposits of out-of-town banks that might be withdrawn and also contribute to instability. There was a conflict between the short-run concern with profitability and the long-run concern with financial stability, the private good with the public good. No one decreed that the New York banks should act responsibly in the public interest; it may or may not have been to their advantage to do so. The problem is a general one in politics and business and centers on who should look out for the public interest.

5

The Critical Stage – When the Bubble Is About To Pop

Changing expectations

The standard model of the sequence of events that leads to financial crises is that a shock triggers an economic expansion that morphs into an economic boom and then euphoria develops; asset prices increase rapidly, much more rapidly than GDP or some other income measure. Then there is a pause in the pace of these increases. A few savvy or lucky investors sell some of their assets to park their speculative gains. The slowing of the increase in asset prices may induce a more cautious approach by others. Distress is likely to follow as asset prices begin to decline. The pattern is biological in its regularity. A panic is likely and then a crash may follow. Lord Overstone, the leading British banker of the middle of the nineteenth century, saw a similar pattern and was quoted with approval by Walter Bagehot: 'quiescence, improvement, confidence, prosperity, excitement, overtrading, CONVULSION [sic], pressure, stagnation, ending in aquiesence'.[1] Overstone identified five stages of euphoria before the financial crisis, or, in his words, the convulsion.

The theory of rational expectations assumes that investors' expectations change more or less instantaneously in response to each shock and that investors immediately see through to the impacts of each shock on the long-run equilibrium prices for real estate and stocks and commodities. In contrast the insight from financial history is that expectations change slowly at some times and rapidly at others as various groups realize – sometimes at different moments and at other times more or less simultaneously – that the current forecasts of prices and values in the distant future differ from earlier views of these same prices and values.

The change in the mind-sets of investors from confidence to pessimism is the source of instability in the credit markets as some borrowers – individuals as well as firms – realize that their indebtedness is too large relative to their incomes. These borrowers adjust to their new perceptions about the future by

reducing their spending so they will have the cash to reduce debt or to increase saving. Some firms may sell divisions and operating units to get the cash to repay loans. The lenders recognize that they have too many risky loans and they seek repayment of outstanding loans from the riskier borrowers and they become reluctant to renew these loans as they mature and they raise the credit standards for new loans.

The period of financial distress may last weeks, months, or even for several years, or it may be concentrated in a few days. The economic downturn that followed the collapse of the US stock market in 1929 continued for four years – until a new, more interventionist government came to power. Japan was in the economic doldrums for more than ten years after stock prices and real estate prices began to decline in January 1990. In contrast South Korea recovered from its 1998 economic shock by the beginning of 1999. The handover of sovereignty in Hong Kong in mid-1997 and its advent as a special economic region of China coincided with the onset of the Asian Financial Crisis; property prices declined by 40 to 50 percent during the next five years. The recession that began in the United States in 2001 was relatively mild and brief, even though stock prices declined by 40 percent from 2000 to 2003.

The United States experienced a much sharper decline in household wealth in 2008 and 2009; prices of residential real estate declined by more than 30 percent and stock prices fell by nearly 50 percent – before a partial recovery. Property prices began to decline at the end of 2006 and economic growth was positive throughout 2007 and in the first half of 2008; the first negative quarter was the third quarter of 2008 – and even then the decline was modest. Then the collapse of Lehman Brothers in September 2008 triggered a panic and a crash. The credit system froze; interest rate spreads surged, and LIBOR increased by 500 basis points relative to the Federal Funds rate whereas the traditional spread had been 10 to 20 basis points. The combination of the more restrictive supplies of credit and the greater cautiousness by households led to a very sharp decline in GDP.

Should a government intervene to moderate the cycle? Government policies play a vital role in the formation of expectations. Can the government head off a financial crisis by dampening the expectations that develop in the euphoria? Should the government seek to moderate the impacts of the decline in prices of stocks and real estate and commodities after the bubble has imploded?

Bear Stearns, Fannie Mae and Freddie Mac, and Lehman Brothers

One of the first signs of adjustment in the United States to the extraordinary increase in home prices and the large excess supply of homes that began in 2002 was a surge in the bankruptcies of mortgage brokers that started in the

last few months of 2006; these firms were 'middlemen' and wholesalers who acquired mortgages from home buyers, after having indicated the terms of the mortgages that they would acquire. Then they sold the mortgages to the investment banks, which would place mortgages with similar terms in a trust that they would use as collateral for issuing mortgage backed securities (MBSs). The investment banks retained the right to 'put' these loans back to the mortgage brokers if the borrowers stopped making their monthly payments within 12 or 18 months. As the borrowers fell behind in their payments, the investment banks returned these mortgage loans to the brokers and asked for the return of their money, and the firms that were unable to repay went bankrupt. The investment banks were stuck with tens of billions of dollars of failed mortgages after the brokers went bankrupt.

The combination of the default by those who had recently borrowed to buy homes and the failure of the mortgage banking firms led to a sharp decline in the supply of credit available for mortgages. The pace of home purchases slackened. The property developers that had built homes in anticipation of future sales experienced a surge in their inventories of unsold properties as home prices declined.

House prices in some of the regional markets peaked toward the end of 2005, and at the national level at the end of 2006. The decline in 2007 of nearly 15 percent led to a dramatic decline in new housing starts.

The first major casualty was Bear Stearns; two of the leveraged hedge funds that it managed tanked, and the investors in these funds lost most of their money. Then there was a run on Countrywide Financial, the largest US mortgage lender, in mid-August 2007. At the same time, there was a run on Northern Rock, which was the largest mortgage lender in Britain. Both firms had been aggressive in their efforts to increase their market share, both had relied on the commercial paper market for more than 30 percent of the money that they used to buy mortgage loans. In effect they had been selling short-term IOUs to firms that had extra cash to get the money to buy more mortgages. Countrywide Financial was rescued by the Bank of America, which initially made a capital investment through the purchases of preferred shares and subsequently acquired Countrywide. Northern Rock was rescued by the Bank of England, which initially provided liquidity; subsequently the British government acquired shares in the firm and became its majority owner.

The next striking development were the runs on Bear Stearns in March 2009, one by the owners of its shares and the other by those who owned fixed-price claims on Bear and were reluctant or unwilling to acquire new claims on Bear as the ones they owned matured. The Federal Reserve facilitated the acquisition of Bear by JPMorganChase by agreeing to accept up to $29 billion of securities of dubious quality that Bear owned.

Fannie Mae and Freddie Mac, the two very large government-sponsored enterprises that carried more than 50 percent of the credit risk attached to US home mortgages, were 'nationalized' by the US Treasury in the second week of September 2008. The owners of their common shares lost virtually all of their money. The owners of their preferred shares also were wiped out; many banks had purchased these preferred shares because that could be considered as part of their required Tier I capital. Thus the bankruptcy of Freddie and Fannie had a major whammy on the capital of other financial institutions.

The next week Lehman Brothers went bankrupt, because the US government had decided that it didn't want to 'save Lehman'. The government officials said that they didn't have the authority to save Lehman. Perhaps. Or perhaps they could have found the authority if they are tried harder. Or they could have provided some temporary financial assistance to enable Lehman to avoid bankruptcy while they acquired the legal authority to save Lehman.

The decision not to 'save Lehman' triggered the most massive credit crisis in the last one hundred years; credit spreads soared. A large number of different firms relied on short-term funding for the money to buy long-term securities, and they found it extremely difficult – and costly – to refinance their maturing IOUs.

The decision not to 'save Lehman' reflected that government officials were concerned by the public outrage about 'bailing out' the rich bankers. They – like officials and analysts in the previous two hundred years – believed that saving a large firm would encourage risky financial behavior ten and twenty and thirty years in the future.

Then on Tuesday 14 September – the day after Lehman failed – the Federal Reserve lent $85 billion to AIG; the loan gave the Fed the option to buy 79.9 percent of AIG's stock. The 'too-big-to-fail' doctrine had been revived. The concern was that if AIG tanked, runs on many other financial institutions would occur, and this dramatic rush for cash would lead to the collapse of the financial system.

The interplay among the decline in real estate prices and bank loan losses was inevitable and predictable, although the identity of the firms that would be pummeled could not have been predicted. However, those firms that had rapidly increased their market share – Countrywide, Northern Rock, Washington Mutual – had done so on the basis of short-term funds borrowed in wholesale markets, and they were more vulnerable to runs than the lenders that relied almost exclusively on deposits for the money to buy mortgages.

The money market funds – established primarily by brokerage firms as an alternative to government-guaranteed deposits – were subject to runs, and the US Treasury hastily extended the umbrella of deposit insurance to these firms.

The major question is how many of these developments were predictable in response to the decline in real estate prices. It was inevitable and hence predictable that the net worth of millions of households would decline as property prices fell. Several millions of these households would have 'upside down mortgages' – they would owe more on their mortgage loans than the market value of their homes. Millions would walk away from these properties, and leave the lenders with losses.

Warnings

One proposal is that a government should make its knowledge or publish its own forecast if it thinks it knows more than the speculators.[2] Thus the government might calm the concerns or fears of investors by making that knowledge publicly available.[3] Many individuals within government have views about the economic and financial outlook but these are often at variance with each other so that developing a 'government view' could occur only if someone – the prime minister, the head of the central bank, the minister of finance – forged a consensus.

The historical record provides little support for the view that statements from government officials have had much of an impact in dampening euphoria. In some cases 'a word to the wise' may be sufficient but in others the warnings are not likely to be adequate to forestall further increases in asset prices. The likelihood that investors and speculators would heed the warnings of a government official when asset prices are increasing at annual rates of 20 to 30 percent a year is not especially high.

The first such warnings on the record were made about 1825. Although many writers have viewed the Bubble Act of June 1720 as a warning by Robert Walpole and King George II against speculation, the primary objective of that legislation was to repress competitors of the South Sea Company because the other bubbles were draining cash that the South Sea Company wanted and needed.[4] The Bubble Act, which had been strengthened in 1749, made both swindles and starting a legitimate business more difficult and was not repealed until the nineteenth century.

The banking authorities began to warn about speculative booms in the nineteenth century. In the spring of 1825, Prime Minister Canning, the Chancellor of the Exchequer, Lord Liverpool, Sir Francis Baring, and W.R. McCulloch in *The Scotsman* warned against the excesses of speculation. The warning contributed to the crisis – but the crisis was probably inevitable. In the panic and crash of December 1825, Lord Liverpool did not rescue the speculators because nine months earlier he had said he would not.[5] In 1837 the President of the Board of Trade, J. Poulett Thomson, excoriated the prevalent spirit of speculation, which differed from that of 1825 in that individuals were investing at home rather than abroad.[6]

In the fall of 1837 the gambling spirit crossed the Channel, and the Belgian and French authorities attempted to repress speculation by forbidding the quotation of the prices of notes and shares of corporations. Their efforts were futile; speculation had gone beyond the narrow framework of the Bourse and non-professionals such as *rentiers* and even 'women and foreigners' were becoming involved. Chambers of Commerce in Liège, Vervier, and Antwerp condemned stock market speculation. The Belgian king refused to charter a proposed bank, the Mutualité Industrielle. Investment slowed as a result of a decline in economic activity rather than in response to the statements from the administration and business establishment.[7] In July 1839, Lamartine in the French Chamber of Deputies spoke against speculation and especially warned about the guarantees of railroad securities.[8]

The only suggestion that official condemnation of speculation may have been effective was from a French observer who commented on the 1857 crisis. In March 1856 the minister of the interior brought legal action against certain swindlers. Emperor Napoleon III congratulated O. de Vallée, the author of *Les Manieurs d'Argent*, which dealt sternly with dubious financial practices. The Senate passed laws. The Bank of France raised the discount rate to 10 percent. Napoleon III published a letter in *Le Moniteur* on 11 December that indicated that the government would provide support only for those catastrophes that were beyond human anticipation. According to d'Ormesson, speculative exuberance declined and the memory of the 1857 crisis reflected a certain glory on French commerce.[9] But Rosenberg concluded that the warnings had been too late.[10]

Restrictions by the Austrian National Bank in 1869 produced a 'great crash' that proved to be a mini-crash compared with the one that followed in 1873.[11] The warnings and revelations of Eduard Lasker, a member of the Diet who in February 1873 had exposed the scandalous connections between the Prussian government and its commerce ministry and the railroad concessionaires, had no significant impact in quelling speculative sentiment.[12]

More timely were warnings issued by *The Economist* in 1888 against commitments to buy *cedulas*, the Argentinean land bonds. In April *The Economist* said that 'the bonds...might...become a very inadequate security. Just at present all real estate at the River Plate commands inflated prices, but the occurrence of financial difficulties might easily render them unsaleable.'[13] Then in May 'A collapse of the "boom" in real estate, which is easily conceivable, would be sure to severely depress the value of the cedulas.'[14] The warnings proved ineffective.

More memorable is the Cassandra-like utterance of Paul Warburg, a partner of Kuhn, Loeb & Co. and one of the founders of the Federal Reserve System, who in February 1929 warned the American public that US stock prices were too high and showed symptoms reminiscent of the 1907 panic; his statement followed a similar statement from the Chairman of the Federal Reserve Board. Investors paused briefly and then stock prices again increased. The

ineffectiveness of Federal Reserve Chairman Greenspan's statement about 'irrational exuberance' in December 1996 about the high level of US stock prices was noted earlier. In August 1999 Greenspan stated that the Fed would consider the level of stock prices when it set interest rates after the discount rate had been increased by 0.25 percent in August. Again the stock market barely noticed.

If prices of real estate and of stocks continue to increase despite these warnings, then it appears that the warnings lack credibility. Economic forecasters may know – or at least think they know – the long-run equilibrium values for real estate and for stocks, but the ability to foretell when market prices will move back toward these long-run average values rather than away from them is modest. Roger Babson sold his clients' stocks in 1928 and he appeared foolish for more than a year as stock prices continued to increase.

The timing issue is complex. If the government authorities want their warnings to be effective, they need to provide their cautionary statements early enough to forestall some of the excesses of the euphoric period and late enough so the statements are credible. One metaphor from a former chairman of the Federal Reserve is that authorities are reluctant to take the 'punch bowl away from the party just as the party is getting going' because of the unfavorable public reactions.

The modern tradition is that the central banks develop their monetary policies to moderate the increases in the consumer price level or some other price level index; 'inflation targeting' became their new mantra. The policy question is whether the central bankers should ignore the increases in the prices of real estate and stocks if they are far above their long-run equilibrium values. Attempting to convince the speculators through statements alone generally has been futile.

Financial distress

Distress is widely discussed in connection with financial crises. The term is imprecise: one meaning is a state of suffering and another is of a hazardous situation. Commercial distress reflects the first definition, financial distress the second. Commercial distress implies that prices and economic activity and profitability have declined and that many mercantile and industrial firms have gone bankrupt – or are on their way to bankruptcy. Financial distress for an individual firm means that it is incurring significant losses and that there is a non-trivial probability that it will not have enough cash to pay the interest due on its debt according to the agreed-upon schedule.[15] The credit panic in the last several months of 2008 led to a sharp decline in automobile sales, and both General Motors and Chrysler became bankrupt. Financial distress for an economy also has a prospective significance and implies the need for

economic adjustments; firms may be on the verge of bankruptcy and banks may need additional capital to compensate for the decline in capital that has resulted from exceptionally large loan losses. The US government developed a $700 billion-plus Troubled Assets Relief Program (TARP) that would make money available to banks and other financial firms. Many investment projects become stalled because the developers cannot obtain the money needed to complete construction.

Other words used to describe the interval between the end of euphoria and the onset of what classic writers called revulsion and discredit (or crash and panic) are *uneasiness, apprehension, tension, stringency, pressure, uncertainty, ominous conditions, fragility.* More colorful expressions include *an ugly drop in the market*[16] or *a thundery atmosphere.*[17] Meteorological metaphors have been used frequently: 'One feels again the oppressive atmosphere that precedes a storm.'[18] And geological metaphors have been used: two years before the panic of 1847 Lord Overstone wrote to his friend G.W. Norman (the grandfather of Montagu Norman, the Governor of the Bank of England in the 1920s): 'We have no crash at present, only a slight premonitory movement of the ground under our feet.'[19] The seismographic metaphor was also used by Michel Chevalier who wrote from America about President Jackson's war against the Second Bank of the United States: 'A general collapse of credit, however short the time it lasts, is more fearful than the most terrible earthquake.'[20] Another French writer noted the 'presentiment of disaster'.[21] A German metaphor spoke of the 'bow being so bent in the fall of 1782 that it threatened to snap'.[22]

Distress is not an easily measured condition for an economy. Investors may have become apprehensive when the values of certain variables diverged significantly from average values; some of these variables include the gold reserve ratios of central banks, the ratio of debt to capital of a large number of firms or individuals, the losses of banks in relation to their capital, the ratios of a country's external debt service payments to its export earnings, and the price-earnings ratios for stocks and the rental rate of return on real estate. Investors may be increasingly aware of the approach to a limit, such as the ceiling on note issues by the Bank of England stipulated in the Bank Act of 1844, the $100 million gold minimum requirement of the US Treasury in 1893, the ceiling on advances from the Bank of France to the French Treasury in 1924, the gold reserve ratio of the Reichsbank under the Dawes Plan in June 1931, or the free gold available to the Federal Reserve System prior to the passage of the Glass-Steagall Act in February 1932. A ratio of external debt to GDP of 60 percent for a country has been viewed as a premonitory indicator by investors; the 'ice is thin' for a country when this ratio is much higher. Similarly a ratio of government debt to GDP of significantly more

than 60 percent is viewed as too high. Limits excite, as one Chancellor of the Exchequer noted in 1857:

> Now when you impose a limit, there is no question that the existence of that limit, provided it makes itself felt at a time of crisis, must increase the alarm. People feel at the moment that a peril presses on them, they begin to calculate how much remains of that fund to which they look for assistance in times of commercial difficulty, and in whatever way you fix the limit, whether by Act of Parliament, or, as Mr. Thomas Tooke [a leader of the Banking School] proposed, by a sort of usage, or, as in France, by the discretion of the Government acting on the Bank of France, there is no doubt that in moments of crisis the limit must aggravate the alarm.[23]

A French official expressed the same idea two decades later when he defended the tradition that the Bank of France should maintain a specie reserve of about one-third of its demand liabilities but without a hard and fast legal requirement: 'A fixed ratio is not required. That would be unwise...the terror of a limit fixed and absolute.'[24] Overshooting of the limit may have psychological significance. In March 1924 although sophisticated bankers knew that a small increase in the French money supply would not be dangerous, the public had come to regard the ceiling on Bank of France advances to the Treasury as an index of economic health. As one minister put it, the French were close to the upper limit of elasticity of confidence in their own currency.[25]

Causes of distress and the symptoms of distress are observed at the same time and include sharply rising interest rates in some or all segments of the capital market, an increase in the interest rates paid by subprime borrowers relative to the interest rates paid by prime borrowers, a sharp depreciation of the currency, an increase in bankruptcies, and an end of price increases in commodities, securities, and real estate. These developments are often related and show that the lenders have become over-extended and are trying to reduce their exposure to risks and especially to large risks.

Financial distress in the nineteenth century was compounded by the arrangements for purchases of newly issued stock that provided for a series of payments by the buyers in response to 'calls' from the issuers of stocks as construction work proceeded. In 1825 and 1847 in Britain and in 1882 in France some of the buyers of the stocks did not have the cash to meet the calls; they may have counted on selling the security at a profit before the next call. Thomas Tooke describes this embarrassment in 1825 as acute because the call for cash payment was immediate and pressing while prospects for earnings on the stock were remote and uncertain.[26] Distress developed in January 1847 when railroad calls amounted to £6.5 million in one month.[27]

The chain-letter aspect of security issues became evident in the finances of the South Sea Company in June, July, and August 1720 with repeated attempts to raise cash through new issues of stock. In 1881 more than 125 new issues with a market value of 5 billion francs were sold in Paris when France's annual savings were estimated at 2 billion francs.[28] Nor was this an era when private companies were going public in large numbers, as in the late 1880s in Britain and the late 1920s in the United States; in both cases there was no need for an increase in savings because the newly issued publicly owned shares were exchanged for privately owned shares.

The end of a period of rising asset prices leads to distress whenever a significant number of investors have based their purchases of these assets on the anticipation that these prices will continue to increase. Some of these investors then may have a 'negative carry' because the interest rates on the funds borrowed to buy the assets exceed the cash income on the assets; these investors anticipated that they would be able to use the increase in the value of the assets as collateral for new loans that would provide them with some of the cash needed for interest payments. When asset prices stop increasing, these investors are shunted into distress mode since they have no ready way to get the cash they need to pay the interest on their indebtedness.

Distress enveloped the market for subprime mortgages toward the end of 2006 and 2007 when real estate prices began to fall. The hallmark of these mortgages is that the homebuyers' equity often was less than ten percent; when prices began to fall, they then had 'upside-down mortgages' – their mortgage indebtedness was greater than the market value of the properties. One response to the distress was 'jingle mail' – to ship the keys to the properties to the lenders.

Distress may arise from an increase in the flow of funds from the country – a bad harvest may require an increase in imports and an increase in interest rates in a major international financial center may attract funds from domestic financial markets. Credit in the domestic credit market might become less readily available – tighter – because of a reduction in the reserves of the banking system.

The money outflow may be potential; there was distress in the London money market in 1872 when the French payment of reparations to Prussia meant that the Reichsbank acquired substantial money market securities denominated in the British pound that could be readily converted into gold. Similarly there was a flow of money to London prior to April 1925 in anticipation that the British pound would appreciate to its pre-war gold parity; once the pound was again pegged to gold, the owners of these deposits had modest incentives to keep them in London. The Bank of England's interest rate policy was constrained by the concern that some of these funds might leave London. The Bank of France acquired large British pound balances in the effort to limit the appreciation of

the French franc after the successful monetary stabilization at the end of 1926; the likelihood that the French might use part or all of these balances to buy gold was a bargaining chip that increased nervousness in London.

The essence of financial distress is loss of confidence. What comes next – belief in the future as imbalances in the economy are corrected, or runs on banks, panics, and the collapse of prices?

The issue is concisely posed by James S. Gibbons:

> Bank officers are not always insensible to alarm when respectable merchants, failing in their best endeavors, are driven into a corner and assume an air of desperation. They know the danger that hangs over the market. Credit is prodigiously extended; the public excitement is wrought up to a high pitch of apprehension, and there need be but a single failure of a 'great house' to explode the 'mighty bubble.' Who knows that it is a bubble? Who knows that the highest point of pressure is not reached today, and that tomorrow the waters will not begin to subside? And then gradually things fall into their old channels, confidence revives and it is proved that there was no bubble to burst after all.[29]

How long does distress last?

Financial distress may subside as in France in 1866 and in Britain in 1873 and 1907 or a panic may follow the distress. In the United States there was great concern over the failed attempt by Bunker Hunt to corner the silver market in 1979, the failure of the Continental Illinois Bank in 1984, and the debacle of Long-Term Capital Management in 1998. There were extended periods of distress after August 1982 over bank loans to Mexico, Brazil, and other developing countries and over loans by the thrift institutions in Louisiana, Oklahoma, and Texas to borrowers who had based their exploration activities on the prospect that the oil price would increase to $80 per barrel. After the collapse of hundreds of US banks and thrift institutions at the end of the 1980s, the US Resolution Trust Corporation (RTC) acquired tens of billions of dollars of real estate that had been the collateral for mortgage loans that were in default. Eventually these properties were sold to the public but the uncertainty about the values of these properties depressed values.

Similarly there was distress in Tokyo for an extended period in the 1990s; in any 'mark to market' test most of the large Japanese banks were bankrupt and yet there were no runs on the banks because depositors understood that they would be made whole by the government if any of the banks closed. The Japanese government's policies toward these failed institutions were a cause of distress: would the government close these failed banks or would the government provide new capital to them on favorable terms?

The sharp decline in stock prices on Monday 19 October 1987 proved to be a correction rather than a panic because it did not spread to other US markets, although there were nearly simultaneous sharp declines in most other national stock markets. Distress lasted for several weeks while investors waited to see whether the decline in stock prices would have significant impacts on other markets.

The debacle of Long-Term Capital Management in the summer of 1998 occurred at the same time as the collapse of the Russian banking system and the ruble. The prospect of the impending disaster in Russia induced changes in interest rate and yield relationships that contributed significantly to the collapse of LTCM. LTCM was an unregulated bank even though it was usually considered a hedge fund. LTCM was considered a 'very smart' financial institution; two Nobel laureates in finance were among its top officers. The firm was much more highly leveraged than traditional banks and most other hedge funds and in addition it had tens of billions of positions in derivatives contracts like futures and options. Initially LTCM had been viewed as a money machine, and extremely clever in taking advantage of small deviations in prices of very similar securities. For example, the thirty-year US Treasury bond was extensively traded but the twenty-nine-year bond was less liquid and not extensively traded, so the interest rate on the twenty-nine-year bond was modestly higher than that on the thirty-year bond. LTCM bought hundreds of millions of dollars worth of the twenty-nine-year bond and sold short more or less the same amount of the thirty-year bond to profit from the interest rate differential. The excess of the interest rate earned on the bond with the twenty-nine-year maturity over the interest rate on the bond with the thirty-year maturity was modest, but the product of this small number and the position of hundreds of millions of dollars was large – and very profitable.

Some of the major banks that were large lenders to LTCM tended to mimic some of its portfolio positions. The positions of LTCM and these major banks in some securities dominated the markets for these securities.

In the spring of 1998, LTCM had a long position in emerging market bonds that it hedged by shorting US Treasury bonds. As investors became increasingly apprehensive about Russia's financial future, the prices of emerging market bonds declined as the contagion effect cascaded. The Federal Reserve responded with greater monetary ease and the prices of US Treasury bonds increased. LTCM lost money on both legs of its hedge, which eroded its capital base. As the prices of these emerging market bonds declined, LTCM was between the proverbial 'rock and a hard place'; if it sold any of its holdings of individual securities, their prices would fall further and its net worth would decline even more rapidly.

The Federal Reserve was concerned that if LTCM failed there would be an extended period of significant uncertainty – distress – in the capital markets

while the firm's positions in futures, options, and other derivatives were unwound and bond prices would fall further. The Fed used its muscle – more precisely the threat of its regulatory muscle – to induce the major banks that were lenders to LTCM to invest their own money in LTCM, and they then acquired 90 percent ownership.

Where does the money go?

The financial wealth of Americans declined from the peak of $74,000 billion before the crisis to $63,000 billion in the second quarter of 2009; most of the decline was in the market value of real estate.

Where did the money come from before it disappeared?

The feature of a bubble is the exceptional and non-sustainable increase in prices of real estate and stocks. Individuals buy the real estate using borrowed money. For every buyer of real estate there is a seller, similarly for every buyer of stocks there is a seller.

These sellers of real estate and stocks park their sales receipts in the banks and in bonds. The money hasn't disappeared, it simply has changed hands.

The crash or panic that follows financial distress may do so immediately, in a matter of weeks, or with a delay of several years. John Law's system peaked in December 1719 and collapsed in May 1720 – five or six months from glory to disaster. In the South Sea Bubble of 1720, the lunatic note sounded clearly at the end of April, the ugly drop in the market occurred in August, and the collapse came in the first days of September. In 1763 distress developed in March while the actual crisis precipitated by the failure of DeNeufville in Amsterdam occurred in July. In 1772 the Bank of England raised its discount rate early in the year; the Ayr Bank curtailed its operations in May, but it was too late. Fordyce absconded on 10 June and the news precipitated a panic in Britain on 22 June; the consequent distress in Amsterdam lasted until the failure of Clifford & Co. in December.

Timing of crises from 1789 to 1815 was dominated by individual apocalyptic events, including the guillotining of Louis XVI in January 1793 (losing one's head always is apocalyptic), the landing of the French army at Fishguard on the southwestern tip of Wales in February 1797, and the penetration of the continental blockade in 1799. The periods of distress on these occasions were short because the panic was virtually immediate. In 1809–10 the setback arose from a tightening of the continental blockade and overtrading in exports to Brazil. Pressure mounted slowly from the middle of 1809, then more rapidly from mid-1810 to the climax of bankruptcies in January 1811.

Calls for further payments on subscriptions to the shares in railroads in January 1847 set the background of tension against which speculation in grain

peaked in May, collapsed in August, and led to panic in November. The crisis of 1866 was the delayed result of the 1864 collapse of cotton prices that had brought panic to France in that year. Britain had two 'critical moments' in 1864, one in January – the real crisis that was related to the collapse in cotton prices – and another in the last quarter.[30] British treatment of the period tends to look more to speculative expansion that affected the discount houses and that started in the previous year, and to a series of firms that resembled the Credit Mobilier and were using the funds received from selling new share issues to buy back their own shares to pump up the investors. W.T.C. King wrote that one Albert Gottheimer used the name Albert Grant to float the imposing Credit Foncier and Mobilier of England, which ultimately achieved a paid-up or rather called-up capital of £1 million.[31] The conversion of the discounting house of Overend, Gurney & Co. to a public company in July 1865 at the peak of the boom and 'dividend race' led to a 100 percent premium on the stock in October. The Bank of England responded by raising its discount rate from 3 percent to 7 percent; the crash did not occur until May 1866. The distress in Britain lasted seven months, from October 1865 to May 1866, while the distress in France lasted nearly thirty months.

Periods of financial stringency and crisis and panic (in the United States) occurred in the autumn when Western banks drew large sums of money from the East to pay for shipments of cereals.'[32] Credit demand peaked in the autumn when banks in the grain-producing areas of the country borrowed large amounts of money from banks in the East that would be used to pay for the purchases and shipments of cereals. Sprague noted that the crisis of 1873 came in September because of the early harvest, that the outbreak of a crisis invariably came as a surprise to the business community and that the crisis of 1873 was not an exception.[33] The seasonal tightness of money was well known and the puzzle is that it was a surprise. The 'excessive tightness' of money from September 1872 to May 1873 caused the railroads to borrow short-term funds rather than issue bonds, which could have been seen as a sign of distress, and then the seasonal tightness precipitated the crash.[34]

Distress may be continuous or it may oscillate in its own rhythm. The crash of the Union Générale in January 1882 was preceded by three separate tense periods, in July, October, and December 1881.[35] The panic of October 1907 was anticipated (although Sprague indicated its exact timing was not foreseeable) and preceded by a 'rich man's panic' in March when Union Pacific stock, the security most widely used as collateral for finance bill operations, dropped 50 points.[36] Markets recovered from this blow and from the failure of an offering of New York City 4 percent bonds in June (only $2 million was tendered for an offering of $29 million of 4 percent bonds) and from the collapse of the copper market in July, and from the $29 million fine levied against the Standard Oil Company for antitrust law violations in August – only to succumb to the failure

of the Knickerbocker Trust Company in October.[37] In 1929 distress lasted from June to the last week in October.

Financial distress in Japan began at the start of the 1990s and continued throughout that decade and into the next. Japanese industrial firms were extremely reluctant to downsize and make the other adjustments that were necessary to bring their costs below their current revenues; in the previous forty years these firms had been able to rely on the banks for the cash to finance their operating losses and investments. Japanese banks were extremely reluctant to stop making new loans, even to enterprises that would be considered bankrupt in any 'mark to market' test; and the regulatory authorities were reluctant to close banks that would be considered bankrupt. Traditionally the risk of financial losses in Japan has been 'socialized'; Japanese society prefers to distribute these losses among taxpayers rather than to impose the costs of adjustment associated with closing failed enterprises on their employees.

Argentina experienced an extended period of distress at the end of the 1990s and in 2000 before its currency collapsed in January 2001. At the end of the 1980s Argentina had suffered through two years of hyperinflation. The incoming government of President Carlos Menem pegged the new Argentinean peso to the US dollar at the rate of one Argentinean peso to one US dollar; at the same time Argentina adopted a currency board arrangement that meant that its central bank would increase the supply of peso liabilities only if its holdings of US dollars increased – more or less a very strict application of the Currency School doctrine. During the 1990s, the tax revenues of the Argentinean government were smaller than its expenditures and the excess of expenditures was financed partly with the receipts from privatization and partly by government borrowing. The hyperinflation of the 1980s had sharply reduced the real value of the Argentinean government debt and so investors purchased Argentinean government bonds denominated in the US dollar because the government looked like a good credit risk. As the ratio of Argentinean government debt to the country's GDP increased, the interest rates that the Argentinean government had to pay to sell new bonds rose. There was a recession in Argentina toward the end of the decade, in part because the US dollar (to which the peso was pegged) was appreciating; and the Argentinean fiscal deficit increased as its tax receipts declined relative to its expenditures. The recession was intensified by the depreciation of the Brazilian real in January 1998. Brazil was Argentina's major trading partner. The policy question was whether Argentina would be able to reduce its fiscal deficit while maintaining the established parity. (The Argentineans do not have an established tradition of paying taxes so tax rates tend to be high and tax collections low, while government salaries tend to be high and the performance of government officials low.) The efforts to raise taxes and cut government expenditures led to a series of political problems; Argentinean citizens were extremely reluctant to pay

more taxes when the economy was doing poorly. The country was moving in slow motion toward a disaster. In the end Argentina both devalued and defaulted on the government debt.

Suppose the monetary authorities tighten credit to raise the cost of speculation. When commodity and asset markets move together, up or down, the direction that monetary policy should take is clear. But when share prices or real estate or both soar while commodity prices are stable or falling the authorities face a dilemma. The Federal Reserve encountered this dilemma in the 1920s; President Benjamin Strong of the New York Fed had agonized over the appropriate policy in 1925 and again in 1927. The dilemma is that the policymakers cannot kill two birds with one stone, or more precisely they cannot achieve two policy targets with one policy instrument, or in what is perhaps a better metaphor, it is difficult to pick off a target if it is standing next to another one that one wants to leave untouched and the weapon is a shotgun rather than a rifle. At the same time, any tightening of credit to dampen the boom in real estate would be likely to stall a remarkable economic expansion.[38]

Fed Chairman Greenspan was concerned that US stock prices were too high or increasing too rapidly when he remarked about 'irrational exuberance' in December 1996. The Fed proved reluctant to raise interest rates to dampen the increase in stock prices because of the negative impacts on economic growth and employment. In 1999 the Fed became concerned (obsessed?) with the Y2K problem, the likelihood that US computer systems would collapse because so many software programs were not designed to recognize the transition to 2000. In the last several months of the year the Fed provided the monetary system with abundant liquidity to forestall problems associated with the end-of-the-millennium transition and in the meantime the money – which had to go somewhere – fed stock market speculation.

Onset of a crisis

Students of logic have discussed the damp squib thrown by A that lands at B's feet which is then thrown from B to C and from C to D and so on only to explode after Y throws it in Z's face. Who is to blame? A, *causa remota*? Or Y, *causa proxima*? *Causa remota* of any crisis is the expansion of credit and speculation while *causa proxima* is some incident that saps the confidence of the system and induces investors to sell commodities, stocks, real estate, bills of exchange, or promissory notes and increase their money holdings. The *causa proxima* may be trivial: a bankruptcy, a suicide, a flight, a revelation of fraud, a refusal of credit to some borrowers, or some change of view that leads a market participant with a large position to sell. Prices fall. Expectations are reversed. The downward price movement accelerates. To the extent that investors have

used borrowed money to finance their purchases of stocks and real estate the decline in prices is likely to lead to calls for more margin or cash and to further liquidation of stocks or real estate. As prices fall further, bank loan losses increase and one or more trading firms, banks, discount houses, or brokerages fail. The credit system appears shaky, and there is a race for money and liquidity.

Identifying the original sellers is difficult. Conspiracy theories abound. One can single out bear speculators like Joseph P. Kennedy, Sr, or Bernard Baruch in 1929; the Protestant-Jewish cartel that allegedly did in Eugène Bontoux in 1882; or Thomas Guy, who liquidated £54,000 of South Sea stock over six weeks between April and June 1720, never selling more than £1000's worth at a time. (He used his fortune to endow Guy's Hospital in London, 'the best memorial of the Bubble'.)[39]

Someone sells. Occasionally it is a foreigner. In 1847, for example, the French (according to one S. Saunders, quoted by Evans) bought up surplus wheat and sent it to Britain in June and July where it was sold at prices much below the then-prevailing market prices, which fell from 96 shillings to 56 shillings per quarter (a quarter is equal to eight bushels) and brought about the bankruptcies of a large number of firms connected with the corn trade.[40] The story is not persuasive. The price of wheat had risen from 46 shillings in August 1846 to 93 shillings in May 1847 because of violent storms that ruined the crop and because of the potato blight in Ireland and on the Continent. The price dropped in July 1847 with favorable weather and the prospect of a good crop. Imports of wheat and flour rose from 2.3 million quarters in 1846 to 4.4 million in 1847, aided by the repeal of the Corn Laws;[41] the 70,000 quarters are a trivial proportion of this sum. In 1846 France had its smallest crop of wheat in 100 years (a problem exacerbated by the potato crop failure); in 1847 the crop was the largest for 100 years. But the condition was general, and British wheat speculation had been excessive.

One view is that the Baring crisis of 1890 was triggered by the sale of Argentinean bonds by German investors, who had stopped buying these bonds two years earlier either because of general uneasiness,[42] or because they were concerned about the instability of the Argentinean currency,[43] or because the domestic boom led them to sell other foreign bonds including Russian bonds.[44] German sales of the Argentinean bonds contributed to distress rather than to a crisis, since British investors then acquired more of the Argentinean bonds. In November 1888 a £3.5 million offering of the Buenos Aires Drainage and Waterworks Company failed, and Baring felt obliged to lend to Argentina through acceptance credits. Declining primary product prices in 1890 meant that the Argentine government did not have the money to repay these loans as they came due. The Baring crisis of November 1890, after two years of distress, resulted from a warning from the Bank of England to Baring Brothers to limit

the level of its acceptances (which stood at £30 million in the summer of 1890), from the crisis in New York in October, and from the maturing of £4 million of acceptances in November.

New information may precipitate a crash, as when it was revealed that the Paris-Lyons-Marseilles railroad would cost 300 million francs instead of the projected 200 million.[45] *Causa remota*, and much more important, were the large balance of payments deficits from extensive imports of railroad materials and, especially, the crop failure of 1846 followed by the glut of 1847. The Granger movement helped precipitate the collapse in the United States in 1873. The Grangers, who in some ways resembled the environmentalists of today, started in the late 1860s and early 1870s as activists for legislation that would control intrastate transportation by prohibiting discriminatory freight rates, establishing regulatory commissions, and setting maximum freight rates.[46] A very large volume of railroad securities had been sold on credit – including a number of 'superfluous and ridiculous' enterprises like the Rockford, Rock Island and St Louis line which had been sold at par and then declined to 6 cents on the dollar as the prospect of local control of freight rates put an end to optimism and triggered sales and then liquidation of these bonds.

One 'accidental' detonator of a crisis has been the sinking of a ship. In 1799, when interest rates ranged between 12 and 14 percent and the price of sugar was 35 percent below the peak before the convoy had broken through the blockade, British merchants shipped £1 million on the frigate *Lutine* destined for Texel in an attempt to assist in the Amsterdam crisis. The ship sank in a storm off the Dutch coast and the hope of alleviating the crisis was dashed.[47] During the 1857 crisis in New York the news that the steamer *Central America*, bound from Panama to New York with a cargo including $2 million in gold, was overdue came at the time of extreme distress in Philadelphia, Cincinnati, and Chicago. Two days later it was learned that the ship had gone down, uninsured.[48]

An accident may precipitate a crisis, but so may action designed to prevent it – or action by the authorities adopted to achieve other objectives. The matter was well put by H.S. Foxwell apropos the crisis of 1808–9:

> To refuse accommodation altogether is always held to be dangerous. To make personal reference is invidious, especially for a National Bank. It is just possible that the Bank might have resorted to the expedient used in 1795–96, I mean the granting of pro rata discounts... [In seeking to contract the circulation] it must have put severe pressure on the market and risked the creation of a panic... The Bank was responsible for the solvency of this crowd of small, ill-managed institutions [country banks], but dared not call them to account, on peril of provoking a general collapse of credit.[49]

Foxwell posed the dilemma neatly. Not to apply discipline will let the credit market expand to dangerous levels; to apply discipline may prick the bubble and induce collapse.

The pinprick

The nature of the bubble is that eventually it will be pricked, and then as with a child's balloon the air may escape suddenly. The bubble in Japanese real estate and stock prices was pricked by the incoming governor of the Bank of Japan who at the beginning of 1990 instructed the banks to limit the growth of real estate loans relative to their total loans. The reduction in the rate of growth of real estate loans meant that some individuals and firms would no longer be able to get enough cash from new loans to pay the interest on old loans, and so they were obliged to sell some real estate. But if it hadn't been this instruction, some other event would have pricked the bubble.

The late 1990s bubble in US stock prices was pricked by the Federal Reserve in 2000 when it sought to withdraw some of the liquidity that it had provided in anticipation of the Y2K problem. The bias of the Fed is to adjust to the anticipated problems by providing the system with more liquidity.

The bubbles in many of the Asian countries burst in 1997 because of the 'contagion effect'. The devaluation of the Thai baht on 2 July 1997 was like a clarion call; each of the Asian countries (except Taiwan and Singapore) had trade deficits that were financed with money borrowed from abroad. Asian firms were eager to borrow dollars because the interest rates generally were significantly lower than the interest rates on loans in their own currencies. When the baht was devalued, the foreign lenders recognized that the Asian countries would no longer be able to maintain the value of their currencies if they were no longer able to borrow foreign money. The money inflow declined and a self-fulfilling prophecy was triggered.

Policies adopted to deal with a crisis often encounter lags. Raising the discount rate in the face of an external drain of cash may induce a return flow. An increase in the discount rate of the Bank of England to 10 percent can 'draw gold from the moon' in the folklore of the City, but how long would it take to produce that result? The issue was debated between the Banking and the Currency schools in the context of the Bank Act of 1844 and the need either to suspend the Act or for the Bank to act as a lender of last resort. In 1825 and again in 1836, speculation in boom conditions led to an outflow of gold and financial stringency. On one interpretation, the boom broke before the Bank of England belatedly raised its discount rate in an effort to reduce its liabilities; thus, a combination of tight money and declining commodity prices produced the crisis and induced the Bank to reverse course and to reduce interest rates.[50] The Banking School believed that the increase in the discount rate produced an immediate inflow of money. The Currency School, on the other hand, had two wings; one believed there would be an immediate return flow and the other, represented by Lord Overstone, thought bank rate would have an impact only after a lag so that a lender of last resort would be required to fill the gap.[51]

Hawtrey pointed to a lag at the level of commercial banks and internal drains, based on backlogs:

> Bankers may take proper steps, but panic because they work slowly: They may have really checked the fundamental danger of the position ... stopped the stress of new orders ... and yet the demand for fresh credits and the drain of cash may go undiminished. The consequence may be a state of panic among the bankers, who, unaware of the cause of the apparent ineffectiveness of the measures they have taken [the working off of the backlog of old orders], despair of saving themselves from failure, call in existing loans regardless of the embarrassment of debtors, and precipitate a series of bankruptcies among their customers and themselves.
>
> The fact is that there is no golden rule for keeping the extension of credits within bounds.[52]

Apart from lags and mistakes in discount policy, the authorities may precipitate a panic by brusque action in the early stages of distress. In the summer of 1836, with credit extended in acceptances drawn by American houses on British joint-stock banks, the Bank of England then refused to discount any bills that carried the name of a joint-stock bank and specifically instructed its Liverpool agent not to rediscount any paper of the three so-called 'W banks' (Wiggins, Wildes, and Wilson) among the seven American banks in Britain, an action that 'seemed vindictive'[53] and led immediately to panic.[54] The Bank of England had to reverse its policies. It had long conferences with the 'W banks' in October, extended credit to them in the first quarter of 1837, but was unable to prevent their failure in June of that year. The Bank acted to dampen the extension of too large an amount of credit. But credit is delicate; expectations can quickly be altered.

Panic in the form of a run on a bank or banks has usually been started by small depositors as happened in the 1980s in Ohio and Maryland and Rhode Island, where some of the state-chartered banks relied on state-chartered insurance arrangements because they charged lower premiums than the FDIC. (All federally chartered banks and thrifts were required to participate in the federal government's deposit insurance programs.) In contrast stock market panic often resulted from selling by big-money insider speculators or institutional investors such as mutual funds, pension funds, and insurance companies, perhaps severally following similar program trading. The run on Franklin National Bank was initiated by other banks, and especially the large money-center banks in New York that refused to take the counterpart of Franklin's forward foreign exchange contracts or to lend it federal funds or to buy Franklin repurchase agreements, except at rates of interest that reflected deep distrust.[55] Similarly the run on the Continental Illinois Bank in 1984 was triggered by the reluctance of other large banks to renew maturing deposits in the federal funds

market and in the offshore deposit market. Smaller depositors were protected by the FDIC deposit insurance. In the stock market meltdown of October 1987, the Fidelity group of mutual funds of Boston was a large seller in the London market before the New York Stock Exchange opened on the nineteenth. These orders were communicated back to New York, where they had accumulated into a mountain of sell orders by the time the market opened for trading. Fidelity's sales were a response to the redemptions by the holders of its mutual funds rather than for its own position, although it may have wanted to raise cash in anticipation of future redemptions – and before stock prices fell further.

The clumsiness of the International Monetary Fund triggered a run on many of the banks in Indonesia during the early days of the Asian crisis in 1997. The IMF induced the Indonesian government to take over and support fifteen of the large private banks, in effect guaranteeing their deposits. But the remaining fifty or so other private banks then were placed in a nether-world position and were subject to runs as the depositors scrambled to withdraw funds before the banks collapsed.

Crashes and panics

A crash is a collapse of the prices of assets, or perhaps the failure of an important firm or bank. A panic, 'a sudden fright without cause' (from the god Pan, known for causing terror), may occur in asset markets or involve a rush from less liquid securities to money or government securities – in the belief that governments do not go bankrupt because they can always print more money. A financial crisis may involve one or both and in either order. The collapse of the stock of the South Sea Company and the Sword Blade Bank almost brought down the Bank of England. The 1929 crash and panic in the New York Stock Market had adverse consequences in both the commodity and real estate markets, and the seizure in the credit markets led to sharp declines in income and employment and output. But there was no panic in the money market; interest rates did not increase because the Federal Reserve was providing liquidity to the market.[56] In 1893 lack of confidence in the ability of the United States to maintain the gold standard under pressure from the silver interests led to money market pressure, bank failures, and downward pressure on prices of securities.[57]

The system is one of positive feedback. The debt-deflation cycles involve a decline in prices of assets and commodities which leads to a reduction in the value of collateral and induces banks to call loans or refuse new ones; firms sell commodities and inventories because their prices are in decline, and the decline in prices causes more and more firms to fail. Households sell securities and firms postpone borrowing and investing and prices fall still further. Further declines in the value of loan collateral lead to further liquidation. The

failure of firms means that the banks incur loan losses and fail. As banks fail, depositors withdraw their money (this was particularly true before deposit insurance).

Deposit withdrawals require more loans to be called, more securities to be sold. Trading reforms, industrial firms, investors, banks in need of ready cash – their riskiest securities cannot be sold at any price, they are forced to sell their best securities and so their prices decline. The banks may carry the firms, corporations, and households that are known to be in trouble for a time in the expectation or hope that prices will increase again and refloat the frail bark of credit. Bank examiners may *in extremis* look the other way as banks continue to value loans and securities at cost rather than at market value, extend loans due, or add to loans of delinquent borrowers so they can pay some of the interest. But when bankruptcy occurs, the nettle of bad loans must be grasped. Prices, solvency, liquidity, and the demand for cash – in German *Bargeld*, in French *numéraire* – are related. Not only banking institutions: as Sprague stated, households, firms, and banks are 'very similar to a row of bricks, the fall of one endangering the stability of the rest'.[58] The metaphor is a cliché, but nonetheless appropriate.

At the height of the panic, money is said to be unavailable. Descriptions are frequently exaggerated, not least about 1825:

> Bankers in Lombard Street called on the Governor [of the Bank of England] on Sunday [after the panic of country banks had reached Pole, Thornton & Co. on 12 December] to warn that if such a house, drawn on by 47 country banks, were allowed to stop, a run would take place on every bank in London.

> It was allowed to stop. A panic seized upon the public, such as had never been witnessed before: everybody begging for money – money – but money was hardly on any condition to be had. 'It was not the character of the security,' observes the *Times*, 'that was considered: but the impossibility of producing money at all.'[59]

This was the occasion when the failure of seventy-three banks brought Britain, according to Huskisson, within twenty-four hours of barter.[60] 'It was, as the Duke said of Waterloo, "a damned nice thing – the nearest run thing you ever saw in your life".'[61] Barter was avoided by exchanging silver for gold with the Bank of France, and by the luck of the Bank of England in finding, just as it ran out of £5 and £10 notes (which were then all it issued), a block of £1 notes left in the vault from 1797. With government approval these were issued on 17 December and 'worked wonders'.[62]

In 1857 New York Central stock went from 93 to 61, Reading from 96 to 36.[63] The price of pork fell from $24 a barrel to $13; flour, from $10 to $5

or $6.[64] In September interest rates rose from 15 to 24 percent as 150 banks in Pennsylvania, Maryland, Rhode Island, and Virginia failed in the last four days of the month. The panic reached a peak in October, when 1415 banks in the United States failed and interest rates rose from 60 to 100 percent per annum – for monies borrowed for a few days.[65] Very high rates of interest, such as 4 percent a day, have sometimes been quoted for a particular kind of loan, as for call money in 1884, when commercial discounts continued at 4.5 to 5 percent a day for first-class endorsed paper;[66] or 5 percent a day at the peak as a premium for cash at the onset of the panic of 1907.[67] Perhaps the apogee of the liquidity squeeze was recorded in 1907, when one bank paid $48 per $1000 for the cash gate receipts of the Harvard-Yale football game.[68] The mild and short recession in 2001 after the massive implosion in US stock prices resulted from the abrupt change in the policy of the Federal Reserve and its rapid and aggressive move to reduce interest rates. The result was a mortgage refinancing boom; millions of individuals refinanced their mortgages at lower interest rates and used some of the cash obtained in the refinancing to buy autos and other consumer durables and to go on vacations. The Fed reduced short-term interest rates to 1 percent and since the inflation rate was nearer 2 percent, real short-term interest rates were negative. The property market boomed; house prices increased sharply in New York, Boston, Washington, Miami, Phoenix, Las Vegas, and Los Angeles. Skeptics wondered whether the deflationary effects of the implosion of the stock price bubble had been largely offset by a bubble in the housing market.

Will the storm subside, the flood crest and fall? Or will boom and crash spread from one market and country to another and the steps taken locally and internationally fail to halt panic and reverse the damage?

6
Euphoria and Paper Wealth

Consider the birthdates of some of the tallest buildings in the world. The Empire State Building in New York City – 1250 feet tall – was started in 1929, at the peak of a bubble. In the late 1980s it seemed like eighty percent of the building cranes used to construct tall structures were in Tokyo. By the mid-1990s many of these cranes had migrated to Shanghai and Beijing, and then they moved to the Persian Gulf. Now the tallest building is the Burj Dubai in the United Arab Emirates, completed in 2010.

World's tallest buildings		
	Year	Height (in feet)
1. Burj Khalifa, Dubai, United Arab Emirates	2010	2717
2. Taipei 101, Taipei, Tawain	2003	1670
3. Shanghai World Financial Center, Shanghai	2008	1614
4. International Commerce Center, Hong Kong	2010	1588
5. John Hancock Center, Chicago	1969	1499
6. Petronas Towers, Kuala Lumpur	1993	1483
7. Nanjing Greenland Financial Center, Nanjing	2009	1480
8. Willis (formerly Sears) Tower, Chicago	1974	1450
9. Trump Interational Tower, Chicago	2009	1389
10. Jin Mao Tower, Shanghai	1999	1380
11. Two International Financial Center, Hong Kong	2003	1362

Tall buildings are as economic as the Pyramids in the Egypt. As the height of a building increases, the share of total floor space needed for elevators and non-rentable space increases in relation to the rentable floor space. Moreover tall buildings need to be 'stronger' since the lower floors must support the weight of more upper floors. (The higher the cost of land, the stronger the economic case for tall buildings – there is an abundance of land on the shores of the Persian Gulf and in Malaysia, if not in Tokyo and Hong Kong.)

These towers of eighty, ninety, or one hundred stories are a visual manifestation of asset price bubbles – and the willingness of governments and private firms to flaunt their wealth by reaching for the sky. A 'mine is bigger' syndrome. So, however, are the number of concert halls and the extensions of art museums and the student unions on college and university campuses. Many of these cultural centers are financed with gifts from wealthy individuals who have become much richer and more generous when asset prices have soared and economic euphoria was a growth stock.

The economic rationale for tall buildings is remindful of the remark of Willie Sutton, America's most famous bank robber, on his choice of profession. 'That's where the money is.' The surge in wealth in a bubble leads to economic behavior that would appear as exceptional – the squandering of wealth – in normal times.

The market for corporate jets soared between 2002 and 2007; there were waiting lists to buy the large Gulfstreams and impatient buyers paid several million dollars to buy the positions in the queue of those with earlier delivery dates.

The association between asset price bubbles and economic euphoria is also strong. One of the best-selling books in Tokyo in the late 1980s was *Japan as Number One*. The World Bank published *The East Asian Miracle* several years before the bubble in real estate prices and stock prices in Thailand and Malaysia and their neighbors imploded. Less has been heard of the New American Economy since the implosion of stock prices and the surge in the US fiscal deficit.

One feedback loop from increases in real estate prices and stock prices to the more rapid growth of GDP is from increases in household wealth to more household spending. Many households have savings or wealth objectives; as the surge in asset prices leads to higher levels of wealth they spend more on consumption goods – including conspicuous homes. (The mirror of increases in consumption spending is that the household saving rate declines.) The second loop is from the increase in stock prices to higher levels of investment spending; firms can raise cash from investors at lower cost and because of the decline in the cost of capital, they undertake new projects that would have been less profitable if this cost had not declined. Thus the 'cost of capital' to a firm varies inversely with the price of its stock; the higher stock prices are, the lower their cost of capital and the larger their investments in plant and equipment.

The cliché is that 'stock prices are a leading indicator' of changes in economic activity. The response is that changes in stock prices have forecast six of the last three recessions. US stock prices began to decline four to six months before the collapse of the economy at the beginning of the 1930s. The Japanese economy began to decline after stock prices started to fall at the beginning of 1990.

The relationship between changes in wealth and changes in spending is symmetrical; when asset prices decline, economic activity slows. During economic expansions business firms borrow more in response to the increase in their market value. Banks relax their loan criteria and increase their loans. When asset prices implode, the banks incur loan losses, and some banks may be so extensively de-capitalized that they are forced to close or to merge with a better capitalized institution or obtain capital from the government.

The strong correlation between the increases in asset prices and economic expansions leads to the question of whether the dominant influence is from changes in asset prices and wealth to the economy or whether instead the dominant influence is from changes in the rate of economic growth to the asset prices.

Albania was one of several former communist countries that experienced a Ponzi-type deposit scheme soon after its transition from a command economy to what was to become a market economy. Financial regulation was virtually non-existent during this transition. Entrepreneurs promised to pay interest at the rate of 30 or 40 percent a month. At such rates, wealth increased rapidly; for example, if the interest rate was 35 percent a month then 1000 leks deposited at the beginning of the year would amount to 64,000 leks at the end of the year. Depositors had every incentive to watch their money grow in the bank rather than to withdraw cash from the bank. Some Albanians dropped out of the active labor force because their incomes from the interest compounding were much larger than their wages and salaries. Others increased their spending because their financial wealth was growing so rapidly. The managers of the deposit arrangement always needed to attract new cash to offset the cash that was being withdrawn to manage the arrangement, in effect for the day-to-day living expenses of the managers.

When the deposit scheme unraveled there were lots of angry Albanians. Economic activity slowed rapidly because households went into a savings mode to compensate for the dramatic decline in their wealth.

Asset price bubbles – at least the large ones – are almost always associated with economic euphoria. In contrast, the bursting of the bubbles leads to a downturn in economic activity and is often associated with the failure of financial institutions, frequently on a massive scale. The failure of these institutions disrupts the channels of credit and thus contributes to the slowdown in economic activity and the sluggishness in the economic recovery.

Tulipmania

The price of Dutch tulips increased by several hundred percent in the autumn of 1636 – and the increases in the prices of the more exotic species of bulbs were even larger. Some analysts, especially those with a strong commitment to

rationality and market efficiency, have questioned whether the use of the term bubble is appropriate to describe the increases in tulip prices. Tulip bulbs are subject to a cobweb-like growth behavior; once planted, a bulb may develop for six to eight months before it begins to bloom – and then each bulb produces many little bulbs.

Not only the exotic varieties of bulbs were affected; ordinary garden-variety tulips such as the Gouda, Switzer, or White Crown that were traded among simple folk at so-called colleges or public houses also soared in price.[1]

The excitement around tulips began in earnest after September 1636, when bulbs were no longer available for examination since they had been planted to bloom during the following spring. Some of the buyers committed to pay for the 'merchandise' that was buried in the ground and that they could not see at the time of purchase. The excited bidding of November and December 1636 and January 1637 was conducted without any specimens in evidence.

The down-payments for the purchase of bulbs were frequently made by a form of barter.[2] Simon Schama, the historian, provided examples; in one case, for a pound of White Crowns (*Witte Croon* in Dutch, sold by weight because of its ordinariness) four cows were paid at the time of purchase and then Fl. 525 was to be paid on delivery (presumably the next June). Other down payments consisted of tracts of land, houses, furniture, silver and gold vessels, paintings, a suit and a coat, a coach and dapple-gray pair; and for a single Viceroy (rare), valued at Fl. 2500, two *lasts* (a measure that varied by commodity and locality) of wheat and four of rye, eight pigs, a dozen sheep, two oxheads of wine, four tons of butter, a thousand pounds of cheese, a bed, some clothing, and a silver beaker.[3]

The changes in the prices of tulips were not isolated from developments in the economy. The Dutch economy had been depressed during the 1620s when war with Spain was resumed after a twelve-year truce, but recovered impressively in the 1630s. The prices of shares in the Amsterdam Chamber of the Dutch East India Company doubled between 1630 and 1639, mostly after early 1636, going from 229 in March 1636 to 412 in August 1639, and then increasing another 20 percent to 500 by 1640. House prices had fallen in the early 1630s but 'shot up' in the middle of the decade. Investment surged in drainage schemes, in the West Indies Company and in canals.[4] Jan de Vries wrote of the *trekschuit*, a system of horse-drawn passenger barge canals which got under way in 1636 and reached a 'fever' in 1640. Construction was undertaken between pairs of towns to make the travel of merchants and officials more dependable than by sailing ships, which depended on the wind. Two lines from Amsterdam to smaller towns were decided on in 1636, and one between Leiden and Delft. Building of the complex network reached a peak in 1659 and 1665, but de Vries connected the project to the tulipmania and to the explosive growth of the Dutch economy between 1622 and 1660.[5]

Jonathan Israel wrote that the tulipmania should be viewed against the background of the general boom and as a mania of 'small-town dealers, tavern-keepers and horticulturalists' while for the most part the wealthy made money in other ways.[6] This perspective undermines one of Peter Garber's points that there could have been no tulipmania because there was no depression once bulb prices declined.[7] The Dutch economy slowed in the 1640s before putting on a tremendous growth spurt from 1650 to 1672, which involved luxury housing, civic buildings, and paintings. However the market for paintings collapsed with the French invasion of 1672.[8] At the height of the boom there was a 'mania' for clocks and clock towers. In Leiden a clock was installed in a tower at the top of the White Gate where at the *trekschuit* station passengers were loaded and unloaded, to assure the punctuality of the personal barges.[9]

The decline in the prices of tulip bulbs led to a decline in economic activity, and the causal connection was that households were less eager spenders as their wealth declined.

The stock market and real estate

Many bubbles in stock markets are related to bubbles in real estate. There are three different connections between these two markets. One is that in many countries, and especially smaller countries and those in the early stages of industrialization, the market value of real estate companies, construction companies, and of firms in other industries closely connected with real estate including banks accounts for a relatively large share of the aggregate stock market valuation. A second connection is that individuals whose net worth has increased sharply as a result of the increase in real estate values buy stocks to diversify their wealth; there aren't many other ways to diversify wealth. The third connection is the mirror-image of the second; the individual investors who have profited extensively from the increase in stock market valuations buy larger and more expensive first homes and they also buy second homes. The changes in the prices of apartments at the Manhattan real estate market are closely tied to the bonuses on Wall Street.

Homer Hoyt's *One Hundred Years of Land Values in Chicago*[10] traced five cycles of boom and relapse in real estate prices to the growth of Chicago. The boom in the US stock market in 1928–29 was linked to increases in prices of raw land, residential sites, and to commercial buildings in both the central business district and in the suburbs. Hoyt quotes from a *Chicago Tribune* editorial of April 1890:

In the ruin of all collapsed booms is to be found the work of men who bought property at prices they knew perfectly well were fictitious, but who

were willing to pay such prices simply because they knew that some still greater fool could be depended on to take the property off their hands and leave them with a profit.[11]

Chicago's reputation for real estate booms was such that Berlin, overindulging in real estate speculation in the euphoria of victory over France in 1870–71, was called 'Chicago on the [river] Spree'.[12] The booms in Berlin and Vienna in 1873 were related to the changes in prices on the New York stock market. One writer claimed that in Chicago in 1871, every other man and every fourth woman had an investment in house lots.[13] The bubbles expanded in parallel until the summer of 1873.

The spread of euphoria from one market to another reflects that when asset prices are increasing at a rapid rate, the widows and orphans climb on the bandwagon. Capital gains can be made without any special skill. When asset prices collapse, shareholders know they are in trouble and must reduce their indebtedness; those investors who have high leverage recognize that their wealth is declining much more rapidly than stock prices and so they sell – or their positions are sold by their brokers and lenders.

Speculators in real estate initially feel no such compunction. Their debts are loans on extended terms, and not callable like day-to-day brokers' loans. They have real assets, not just paper claims. Most choose to wait for the recovery that they think is just over the horizon.

The economic downturn leads to a drying up of the demand for real estate investments. Taxes and interest on loans, however, continue without interruption. Hoyt wrote that slowly but inexorably the speculators in real estate are ground down. The lenders to the real estate speculators, and especially the bank lenders, incur large loan losses. One hundred and sixty-three of 200 banks in Chicago in 1933 suspended payment. Real estate loans in default, not failed stockbrokers' accounts, were the largest single element in the failure of 4800 banks from 1930 to 1933.[14]

Hoyt's analysis of the relation between the stock market and the real estate market can be readily applied to the Japan of the 1990s. The large decline in property prices meant that many borrowers defaulted on their loans. The bank loans to credit cooperatives and other types of financial firms declined sharply in value because these lenders also had made real estate loans that often were underwater. Finally the decline in real estate values led to a sharp decline in banks' capital since they owned large amounts of real estate.

The stock market's troubles of October 1987 cleared up brilliantly after the monetary authorities rapidly increased bank liquidity to forestall a shortage of credit. Margin requirements of 50 percent helped. But the agony in real estate was drawn out. Construction slowed as buildings were completed, but new starts were abandoned or postponed. Vacancy rates in office buildings rose

sharply, varying according to location, whether downtown, midtown, or in the 'edge cities' built in the suburbs during the 1980s boom.

Rockefeller Center Properties, Inc., had a $1.3 billion mortgage on the Rockefeller Center in midtown Manhattan after the complex had been sold to Mitsui Real Estate. The mortgage was held in a Real Estate Investment Trust (REIT). In 1987 the trustees sought to increase the income of the trust by using the cash from short-term loans to buy back bonds that were selling at a discount. The gains were paid out as dividends. In 1989 as the deterioration in the real estate market progressed, the trustees borrowed, using a letter of credit to get the cash to pay off the short-term debt. The president of the REIT said, 'It was a prudent thing to do at the time.'[15] Hoyt's analysis suggested that there would be an extended downturn in real estate values following a stock market crash. After prolonged agony, the REIT crashed.

The story of the collapse of the bubble in the Japanese real estate market in the 1990s begins in the early 1950s when GDP began to increase rapidly in both nominal and real terms from the sharply depressed values immediately after World War II. (Japanese per capita income returned to the 1940 level in 1951.) Exports increased rapidly, and the composition of exports shifted from cheap toys and textiles to bicycles and motorcycles and then to steel and automobiles and electronics. The government began to deregulate financial controls during the first half of the 1980s, and extensive efforts by the Bank of Japan to limit the appreciation of the yen in the second half of the 1980s led to rapid growth in the supplies of money and of credit.

Real estate prices increased steadily, although with substantial year-on-year variability. Because of financial regulation, the nominal interest had been below the inflation rates in the 1950s, the 1960s, and the 1970s; the real rate of return on bank deposits and other fixed price assets often was negative. The price index for residential real estate in six large cities started at 100 in 1955 and reached 4100 in the mid-1970s and 5800 in 1980; the owners of real estate were one of the few groups with a positive real rate of return. During the 1980s the price of real estate increased by a factor of five to six.[16] At its peak, the value of real estate in Japan was twice the value of real estate in the United States; and the ratio of the value of real estate to GDP in Japan was four times higher.[17]

The Nikkei stock market index, which started at 100 in May 1949, reached 6000 by the early 1980s. Stock prices surged in the second half of the 1980s and nearly reached 40,000 at the end of 1989. The volume of shares traded did not quite keep pace, going from 120 billion shares in 1983 to 280 billion in 1989.[18]

The increase in the real estate prices fed the boom in stock prices. Many of the firms listed on the Tokyo Stock Exchange were real estate companies that owned substantial property in central Tokyo and Osaka and Nagoya and a few other major cities. The boom in real estate prices and financial deregulation led

to a surge in construction activity. Banks owned large amounts of real estate and stocks so increases in their values led to increases in the value of bank stocks. Banks usually required that borrowers pledge real estate as collateral; the increases in the value of real estate meant that the value of the collateral increased, and the banks were eager to make more loans because they wanted to increase their size – their total footings – relative to other Japanese banks and to banks in the United States and Europe. Industrial firms were increasingly eager to borrow to buy real estate, since the profit rate on real estate investments was many times higher than the profit rates from producing steel and automobiles and TVs.

The rapid expansion of bank loans was facilitated by financial deregulation, which was partly a response to pressure from the US government. Officials in the US were motivated by the unevenness of the regulatory framework, since US firms found many regulations impeded their expansion in the Tokyo markets while Japanese firms found it much easier to expand in the United States. Moreover the US Treasury wanted Japanese financial institutions to buy US government securities as the US fiscal deficit increased.

Deregulation proceeded slowly and deliberately.[19] The deregulation of interest rates paid by banks on large deposits was staggered, with the minimum reduced by steps from ¥1 billion (for terms of three months to two years) to ¥500 million and ¥300 million in 1986, ¥100 million (and one month) in 1987, ¥50 million and then ¥30 million in 1988, and ¥10 million in 1989.[20] In the early stages of this process, the Bank of Japan reduced its discount rate from 5.5 percent in 1982 to 5 percent in 1983, 3.5 percent at the beginning of 1986 and 2.5 percent a year later. The 1986 reduction was taken simultaneously – with similar action by the Federal Reserve and the Bundesbank. In reverse, however, first the United States in the middle of 1987 and Germany in 1988 began increasing interest rates. The Bank of Japan waited until a new governor, Yasuki Mieno, took over in December 1989, and then restricted increases in credit for real estate. The crash began in January 1990 and became more intense when it became known that some major banks had made good losses on loans to favored customers and hidden these transactions by imaginative accounting.[21]

The euphoria in Japan was evident in many ways. There was a boom in corporate investments in plant and equipment. The vision in Herman Kahn's *The Emerging Japanese Superstate: Challenge and Response*, published in 1970, appeared to be becoming reality.[22] Japanese firms were preparing for a glorious global future. A construction boom followed: there was a real estate construction boom; several hundred new golf courses were built. A new office block next to the Tokyo railroad station was named 'The Pacific Century Building'.

Stock prices peaked on the last trading day of 1989 and declined by 30 percent in 1990. The trough for stock prices occurred in 2002 at a level that

was a bit more than 20 percent of their peak values. Real estate prices fell more slowly but almost as extensively.

The result of the decline in asset values was that many financial institutions were de-capitalized and remained in business only because of the implicit support of the government. A few were allowed (or forced) to close, although no depositor incurred a loss. The banks became owners of thousands of French paintings. Several hundred golf courses went bankrupt.

Economic growth plummeted. The inflation rate began to fall and then ten years later the price level began to fall. There was no way to recover the gold leaf that had been sprinkled on the desserts. The failures of firms meant that the banks took over title to the properties and sold them, which put further downward pressure on property prices which complicated the business plans of other firms. There was a downward spiral and concern about a debt deflation trap.[23] Commercial and industrial enterprises were going bankrupt at a steady rate of 1000 a month. Three large credit unions were rescued by the government. The capital adequacy problems of the banks and insurance companies were compounded by losses on their portfolios of foreign assets. Two economic experts on Japan have characterized the country's problems for the next decade as 'debt, deflation, default, demography and deregulation'.[24]

Commodity prices, asset prices, and monetary policy

The reduction in the Bank of Japan's discount rate, especially after 1986, which touched off the bubble, was stimulated by pressures from the United States and other industrialized countries and rationalized because the price level in Japan was steady. The appreciation of the yen from 240 to the US dollar in 1985 to 130 in 1988 led to downward pressure on prices of goods and services.[25]

A major question is whether central bankers should be concerned with asset prices. Most central bankers choose stability of the goods price level as the target of monetary policy,[26] whether it be wholesale prices, the consumer price index, or the gross domestic product deflator is not a critical issue. Most recently the policy mantra has been inflation targeting – central banks aim to achieve an inflation rate no higher than 2 percent. If, however, the implosion of a bubble in stocks and/or real estate leads to a significant decline in bank solvency, should the central banks be concerned with asset prices? In one view, asset prices should be incorporated into the general price level because, in a world of efficient markets, they forecast future prices and consumption.[27] But this view assumes that asset prices are determined by the economic fundamentals and are not affected by the herd behavior that leads to a bubble.

Central bankers traditionally have not been reluctant to raise interest rates to prevent an increase in the inflation rate. They are exceedingly reluctant to deal with asset price bubbles or even to recognize that they exist or could have

existed – although after 2008 they are well recognized. The decline of 40 percent in US stock prices from 2000 to 2003 is evidence that there had been a bubble in stocks. Similarly the decline of 30 percent in the price of US residential real estate between 2006 and 2010 indicates that there had been a bubble in property prices, which had been especially sharp in Arizona, Florida, and California. One question is why those outside the Federal Reserve found it easier to recognize that the increase in asset prices in both episodes was nonsustainable. And in that sense the much younger Federal Reserve of the 1920s seems more heroic in its statements about asset price developments.

7
Bernie Madoff: Frauds, Swindles, and the Credit Cycle

Charlie Ponzi is the most famous banker in American history, immortalized in the term 'Ponzi scheme'. Ponzi owned a small firm that sold deposits in one of the Boston suburbs in the early 1920s; he promised to pay the buyers of these IOUs 45 percent interest a year at a time when the traditional banks were paying two or three percent. Ponzi's operation was straightforward – he used the money that he received from the sale of deposits on Wednesday to pay the interest to those who had bought the deposits on Monday. The interest rate that he was paying was so high that most of those who bought the deposits on Monday were happy to keep their money with Ponzi, so they could earn 'interest on the interest'. But eventually some of the newly rich depositors decided to withdraw some of their money, and unfortunately, the money wasn't there. Ponzi went to jail. His scheme flourished for eighteen months.

Bernie Madoff managed the largest Ponzi scheme in American history; at the time of the collapse in December 2008 the liabilities were reported to be $50 billion or $64 billion – although the most recent estimates suggest that the losses will probably be in the range of $20 billion to $25 billion. (One measurement problem is deciding whether the losses should include only the amounts of the initial deposits that individuals made with Madoff, or whether these losses should also include the 'investment income' that they had earned over the years. Another involves whether the cash withdrawn from the accounts should be included.) Madoff had run the scheme for twenty years, so he may hold the world's record for both the largest and the most long-lived Ponzi scheme.

Every Ponzi scheme has a 'story' – an explanation for why the firm earns such high returns. Ponzi himself said he was arbitraging international postal reply coupons. (You don't really need to know the details.) Madoff said he had a 'split strike option strategy'.

Madoff's scheme differed from Ponzi's in one other important dimension – the rate of return on the funds entrusted to Madoff were in the range of 10 to

12 percent a year – a bit higher than stocks, and much higher than on bonds. Madoff's returns were steady but not spectacular.

Ponzi was flamboyant. Madoff in contrast was understated, and he created the impression that he accepted money to invest only from a few close friends and their close friends. Subsequent investigation showed that he had more than 4500 close friends, and that he relied on at least four or five 'feeder firms' to bring him the money. The payment arrangements to these feeders varied, they appeared to get one percent of the money up front and then another one percent each year the money stayed with Madoff. Subsequently the attorneys responsible for cleaning up the mess indicated that some of these feeder firms received rates of return of 40 or 50 percent a year on their own money that they invested with Madoff.

Investigation by the court-appointed attorney to seek funds for Madoff's clients showed that Madoff rarely traded any securities. In 2008 clients withdrew $11 billion from their accounts. Madoff went out of business when there was no more money to withdraw. Most of the clients probably withdrew the money because they were losing money on most of their other investments.

Ponzi was in jail for about two years and then he moved to Florida to sell real estate during the mid-1920s bubble around Miami. Madoff will be in jail until the end of his time on earth.

Ponzi and Madoff shared one attribute – they both traded in mis-information. Both lied about the rates of return that their clients would be earning. Much of the corruption in financial markets involves mis-representation about the values of inventories, the values of investments, and profit rates. But some corruption involves the theft of information and the efforts of outsiders to obtain information from insiders before that informa-tion is publicly available.

The implosion of a bubble always leads to the discovery of fraud and swin-dles. As long as asset prices are rising, investors are content and often are happy to borrow against the increase in their wealth if they need the money for their living expenses. When asset prices decline, some of them sell securities to get the cash for living expenses. And they generally sell their most liquid securities because they can't bear to take the large losses associated with selling illiquid assets.

Ponzi schemes tend to be personal, and the basic crime is that the promot-ers – Madoff excepted – almost always make promises about exceptionally high rates of returns from day one. Corporate chicanery involves false state-ments about the value of inventories and of other assets, and the profit rate with the difference that the fraud begins long after the firm has been estab-lished. Information is withheld or dissembled. Enron was the poster-child of

corporate fraud in the 1990s dot.com boom. Enron had been in business for many years – the fraud appears to have begun in the mid-1990s, when the firm had a massive need of money to finance its expansion. It then began its tumble into bankruptcy within a few months of the peak in US stock prices in 2000.

Measuring the supply of corruption is not rocket science. What is 'measured' is the corruption that rises to the surface; the corruption that remains undiscovered obviously cannot be measured. Corruption is discovered when money becomes tight and credit is less readily available, and when asset prices decline. Corruption increases in a pro-cyclical way, much like the supply of credit. Soon after a recession begins the loans to firms that were fueling their growth with cheap credit decline as the lenders became more cautious. In the absence of more credit, the fraud sprouts from the corporate edifice like mushrooms in a soggy forest.

Most fraudulent behavior is illegal, but some hovers on a fuzzy legal borderline. Should the award of options to senior management and employees be considered a cost, like wages, or should the award be buried in a footnote so that costs and profits are not affected? The answer affects how rapidly profits increase and presumably how rapidly the stock prices will increase. Should Henry Blodgett and Mary Meeker and Jack Grubman, the gurus of the telecom firms and dot.coms in the 1990s, have been obliged to inform the investing public of the rationale for their forecasts for the increases in the prices of individual stocks, or was it sufficient for them to announce the price targets for each firm at the end of the next six and twelve months – more or less like touts at the racetrack? (See 'Shills, touts and thieves', below, p. 137) Should the governmental authorities establish 'truth police' to prevent the investing public from being misled, or should this public be on its own in determining which statements made by the stock sales personnel are not strictly true? Certain business practices are legal, yet those engaged in such activities would be reluctant or embarrassed to have their transactions reported on the front pages of the *New York Times* or the *Wall Street Journal* or the *Chicago Tribune* or the London *Daily Telegraph* because the sunshine would awaken the victims to the con. Consider Enron, MCIWorldCom, Adelphia, Tyco, HealthSouth, Global Crossing – the stars of some of the financial excesses of the 1990s stock price bubble. Much of the fraudulent behavior initially had occurred in the mania phase as stock prices were increasing but was obscured in the froth of the bubble; the high-risk borrowers were able to refinance their maturing loans because the lenders were eager to increase their total loans. The investment banks believed in *caveat emptor*; in their view, customers old enough to vote should be capable of looking out for their own financial interests.

The theft of information and insider trading

Information is valuable, since it changes relative prices. After World War II, several US Congressmen got rich – or at least richer – because they just happened to buy some land near the sites where new exit and entry ramps would be built for the Interstate Highway System. Information about the size of the wheat crop – actually, projections of the information – lead to changes in price of wheat, so it would be valuable to have access to these projections before they become generally available. Information about the quarterly earnings of firms, especially when the change is large, often lead to sharp changes in stock prices. Mergers of firms, especially the takeover or purchase of one firm by another, often lead to sharp increases in the prices of the shares of the firm that is in play.

Corruption is multi-faceted. One form involves breaking laws, perhaps with the connivance of government officials. A great deal of corruption involves the misuse of information; someone may provide misleading information about profits, dividends, and possible mergers. Information may be stolen, and there are laws against insider trading, where well-connected individuals may seek to obtain information before it is generally available.

In this context, should Goldman Sachs have told some of the buyers of mortgage-related securities in 2006 and 2007 that the suggestions for the design of these securities had come from the firm that wanted to short them, or were the buyers regarded as sufficiently knowledgeable to conclude that someone believed that the prices of these bonds would decline? (Goldman agreed to pay more than $500 million to the Securities and Exchange Commission to keep its name out of the papers – about $250 million will go to the German bank that bought these bonds.)

Large CPA (certified public accounting) firms had been established to protect investors from miscounts of the numbers of beans and widgets that corporate firms report; the CPAs are there to verify the bean count. Arthur Andersen, at one time one of the 'Big Five' global accounting firms and the one with a reputation for paying exceptional attention to ethical issues, was 'captured' by several of the firms that it was auditing – firms that were paying its bills – and colluded in misleading investors. Did the law firms that provided legal advice to Enron and MCIWorldCom and their ilk discharge their obligations to investors as well as to their clients – or were the lawyers, who may have had billing rates of $800 an hour, also seduced by their clients?

The banks and the investment banks were on the front pages of the news-papers when property prices declined in 2007 and 2008. The Royal Bank of Scotland together with Santander Bank of Spain and Fortis of the Netherlands and Belgium bought ABN-AMRO for $100 billion in what has proved to be one of the most costly business decisions of the decade. (The smartest business decision of the decade was to sell ABN-AMRO to these eager buyers.) While there

was no obvious fraud, there was a lot of delusional grandiosity on the part of the buyers.

AIG (the acronym for the American International Group), then the world's largest insurance company, is the 'poster child' of the 2008 collapse because of the massive amount of financial assistance that it received from the US government to avoid bankruptcy. Most of its problems stemmed from the sale of credit default swaps (CDSs), which are a form of insurance purchased by the owners of bonds to protect themselves against loss if the firms that have issued the bonds go bankrupt. (These swaps are similar in concept to private mortgage insurance; the mortgage lenders require that the borrowers with less than 20 percent of the cash needed for the traditional down-payment buy insurance to protect themselves from loss if the borrowers default.) The buyers of the CDSs want the higher interest rates attached to the riskier securities but want to avoid the default risk attached to these securities.

A lot of the fraud in the mortgage market in the housing boom between 2003 and 2007 occurred at the 'retail level' – borrowers lied about their incomes and their credit histories, and the mortgage brokers knew that the borrowers were lying, but hey, 'It's business'. Several senior managers of Bear Stearns were indicted for misleading investors about the quality of the assets in the hedge funds that they managed – but the jury voted 'Not Guilty'.

The classes of swindles and corruption

Swindles, fraudulent behavior, defalcations, and elaborate hustles are part of life in market economies, more so in some countries than in others. Many of the swindles involve supplying misleading information about inventories and profits. Occasionally the swindle involves the theft of information or news – when the presumption is that in a market economy each investor should have the same information at the same time, when some individuals jump the queue it is a form of fraud.

Transparency International publishes an annual index of corruption; Finland has a lock on the number one position in terms of virtue and Iceland was close behind until the implosion of its massive stock price bubble in 2008. (A subsequent investigation revealed that the banks had been looted by their owners.) For many years Bangladesh, the Democratic Republic of the Congo, and Nigeria were considered the most corrupt. The United States is much nearer the top of this hit parade although it may have slipped several places in the rankings because of the extensive fraud and deception that occurred in corporate America during the stock price bubble of the 1990s and the real estate bubble ten years later.

A traditional form of swindling involves overstating the value of inventories. The McKesson Robbins scandal of the late 1930s involved the use of forged

warehouse receipts as collateral for loans. Billie Sol Estes, the Texas plunger of the 1960s, falsified the number of fertilizer tanks he had under lease and borrowed against the much larger fictitious number. Tino De Angelis stung American Express and other large lenders by using tanks of 'salad oil' in the 1960s as collateral for loans; Tino knew that that oil was less dense than water and he floated a six-inch layer of salad oil on top of twenty feet of water.[1]

Swindles that involve falsified statements about the value of inventories can be tested when the promises are made. Eventually the lenders wised up to Billie Sol Estes; someone went out and counted the fertilizer tanks. The lenders can be taken in by the falsified values of the collateral offered by the borrowers, and initially the lender's accountants don't catch the deception. Swindles in financial markets may involve statements about the growth of corporate earnings or about the 'warranted' prices of shares of individual firms. Corporations 'manage' their earnings so they appear less variable from one quarter to the next, because investors will pay higher prices for the stocks of firms with stable earnings. A typical statement is that the 'price of Amazon stock' will climb to $400 a share by 4 July; or the wording might be 'Our price target is $400 a share'. Another statement might be 'Corporate profits will increase at the rate of 15 percent a year for the next five years'. Some of the swindles in the financial markets involve 'excessive optimism' about the earnings of firms or future stock prices that those making the statements know are not likely to be true.

Wall Street makes a lot of its money selling stocks, and it prospers by having 'stars' who are paid – very well paid – to make statements that the prices of the shares of individual firms will increase; they're like the shills in front of the side-shows at carnivals who persuade the public to buy tickets to see the sword swallowers and the hermaphrodites. Stocks increase in price twice as often as they fall, so that the odds are that the 'market strategists' will be right twice as often as they are wrong – although being right on the direction is different from making the right calls on stock prices. The market strategists typically are reluctant to indicate that stock prices as a group will decline and very rarely suggest that the price of the stock of individual firms will decline (because the top executives of those firms would become furious and threaten never to bring any underwriting business to the investment bank ever again). If share prices decline when the pitch has been that they will increase and the touts become an embarrassment to their employers, well they've been well paid and they're expendable and, hey, it's business. It's not hard to find replacements.

Corruption can't be measured unless an economy or a society has laws or norms or rules that distinguish permissible from illegal or immoral behavior. Anything would be permissible and acceptable in a society without rules or norms, corruption would not occur because there would be no perceptible boundary between acceptable and non-acceptable behavior. Virtually every

society has rules or norms – these codes about what is and what isn't acceptable behavior were adopted to reduce the costs of doing business.

Laws differ among countries, practices that are legal in some countries may be illegal in others. Moreover the laws and the norms about non-acceptable behavior within a country change over time; some financial practices that were legal in the United States in the 1870s were illegal a hundred years later. Despite these differences across countries and over the years, there is a broader, more universal sense of acceptable financial behavior that is based on the Eighth Commandment: 'Thou shalt not steal'.

Corrupt behavior is part of virtually every economy. The number of transactions that appear to overstep both moral and legal norms increases in euphoric periods like the 1990s. Paradoxically, increases in personal wealth as the prices of stocks and real estate and commodities increase at 30 or 40 percent a year for several years seem to trigger an increase in fraudulent behavior by those who want even more rapid increases in their wealth. Some individuals wish to keep up with the Joneses, and they may blur the truth and cut a few corners. Some entrepreneurs and managers may skate close to the edge of fraudulent behavior because of an apparent increase in the reward-risk ratio; the potential increase in their wealth from bending the rules and deceiving the public may seem extremely large relative to the risk of being caught and fined and exposed to public embarrassment. Some may have calculated that they can make a big fortune because the likelihood that they will be detected is low; they may still get to keep half if they're caught. The odds on going to jail are low, and the prisons for white-collar crime are like modest country clubs with drab clothing and hard-surface tennis courts.

Crash and panic, with their motto of *sauve qui peut*, induce many to cheat in the effort to forestall bankruptcy or some other financial disaster. A little cheating today may avert a catastrophe tomorrow. When the boom ends and the losses become apparent, some individuals will make a big bet in the hope that a successful outcome will provide an escape from what otherwise would be a disaster.

Nick Leeson was a modest functionary – one of five or six employees – in the Singapore office of Baring Brothers, the venerable London merchant bank. Leeson traded options on stocks and especially options on the Nikkei, the primary stock price index in Tokyo. The London office of Barings had set a limit on Leeson's position, the maximum amount of the firm's capital that he could risk. Leeson bought and sold call and put options on the Nikkei; the purchase of a call option was a bet that Japanese stocks would increase in price, and the purchase of a put option was a bet that Japanese stocks would decline in price. The sale of a call option was a bet that the Nikkei would not increase significantly in price, and the sale of a put option was a bet that the Nikkei would not decrease significantly. When Leeson bought either call options or

put options, he had to pay a premium to the sellers of the options who were acquiring the price risk; when he sold options, he earned the premium because he was acquiring the price risk.

Someone in Leeson's office apparently made an error in a transaction which led to a loss in his trading account. Rather than acknowledge the error to the head office of Barings, Leeson sold some put options on the Nikkei; his plan was to use the premium income to offset the trading error loss. However, the Kobe earthquake led to a sharp decline in the price of Tokyo stocks, which meant that he incurred a loss on the options contract that was much larger than the premium income. Leeson then 'doubled his bet' in the hope that a profit on the new contract would offset the previous losses, the position would be closed and no one in the head office would be wiser. Unfortunately for Leeson he lost on the second bet. Each time that his position incurred a loss, he made another 'double up' bet in the hope that it was his turn to win. So it went, until the cumulative losses in his trading account were more or less equal to all of Baring's capital. Leeson probably made four or five losing bets in a row. Consider the probability that he could have made five successive losing calls on the traditional coin-flip – a run of five heads in a row is a 1 in 32 chance. If he had won just one of the bets, his misadventure would not have made the front pages of the newspapers and he would not have spent two years in a Singapore jail.

The gallery of rogue traders includes John Rusnak of the Baltimore office of the Allied Irish Bank who lost $750 million trading currencies before the head office in Dublin realized how large the losses were. Yasuo Hamanaka of the New York office of Sumitomo Bank lost several billion dollars trading in the copper market. Five traders for the National Australia Bank in Sydney lost several hundred million dollars trading currencies.

Leeson and Rusnak and the Aussies made an extended series of losing bets before being discovered. Their problems made the headlines because they ran out of cash to pay off on the losing bets before they had one winning bet.

The likelihood is high – very high – that there have been other rogue traders who started like Leeson and Rusnak and incurred losses on two or three or four of their 'double or nothing' bets before winning again. The bank's capital was not affected and their illegal and fraudulent behavior went undetected.

The approach toward financial fraud in this chapter is descriptive and anecdotal. *Guinness World Records* does not yet have a chapter on the magnitude of financial swindles and fraud. Financial chicanery occurred in the US economic expansion following the Civil War, and fraud was widespread in the Gilded Age of the 1880s. In the late 1920s a few bankers continued to sell the bonds of several Latin American countries after they had been told that the countries had stopped paying interest.

BCCI – the Bank of Credit and Commerce International – was a major financial beneficiary of the oil price shocks of the 1970s. BCCI had been chartered in Pakistan; the firm then established a network of branches and subsidiaries throughout the Middle East and in London and subsequently in some of the major cities in Europe and Africa. BCCI prospered following the surges in the oil prices in the 1970s; the Middle East was awash with money and some of its early depositors were wealthy sheikhs from the Persian Gulf. BCCI's subtle appeal was to Muslim depositors. BCCI made large loans to rulers for political reasons, which eased its regulatory problems and helped it expand.

BCCI may have been one of the largest-ever Ponzi schemes in the pre-Madoff days. Non-performing loans were refinanced, in effect delaying the recognition of losses. Some of BCCI's loans to a shipping magnate went sour, and the bank used fraudulent accounting to hide the losses; the auditors were bribed. BCCI dealt extensively in options to make good the loan losses, and the firm recognized 'in the money' options at their market value and valued 'out of the money' options, on which it had losses, at zero. BCCI continued to expand as long as it could sell its deposits to its 'captive clientele'.

When the bubble imploded in Japan at the beginning of the 1990s, the large banks headquartered in Tokyo and Osaka incurred massive losses on their loans that had been made to finance the purchases of real estate and stocks. Some of the regional banks in Japan had incurred even larger loan losses. Many of their loans for golf courses, hotels, and amusement parks were undertaken to promote local economies. The tradition that loans would be repaid was not well established in Japan; instead the practice was that the borrowers would use the cash from new loans to pay the interest on the outstanding loans so that their indebtedness would increase at the rate of interest. This practice of 'evergreen finance' was safe as long as the rate of increase in property values was higher than the interest rate.

Once real estate prices began to decline some of the hanky-panky in the loan approval process became known. A woman who owned a small restaurant in Osaka had borrowed the yen equivalent of several billion dollars from the local branch of Sumitomo Bank; she had a 'friendly relationship' with the local banker. Some of the real estate appraisers for the banks had been bribed – or perhaps frightened – by the *yakuza*, who had concluded that the low-risk way to rob banks was to secure loans based on highly inflated appraisal values of the real estate that was pledged as collateral. The senior officers of several of the large regional banks made loans to real estate developers so that they could buy land that was owned by the bank officials. Several senior officials of the august Ministry of Finance were entertained at a 'pantyless shabu-shabu restaurant' that had a mirrored floor; the suspicion was that these officials provided advance information to their hosts about impending changes in financial regulations.

When the real estate bubbles in Thailand and Malaysia and the other nearby Asian countries imploded and the banks incurred large losses, 'crony capitalism' suddenly appeared – favored treatment for certain borrowers that somehow had not been evident in the earlier years of exuberant economic growth. Indonesia had been a 'Suharto family business' for more than thirty years – and remarkably successful measured by the increase in its GDP. When Indonesia was doing well, the banks were happy to make loans to the business enterprises headed by President Suharto's children – indeed some of the banks were headed by his children. The banks were not especially concerned with whether the projects that would be financed with these loans would be profitable.

One of the persistent US financial headlines in the early 1980s was the impending financial disaster of the US savings and loan associations and the US mutual savings banks – thrift institutions that had been established to help Americans of limited means finance the purchase of their first home. The thrifts used the funds obtained from the sale of deposits with short-term maturities to buy fixed interest rate mortgages with long-term maturities. Because the maturities were not matched, the thrifts acquired a 'transformation risk' (sometimes called 'duration risk') that short-term interest rates could increase relative to long-term interest rates and the excess of the interest rates they earned on their loans over the interest they paid on their deposits would decline, and perhaps even become negative. The thrifts had lived with this risk for fifty years because the interest rates on short-term deposits had been regulated by US government agencies to limit 'excessive' price competition among the banks and thrifts for deposits. In the second half of the 1970s, however, interest rates on US-dollar securities surged, and many individuals withdrew their money from the thrifts so they could buy US Treasury bills and money market funds that offered higher interest rates than the thrifts were allowed to pay.

The conundrum for the thrifts was that if they raised interest rates on their deposits to levels that were competitive with interest rates on US Treasury bills (assuming that they could get the regulators to raise the interest rate ceilings) then their interest payments would be higher than the interest rates they were earning on their mortgages and their capital would erode and eventually disappear. If instead they sold mortgages to reduce the excess of their interest payments over their interest income, they would incur large and immediate losses because the sale prices of the mortgages would be below – probably far below – the prices they had paid for them.

Most thrifts decided to pay the higher interest rates on deposits; slow death was preferable. Each thrift could estimate the monthly rate of depletion of its capital and project the date when its capital would be fully exhausted.

Hundreds, then thousands of these thrift institutions failed. Initially a failed thrift was taken over by the Federal Savings and Loan Insurance Corporation or the Federal Deposit Insurance Corporation, the US government's deposit insurance guarantee agencies. These government agencies would sell the good loans of the failed thrifts and use the cash from these loan sales and their accumulated reserves to pay the depositors 100 cents for each dollar of deposits.

The accumulated reserves and the capital of both the FSLIC and the FDIC, which had been built up over nearly fifty years, soon were exhausted. Then the deposit insurance agencies could no longer afford to close those thrifts with negative net worth since they didn't have enough money to honor their deposit insurance guarantee; they needed a subterfuge to enable the failed thrifts to remain open. They discovered the policy of 'forbearance', which allowed these bankrupt thrifts to continue in business.

Inevitably the losses of the FSLIC and the FDIC would have to be made good, either by a transfer of taxpayer money from the US Treasury or by increases in the deposit insurance premiums charged to the surviving banks and thrifts. These banks and thrifts were vehemently opposed to paying higher premiums to cover the losses of their failed competitors.

Several members of the US Congress stalled the efforts to obtain funds from the US Treasury so the FSLIC and the FDIC could pay off the depositors and close the failed institutions; these elected officials wanted to use this financial disaster to force a deregulation of the financial services industry. Some of the failed thrifts were 're-capitalized' as 'phoenix institutions'; one failed thrift would acquire another and the asset side of the balance sheet of the acquiring institution would be credited with a large entry for 'good will' and there would be a corresponding increase in its capital. The idea – more precisely the hope – was that over the next twenty or thirty years the profits of the acquiring institutions would be sufficient so they could amortize the good will and their negative net worth.

Some entrepreneurs had the bright idea that the way to save the failed thrifts was to help them rapidly increase their deposits and loans so that their profits on new mortgages and loans would be larger than their losses on the older, low interest rate mortgages. In 1982 the US Congress reduced the regulations applied to the thrifts, which meant they would be allowed to buy almost any type of security. At the same time the ceiling on the maximum insured deposit that any individual might have in a single thrift was increased from $40,000 to $100,000. (However, individuals could easily circumvent this ceiling by acquiring deposits under slightly different versions of the same name; Mr Jones might have an insured deposit of $100,000, Mrs Jones might have an insured deposit of $100,000, and Mr and Mrs Jones together might

have a third insured account of $100,000. And the Jones family could buy separate deposit accounts for each child as well as a series of deposits in joint child accounts.)

These failed thrifts were eager to buy any loan or security that offered relatively high interest rates – and so they became one of the 'natural markets' for junk bonds. Initially the supply of junk bonds had been limited to those of the 'fallen angels'; bonds that had lost their investment grade rating because the firms that had issued them had fallen on tough times. The loss of the investment grade rating meant that some financial institutions could no longer hold these bonds, so the interest rates increased sharply.

Michael Milken of Drexel Burnham Lambert developed some market innovations that greatly increased both the demand for and the supply of junk bonds. The decline in interest rates and the increase in the rate of economic growth of the 1980s created an encouraging environment for the growth of the junk bond market. The sales pitch was that the excess of the interest rates on the junk bonds over the interest rates on investment grade bonds was much larger than the losses the owners of its bonds would incur when one of these borrowers failed.

Milken provided the financing for some of his friends and associates to gain control of thrift institutions and insurance companies and other firms that would be 'natural' buyers of the junk bonds. Milken provided 'comfort letters' to entrepreneurs contemplating a takeover of an established firm, assuring them that Drexel could raise the money they would need. Once his friends had ownership of these industrial companies, the firms would issue junk bonds that would be underwritten by Drexel, and Milken would place these bonds with the thrifts and the insurance companies that were controlled by his friends. Drexel established its own mutual funds that would buy the junk bonds that it had underwritten.

Milken had a money machine. Drexel earned underwriting fees when its client firms issued new junk bonds, fees for selling the bonds to the mutual funds, sales fees for selling the shares in the mutual funds to the American public, and management fees for operating the mutual funds.

Merrill Lynch – ubiquitous Merrill – began to broker deposits for thrifts; the thrifts in California and the Southwest began to offer much higher interest rates and Merrill's army of thousands of brokers became the channel for moving money from around the country to the thrifts controlled by Milken's friends. The deposits that the thrifts were selling were guaranteed by the US Treasury – which was all that the buyers of these deposits wanted to know.

Few of the corporate raiders financed by Milken had much industrial experience. They used Uncle Sam's money – money obtained from the sale of deposits guaranteed by the US government – to acquire more than fifty firms. They often 'overpaid' for these firms, but then they were paying with Sam's money.

Many of the firms were unable to earn enough to pay the interest on the outstanding junk bonds. Not to worry, these firms issued some new securities that relieved them of the obligation to pay the interest on their outstanding bonds and the Milken-friendly thrifts bought these securities too. The money machine continued to work as long as the junk bond market remained vibrant. (Is this reminiscent of Ponzi finance?) However, the junk bond market collapsed at the end of the 1980s, after Federal regulators changed the rules so that the thrifts could no longer buy these bonds. The liquidity disappeared from the junk bond market; the prices of these bonds declined sharply, and the bondholders incurred large losses. Most of the junk bonds that came to the market in the 1980s had been underwritten by Drexel, and about half of these junk bond issues went into default. Drexel incurred large losses on its inventories of these bonds and went bankrupt in 1992.

In the end the debacle of the thrifts cost the Federal government more than $100 billion. If the US Congress had supplied the deposit insurance agencies with the money to close the failed thrifts in the early 1980s when their capital was depleted, the cost to the American taxpayers would have been in the range of $20–$30 billion. A large part of the excess of the actual costs over the costs that would have been incurred if the failed thrifts had been closed in a timely manner resulted from the costs of closing down failed thrifts that had become major buyers of junk bonds.

Milken and his family became billionaires and he probably remained one even after paying a fine or penalty of $550 million and spending thirty months in a federal 'country club' (minimum-security prison).

Fact and fiction about junk bonds

Corporate takeovers and junk bonds led to an interesting literature. Consider the titles of both the fiction and the non-fiction. Tom Wolfe's *Bonfire of the Vanities* is a remarkable description of the values of New York's financial elite. *Predator's Ball* by Connie Bruck is a description of an annual party for the buyers and sellers of junk bonds. *Barbarians at the Gate* is a tale about the would-be takeover of RJR Nabisco; it is hard to decide whether the would-be acquirers or the target was less attractive. The title of James Stewart's *Den of Thieves* offers a clue to the story of Milken and his friends. Ben Stein's *A License to Steal* provides a record of large corporate failures in the 1980s and the 1990s and the number that had had their securities underwritten by Drexel Burnham Lambert.

Swindles and theft in the 1990s

There was a bumper crop of scams, swindles, and frauds in the United States in the second half of the 1990s. As outlined earlier, Enron was at the top of the hit parade of firms involved in the fraud and chicanery associated with the stock

market boom; the firm was born from the merger of two regulated natural gas pipelines. Enron had two 'arms'; 'Enron Heavy' made massive investments in an electrical generating plant in India and in water systems in Britain and in Mexico. 'Enron Light' began to expand rapidly in the production and trading of electricity and natural gas and broadband and anything else that could be traded in a wholesale market; the press hyped Enron as the single most important firm in the move to deregulate the market for electricity. The rapid expansion of both arms of Enron required large amounts of money for investment in plant and equipment, trading facilities, and software. Enron sold a lot of bonds and provided substantial revenues for its primary investment bankers, Merrill Lynch and Salomon Smith Barney. At its peak, the market capitalization of the firm was $250 billion, the market value of its stock at $100 a share was over $200 billion and the market value of its publicly owned bonds was $40 billion.

Soon after US stock prices began to fall sharply in 2000, Enron – at one time touted (by its own public relations flacks) as the seventh largest firm in the United States and described by *Fortune* magazine as one of the most innovative firms in the country – filed for bankruptcy. The market value of the stock was zero, and the market value of the bonds declined sharply.

One intriguing issue is when the top managers at Enron came to the proverbial 'fork in the road' and started on their elaborate path of financial chicanery. A large part of the total compensation of the senior managers of Enron would come from selling the stock options they had received from the company as incentives. The more rapid the rate of growth of Enron's profits, the higher the price of Enron stock; the more valuable the options, the wealthier the owners of the options. Enron's senior officers had a powerful reason to keep profits growing, since their individual wealth was geared to the stock price.

Enron was responding to the challenges presented by the stock market analysts on Wall Street. Quarter by quarter, the analysts predicted the earnings per share to the penny. There were plenty of examples of firms that failed to make their earnings estimates; their stock prices declined by 10 or 20 percent. The chief financial officers of Enron, as well as of other companies, had a strong incentive to 'smooth' earnings so they met the estimates of the Wall Street analysts.

Enron had entered into sale-and-leaseback arrangements with Merrill Lynch and JP Morgan for some electrical generating barges in Nigeria. Merrill and Morgan paid above-market prices for the barges so that Enron could realize a profit from their sale, which would contribute to Enron's reported profits for the year. Merrill and Morgan are not charities; because they paid above market prices for barges, the annual lease payments that they were charging were correspondingly higher. From Enron's point of view, the increase in profits in year one would be at the expense of profits in the next few years.

Enron was scrambling to increase this year's profits regardless of the negative impact on next year's profits. Next year's problem in growing earnings could be solved next year.

Enron had placed very high values on some esoteric futures contracts that were due to mature in five and ten years and which had no ready-reference market prices. Enron increased the value placed on these contracts from one year to the next and the increase in these values contributed to Enron's reported profits; a 1990s counterpart to Billie Sol Estes exaggeration of the number of fertilizer tanks that he had leased.

Enron's financial finagling led to increases in its reported profits of more than $1 billion at a time when the firm was concealing substantial debt, which had been buried in off-balance sheet financing partnerships, so-called special financing facilities or vehicles (SFVs). Many banks and industrial firms used these SFVs to remove debt from their balance sheets so they would be better positioned to increase the amounts they could borrow. The accounting rules provided that these SFVs could be considered independent entities as long as 3 percent of the equity in the SFV was owned by 'unrelated individuals'. Enron owned 97 percent of the partnerships and some of the individuals who owned the remaining 3 percent were senior employees and some were the firm's bankers. These bankers were eager to lend to and invest in these partnerships because Enron was a source of large underwriting income. These partnerships had names taken from *Star Wars* – JEDI, Chewco, and so on. The partnerships would borrow from the banks and other lenders as if they were independent from Enron and then invest the funds with Enron.

Enron used some of the cash obtained from the SFV's borrowings to support the price of its own stock. That is more or less a Nick Leeson go-for-broke strategy; if the stock price should fall, then the value of the partnerships would decline and they would be 'under water'. But that is the traditional practice and it had been used by American railroads in the nineteenth century.

The collapse of Enron led to the demise of Arthur Andersen, formerly the most respected of the large US accounting firms – although in the previous few years Andersen had been sued by the creditors of the Baptist Hospital of Arizona, Waste Management and several other audit clients that had gone belly-up. Andersen was accused of shredding documents after the Securities and Exchange Commission had started its investigation into Enron's finances.

Andersen had been selling $2 million a year of auditing services to Enron and $25 million of consulting services. The implication was that Andersen's desire to retain the consulting income clouded its judgment about the appropriateness of how Enron should measure its profits – and that's a charitable statement. Andersen was convicted of obstruction of justice in June 2002; hundreds of its clients did not want to be associated with a firm with a tarnished reputation, and the firm folded. Andersen was only one of the outside groups

that had been co-opted by Enron; its board of directors had multiple conflicts of interest since some members received consulting income from the firm. Enron had established several groups of advisers, usually from the media, on a $25,000 a year retainer for one meeting a year. Enron's chair, Kenneth Lay ('Kenny boy' in some political circles in Washington) was also a major contributor to politicians.

Criminal charges were filed against more than thirty Enron officials. There were three principal types of financial chicanery; the general thrust was to overstate income and understate the growth of indebtedness. Some of the top officials also under-reported their incomes to the Internal Revenue Service. Ken Lay and others were charged with providing misleading information about the financial well-being of the firm. Five were found guilty after trials and one was acquitted; fifteen pleaded guilty.

Jeffrey Skilling, the number two at Enron, was charged with thirty-five violations of the law, including conspiracy, securities fraud, wire fraud and insider trading and received a long prison sentence (24 years 4 months) in 2006. Andrew Fastow, the chief financial officer, pleaded guilty to the conspiracy to commit securities fraud and was imprisoned for a minimum of ten years without the possibility of parole; Leah Fastow, his wife, pleaded guilty to tax fraud and was jailed for six months. Ben Glisan, the corporate treasurer, pleaded guilty to wire and securities fraud and received a five-year sentence. Michael Kopper, a finance executive, pleaded guilty to fraud and money-laundering. The former directors of Enron agreed to a $168 million settlement of a lawsuit brought by the shareholders; $13 million of this amount came from their own pockets and the rest from the proceeds of an insurance policy. Lehman Brothers paid $222 million as part of this settlement and Bank of America paid $69 million.

WorldCom had been one of the most rapidly growing telecommunications firms in the 1990s, with more than sixty acquisitions. Bernie Ebbers, a Canadian-born former high school gym teacher and basketball coach in Jackson, Mississippi, was the maestro of its rapid expansion. WorldCom's earnings per share increased at a rapid rate which meant that the ratio of its stock price to its earnings was higher than the stock prices of the firms that were being acquired. WorldCom paid for its acquisitions by exchanging newly issued shares in WorldCom for the shares in the firms being acquired.

WorldCom's earnings per share were increasing at a rapid rate because it was 'buying' the earnings of firms that had lower stock prices. (WorldCom could not afford to buy a firm with a stock price higher than its own because that would lead to decline in the rate of growth of its earnings.) Ebbers was riding the proverbial tiger: to keep WorldCom's stock price high, the firm had to continue to acquire other firms that had lower stock prices. As these acquisitions led to an increase in the WorldCom's size, the only way it could maintain the

rapid growth of earnings was to acquire larger and larger firms. The problem – the generic problem for firms that grow their earnings per share by acquisition – is that the number of firms that remained attractive takeover targets was declining as WorldCom became larger.

WorldCom's last acquisition was MCI, one of the most innovative firms in the US telecommunications industry; through the use of microwave towers it had broken AT&T's monopoly on long-distance phone services. MCI was so large that the merged firm was re-named MCIWorldCom. In the effort to maintain the growth of its earnings, MCIWorldCom then attempted a merger with US Sprint, but the proposal was blocked by the US regulatory authorities.

Soon thereafter MCIWorldCom filed for bankruptcy – at the time the largest bankruptcy in US history. Subsequent investigation revealed that the chief financial officers in the firm had overstated earnings by claiming that $4.8 billion of everyday expenses were 'investments' – eventually the fudged numbers amounted to $10 billion. The major reason for this accounting fraud was to maintain the growth of earnings to support the stock price. Two of WorldCom's senior officials were arrested and charged, pleaded guilty and went to prison.

Later it became known that MCI had lent Bernie Ebbers more than $400 million in off-the-books loans. Ebbers had used his WorldCom stock as collateral for the loan – and he used some of the cash obtained from the loan to buy more WorldCom stock. Ebbers flunked the first lesson in investing – diversification of assets.

Enron and WorldCom had become victims of their own success. The stock market analysts on Wall Street had developed the practice of forecasting each firm's quarterly earnings. If the firm failed to meet the earnings targets, the stock price would decline by 15 or 20 percent. A big hit. So the senior financial officials began to play the game, in two ways. If the prospective earnings were likely to be much higher than anticipated, they had an incentive to prepay expenses or delay receipts, in effect postponing the recognition of earnings. The rationale was that the Wall Street analysts would raise the bar, and what might have been a one-time surge in earnings could become the new and higher benchmark. Conversely if earnings reported by the firm came in below estimates, they had a strong incentive to 'borrow' earnings from the future by delaying payments or advancing receipts. The problem would become more intense the next quarter.

Bernie Ebbers' trial on the charge of misleading the public about the financial condition of the firm began in January 2005. His lawyers claimed he was misinformed by Scott Sullivan, WorldCom's chief financial officer, and that Sullivan claimed that Ebbers directed the fraud to obtain a reduction in his own prison time. Ten of the former directors of WorldCom agreed to pay

$18 million from their own funds as part of a $54 million settlement with the shareholders. Ebbers received a 25-year prison sentence.

Dennis Kozlowski, head of Tyco, a $200 billion dollar conglomerate, and Mark Swartz, its chief financial officer, were accused by the Federal government of looting Tyco of hundreds of millions of dollars in two ways: they awarded themselves options to buy Tyco stock without the approval of Tyco's directors, and they tapped into Tyco's treasury to pay their personal living expenses. The Feds made a big deal of the $6000 shower curtain and the $2 million birthday party for the second Mrs Kozlowski in Sardinia; Tyco picked up 50 percent of the costs of the party. The press reported that one of the highlights of the party was an ice sculpture reproduction of Michelangelo's *David* that dispensed vodka from a vital lower body part.

Tyco had acquired several hundred firms in a wide variety of industries, mostly paid for with its stock but occasionally with the cash obtained by selling bonds and by borrowing from banks. Tyco massaged the earnings of the firms that it was about to acquire. After these firms had agreed to the acquisition, their earnings would be squeezed or otherwise temporarily depressed; then earnings would surge when they were integrated into the Tyco family, which contributed to the rapid growth in Tyco's earnings.

Tyco was also sensitive about its costs and especially its taxes; the firm's operating headquarters were in New Hampshire (a state with neither an income tax nor a sales tax) and its legal headquarters was in Bermuda, which enabled the firm to avoid paying US corporate income taxes. The State of New York alleged that Kozlowski had failed to pay the appropriate sales tax of $1 million on paintings that he purchased in New York City by asserting that the paintings would be shipped to New Hampshire; the empty crates that had contained the paintings were shipped there after the paintings had been removed to Kozlowski's New York apartment.

Kozlowski was found guilty at a second trial, the first trial ending with a hung jury. Tyco, unlike Enron and MCIWorldCom, did not go bankrupt; the fraud was personal not corporate.

The Rigas family was charged by the Federal government with looting Adelphia Communications, the sixth largest cable system in the United States, of $2.3 billion; the family's transgressions combined elements of the misbehavior of both Enron and WorldCom. The family collected more than $3 billion in off-balance sheet loans; the company inflated capital expenses and hid debt. Adelphia had started as a small family-owned firm; and the boundary between personal accounts and business accounts is often blurred in such situations. Many of the borrowings by the firm were guaranteed by the family, and many of the Rigas family borrowings were guaranteed by the firm. The family borrowed in large part to use the cash to support the firm's stock. The former

director of finance and the former vice president pleaded guilty. Two of the Rigas family members were convicted.

The founder and chair of HealthSouth, Richard Scrushy, was charged with an accounting fraud of nearly $3 billion; a trial began in January 2005, in which he was acquitted, although there was agreement that fraud was committed. Scrushy claimed that the fraud was committed by his subordinates and that they suggested that he managed the fraud so their own prison time would be reduced. At a second trial he was found guilty on a range of charges and jailed; later, in a civil trial, he was ordered to pay a $2.86 billion penalty, perhaps the largest in US history, according to the *Wall Street Journal's* report.

Sam Waksal, the founder and promoter of ImClone, was committed to a federal prison after pleading guilty to six federal charges. Waksal had told his father and his daughter to sell their shares in the firm because an adverse statement from the Food and Drug Administration would be released the next day and lead to a decline in the stock price.

Martha Stewart's five-month stay in a federal prison may have been related to Sam Waksal's decision to sell shares in ImClone. Stewart sold her shares but denied that she had received any information from Waksal. An interesting coincidence. The Federal government indicted Stewart on an 'obstruction of justice' charge. She was found guilty and went to jail.

Richard Grasso, the chairman and chief executive officer of the New York Stock Exchange, made it to the nation's front pages with the news that his retirement benefit package would be $150 million. The NYSE is owned by its members and both provides them with a trading floor and regulates their trading practices, as well as being, formally, a not-for profit institution. The Exchange is the arbiter of permissible trading practices. Most of the directors of the exchange are senior officers of the firms that are regulated by the exchange – or perhaps more precisely, are supposed to be regulated by the exchange. Grasso's retirement compensation package seemed high – very high – relative to the revenues of the Exchange (a $28m surplus the previous year) and to the retirement benefits packages of other leaders on Wall Street. The suspicion was that Grasso was soft on the regulatory front because he did not want to offend the directors who determined his salary and bonuses. After an uproar and pressure from pension fund investors Grasso resigned. Subsequently Grasso was sued by the New York Attorney-General over his compensation package.

Global Crossing swapped network capacity with other carriers to inflate its revenues. The company shredded documents and went bankrupt, but before it did, its founder and Chairman Gary Winnick received $800 million from the sales of his shares. He later paid substantial compensation in a class-action lawsuit by former shareholders, without, however, admitting wrongdoing.

Shell Oil agreed to pay a fine of $150 million to the US and the British regulatory authorities for overstating the volume of oil in its reserves in the ground. It's harder to quantify underground oil reserves than to count Billie Sol Este's fertilizer tanks. Several Shell officials were bounced from the firm, but none went to jail.

Mutual fund scandals of 2003

In 2003 a number of the large US mutual fund families – maybe half of the twenty largest firms that manage mutual funds – were accused of providing exceptional trading privileges to several hedge funds, which enabled them to earn large profits at the expense of the other owners of shares in the mutual funds. Some of these transactions were legal, most were not, and all violated the implicit contract between the mutual fund and its shareholders that each shareholder would be treated equally. Each US mutual fund is required to file a contract with the US Securities and Exchange Commission that stipulates its own rules for transactions – a description of what is permissible behavior and what isn't. A feature of most such contracts is that investors will not be allowed to make 'in-and-out' transactions – to sell the shares in the fund a day or two after the shares were purchased. The contract generally provides that if an investor does undertake an in-and-out transaction, the investor will be charged an additional 1 percent or the request to sell may be delayed. Another standard boiler-plate paragraph in the offering agreement is that the officers of the fund will not buy shares in a firm if the fund is buying shares in the same firm – or that if the officers buy shares in the firm, they will not 'front-run', that is, buy the shares in a firm for their own private accounts before the mutual fund buys the shares in the same company. Moreover the funds' standard practice was to disclose information about the securities that each fund owned only at the end of the quarter and not to provide any information on purchases and sales of individual securities between these end-of-quarter dates.

A number of hedge funds engaged in market timing transactions with the mutual funds; the hedge funds would buy the shares in the mutual fund if they thought it likely that the news would lead to increases in the prices of the stocks owned by the mutual funds and increases in the funds' net asset value per share. The hedge funds would trade on 'stale prices', like shooting fish in the proverbial barrel; the opportunity to trade on stale prices arose because some of the mutual funds owned foreign securities, and the markets in which these foreign securities were traded closed before the US market closed. For example, the Japanese market closed on Tuesday before the US market opened for trading on the same Tuesday. The mutual funds would inform some of the hedge funds about the stocks they were buying and selling during the quarter, although they would not provide the same information to other shareholders. Some mutual fund managers would trade the shares of their own funds on an in-and-out basis; some would front-run the purchases of the mutual fund.

Why did the mutual funds break their commitments to their shareholders?

Because a lot of business is quid pro quo, and the hedge funds might purchase shares in some mutual funds owned by one of the mutual fund management companies that needed a bit of a boost.

Canary Capital, a family-owned hedge fund, agreed to pay restitution of $20 million to the mutual funds and a penalty of $10 million to the Securities and Exchange Commission for its transgressions. Massachusetts Financial Services, the oldest US mutual fund, agreed to a $200 million settlement. Alliance Capital paid a $250 million penalty and to cut its fees by $350 million over several years. The founder and the major owner of the Strong Funds in Milwaukee resigned his management positions, sold his ownership interest and paid a penalty in the tens of millions. Morgan Stanley instructed its brokers to hustle its own mutual funds rather than funds in general in a form of payola; the clients of Morgan Stanley were not made aware that the advice of the brokers was not impartial.

The increase in the number of swindles and the scope of fraudulent behavior has attracted the economic theorists. Many of the swindles – often the low class ones – involve 'Ponzi finance', a pattern of financial transactions when a firm's interest payments are larger than its cash flows from operations.[2] Another theorist noted that those borrowers who can set the interest rates on their personal debts engage in Ponzi finance and use the cash obtained from new loans to make the debt service payments on their outstanding loans.[3]

Most of the entrepreneurs who establish a Ponzi finance operation appear to know and understand what they are doing; they live high for a few months or years before the authorities catch up with them. Occasionally an innocent becomes involved and seems unable to understand the pattern of finance.

Shills, touts, and thieves

If you're at the carnival or the circus or the county fair, you know that the shill for the sideshow is likely to engage in some exaggeration about the attractions of those that you might see if you pay the six bucks entry fee. If you are at the racetrack, you know that the touts that want to sell you their list of winners are not likely to pass the boy scouts' code of honor when they describe how many winners they have picked in the last several days and the last week – somehow your intuition suggests that if they were half as successful as they claim they are, they would not be hustling their tip sheets. But what interpretation should you place on the statements of the touts for Merrill and Morgan Stanley and Credit Suisse when they indicate their price targets for the shares of individual firms?

In a good year in the late 1990s Henry Blodgett earned $10 million, Mary Meeker $15 million and Jack Grubman $20 million. Henry, Mary and Jack were the apostles of the dot.coms and the telecommunications firms in the bubble years of the 1990s. They earned these rock-star incomes because they brought a lot of underwriting and trading business to their Wall Street employers. There is no rule of thumb that relates their incomes to the profits they generated for

their employers, but it's a safe bet that their employers had calculated that the profits they produced were at least three to five times larger than their incomes.

Henry Blodgett of Merrill Lynch became famous for setting price targets for the stocks of individual dot.com firms, and some of these targets were achieved; in the heyday of the bull market these achievements could be seen as testimony to his success as a forecaster, or to a self-fulfilling prophecy. Henry got a lot of attention because in some internal e-mails within Merrill he was deriding the attractiveness of some of the same stocks that he was promoting to the American public. Henry agreed to leave the security business never to return. But he kept the money.

Jack Grubman of Salomon Smith Barney, then part of the Citigroup family, became infamous because he changed his recommendation on AT&T stock after a polite request from his boss Sandy Weill. The apparent *quid pro quo* was that Jack wanted Sandy's help in getting Jack's twin daughters admitted to the kindergarten class at the 92nd Street YMHA. About the same time Citigroup made a contribution of $1m to the '92Y', Jack left Salomon Smith Barney with a $20 million golden handshake. Jack paid a penalty of $20 million and agreed not to work in the securities business again.

Parmalat (the name combines that of the city of Parma, Italy and *latte* or milk) the Italian dairy and food products company headquartered in Parma, used fabricated certificates of deposits to overstate its assets by $4 billion; the fraudulent CDs were prepared by superimposing one document on another in a copying machine. This fraud – the deception of the investing public and the investment banks – continued for more than ten years.

Boiler shops are another form of a swindle. Robert Brennan of First Jersey Securities operated a series of high-profile boiler shops. The sales brokers in these shops specialize in making cold calls to individuals about a stock, say Shazam Rockets. Shazam's share price is usually low, perhaps between $2 and $5 a share, and initially many of the shares in Shazam are owned by insiders – those who own the boiler shop. The insiders trade with each other and manage increases in the price of the stock say from $2 to $3 a share, and they then begin to make their cold calls, pointing to the 50 percent increase in the price of Shazam's stock in the last six weeks; they've learned that the gullible are much more willing to buy a stock whose price has increased rapidly. They also know that the unwashed investors have a preference for low-price stocks.

Swindles differ from ordinary robbery in that they abuse a trust. Daniel Defoe thought the stock-jobber cheat 10,000 times worse than the highwayman because the swindler robbed people he knew – often his friends and relatives – and ran no physical 'risque'.[4]

Swindles should be distinguished from bribery, both of government officials and of the employees of one business by those of another. These illegal and/ or immoral transactions involve both misrepresentation and the violation of an explicit or implicit trust between particular groups. Was Arthur Andersen corrupted by the $25 million a year in consulting fees from Enron? Was Bernie Ebbers bribed by Jack Grubman because Jack arranged for Bernie to be able to buy shares in twenty or thirty firms on the days when these stocks would be sold to the public for the first time? In the bubble euphoria, the odds were extremely high that the share prices would double on the first day of trading. Or did Bernie bribe Jack Grubman with the implicit threat that unless he was provided with favored treatment in buying newly issued stock he would take WorldCom's underwriting business across the street to Merrill Lynch or Morgan Stanley? The rules of the stock exchanges and the rules of the futures exchanges are like a 'code of purity' designed to instill confidence in the public that everyone will be treated fairly – but the rules are really designed to protect some of the members of the exchanges from the adverse consequences for the industry of recognition of cheating by others.

The National Association of Security Dealers (NASD) pursues disciplinary actions against those of its members who have violated the rules. In recent years, hundreds of its members have been at the receiving end of disciplinary actions.

Corruption has been discussed historically by Jacob van Klavaren, who was especially interested in corruption as a form of market transaction that facilitated getting things done that were profitable but forbidden, as in black markets, or simply dishonest, such as the looting of India by Clive or Hastings.[5] Van Klavaren also discussed the systematic embezzlement from the Royal African Company and the East India Company by insiders who skimmed the profits due to stockholders through contracts with companies that they controlled. In a similar manner, the Credit Mobilier in the United States in 1873 diverted profits from Union Pacific stockholders to the inside group run by Oakes Ames, a congressman from Massachusetts, and his congressional and business cronies. Drew, Fisk, and Gould milked the Erie Railroad in similar fashion.[6]

Swindling in financial markets is many-faceted – the directors swindle the stockholders, the senior managers swindle the directors, the security underwriters swindle both the owners of the firms that they are bringing to the public and the stockholders, borrowers swindle their bank lenders, and one group of employees may swindle another. Some swindlers issue fraudulent bills of exchange to the cost of those who own these bills when the fraud is discovered and who only then discover that they are holding counterfeit securities.

The line separating moral from immoral acts is now less fuzzy than it once was. One way that modernity is distinguished from backwardness is morality. In early stages of economic development, codes provided for honor and trust only within the family. Nepotism was efficient in these circumstances since strangers virtually had a license to steal. In 1720 a firm could buy a man's services but not his loyalty; to be a clerk was an invitation to start a new and competitive business at a time when embezzlement and fraudulent conversion were not regarded as crimes.[7] Lines between business and theft, commerce and piracy were not precise.[8] Hammond noted that it was not until 1799 that the borrowing of bank funds by bank officials was definitively ruled illegal,[9] although the 1720 House of Commons investigation of the South Sea Bubble ruled that the directors of the South Sea Company, having been guilty of a breach of trust in lending money of the company on its own stock, should use their own wealth to make good investor losses.[10]

A financial journalist writing a preface to a fictionalized biography of Ponzi drew a parallel between the 1920s and the 1970s and suggested that swindles were a product of inflation; his view was that when increases in the cost of living pinched family budgets, the heads of some households took additional risks to increase their incomes.[11] A somewhat different view held that destabilizing speculation in the absence of 'avoidable ignorance' is gambling, which provides utility to the participants even when they know they are likely to lose, in much the same way as participating in a lottery.[12]

Whether swindles should be included in the category of 'avoidable ignorance' is debatable. Cynics may share the belief of W.C. Fields that 'You can't cheat an honest man' and conclude that victims of swindles mainly have themselves to blame. *Mundus vult decipi – ergo decipitatur*: 'the world wants to be deceived, let it therefore be deceived'.[13] Some psychiatrists believe that the swindler and his victims are bound together in a symbiotic, love-hate relationship that provides satisfaction to both.

The theft of information

In 2009, Raj Rajaratnam, founder of the Galleon Fund group, was arrested by the FBI and indicted by the Securities and Exchange Commission for the use of 'inside information' – before the information was available to the public. His fund traded on that information and realized exceptional profits. Rajaratnam was charged with fourteen counts of securities fraud and pleaded not guilty; he asserted that the profits reflected his research. The government charged 26 people with illegal behavior; 21, including several who were traders for Galleon, have pleaded guilty. Some of the inside information involved corporate acquisitions, some involved large changes in corporate earnings. The jury was still deliberating when this edition went to press.

Fraud and euphoria

Fraudulent behavior increases in economic booms. Fortunes are made in a boom, individuals become greedy for a share of the increase in wealth and swindlers come forward to exploit that greed. The number of sheep waiting to be shorn increases in booms and an increasing number offer themselves as sacrifices to the swindlers. 'There's a sucker born every minute.' In *Little Dorrit*, Ferdinand Barnacle of the Circumlocution Office tells Arthur Clennam, who had hoped that the exposure of Mr Merdle's swindles would serve as a warning to dupes that 'the next man who has as large a capacity and as genuine a taste for swindling will succeed as well'.

Greed also induces some amateurs to commit fraud, embezzlement, defalcation, and similar misfeasance. The demise of Overend, Gurney and Company, the well-established 'Corner House', in 1866, after the original partners had retired and the firm had gone public, was brought about by a pleasure-loving gallant inside the firm who had appointed an outsider as adviser for £5000 a year, paid in advance and returned to the insider. D.W. Chapman, the insider, kept ten horses and entertained lavishly at Prince's Gate, Hyde Park. His outside adviser, Edward Watkins Edwards, a former accountant, recommended that many new activities be added to the bread-and-butter discount business: speculation in grain, production of iron, shipbuilding, shipping, railroad finance. The firm became 'partners in every sort of lock-up and speculative business' and Edwards drew commissions on each. By the end of 1860 the firm was losing £500,000 a year net even though it was earning £200,000 a year on its discount business. The bubble was pricked by the failure of an unrelated firm of railway contractors, Watson, Overend and Company.[14]

From the world of fiction – in this case, *Melmoth réconcilié*, by Honoré de Balzac – the comparable figure to Chapman is Castanier, a bank cashier, whose mistress, Mme Aquilinia de la Garde, has expensive tastes in silver, linen, crystal, and rugs, passions that prove his undoing. For a time he survives by issuing promissory notes. At the point of no return, when he finally calculates his debts, he might have been saved by leaving Mme de la Garde, but he cannot give her up. Finally, because of the impossibility of continuing his financial maneuvers, given the growth of his debts and the very large interest payments it is clear that he is bankrupt. But he prefers fraud to honest bankruptcy and dips into the bank's till.[15]

Swindling increases in economic booms because greed appears to grow more rapidly than wealth; it's as if the increase in wealth triggers an increase in greed. Dennis Kozlowski was one of the richest people in America; yet the money to pay for the $6000 shower curtain and the other furnishings in his apartment were taken from Tyco. Swindling also increases in times of financial distress as a result of a taut credit system which induces declines in asset prices; at that

stage the fraudulent behavior is undertaken to avoid a financial disaster. Ponzi resisted the suggestions of his associates that he should take the money and run and they in their turn swindled him.[16] The London banker Henry Fauntleroy forged conveyances of estates to use as collateral for loans. These men served as models for Augustus Melmotte, the swindler of Trollope's *The Way We Live Now*, who forged both a conveyance and a deed when the price of his Mexican railroad stock declined and he could no longer sell stock to raise cash.[17] John Blunt of the South Sea Company, Eugène Bontoux of the Union Générale, Jacob Wasserman of the Darmstäderund Nationalbank (Danatbank), and the directors of the Credit Anstalt all bought the shares of their firms in the open market to support their prices in the hope that they might sell more shares at a later date. A bank that buys its own stock to keep the price high reduces its own liquidity since the ratio of its cash holdings to its deposits declines as it pays out cash to obtain the stock. In 1720 the Bank of England borrowed using its own stock as collateral. Clapham noted that the Bank of England did not penetrate into the far wilder and 'absolutely dishonest' finance of the South Sea Company.[18]

The revelation of a swindle or embezzlement increases distress which in turn often precipitates a crash and panic. In 1772 Alexander Fordyce absconded from London to the Continent, leaving his associates to meet obligations of £550,000, largely in dubious acceptances of the Ayr Bank, if they could – but they could not. Fordyce had personally been short of East India stock, whose price had risen enough to wipe him out.[19] On 24 August 1857, it became known that a cashier in the New York office of the Ohio Life Insurance and Trust Company had embezzled almost all the assets of that highly reputed enterprise to sustain his stock market transactions; the news triggered a series of failures that reverberated in Liverpool, London, Paris, Hamburg, and Stockholm.[20] A kind of mid-nineteenth-century version of Nick Leeson.

In September 1929 the Hatry empire in London, a set of investment trusts and operating companies in photographic supplies, cameras, slot machines, and small loans, collapsed. Clarence Hatry wanted to expand into the steel business, but was caught using fraudulent collateral in an attempt to borrow £8 million to buy United Steel; his failure led to tightening of the British money market, withdrawal of call loans from the New York market, and a topping out of the stock market.

Bubbles and swindles

Some bubbles are swindles, some are not. The Mississippi Bubble was not a swindle; the South Sea Bubble was. A bubble generally starts with an apparently legitimate or at least legal purpose. What became the Mississippi Bubble initially started as the Compagnie d'Occident, to which the Law system added the farming-out of national tax collections. John Law owned about one-third of the Place Vendôme and other valuable real estate in Paris and at least a dozen

magnificent rural estates. His activities were not a swindle, but his financial demise reflected a mistake that was based on two fallacies: (1) that stocks and bonds were money and (2) that issuing more money as demand increased was not inflationary.[21]

In the South Sea Bubble, the monopoly of trade in the South Atlantic was purely incidental.[22] Very quickly, the consolidation of British government debt overwhelmed the South Atlantic trade aspects of the enterprise, and stock-jobbing overwhelmed government debt shortly thereafter. John Blunt and his insiders sought to profit on stock issued to themselves against loans secured by the stock as collateral. As they realized the capital gains from the increases in the price of the stock, they used the cash to buy estates; Blunt had six contracts to buy estates at the time of the collapse and a man named Surman had *four* contracts to buy real estate on which he owed £100,000. To get the cash to pay profits, the South Sea Company both needed to increase its capital and to have the price of its stock increase continuously. And it needed both increases at an accelerating rate as in a chain letter or in a Ponzi scheme.[23]

History has given less immortality to certain of Ponzi's forerunners. The former Munich actress Spitzeder paid 20 percent a year interest on the funds received from Bavarian farmers. She received 3 million gulden from these farmers. She and her helpers drew long jail sentences at the end of 1872. Placht, a dismissed officer, promised to pay 40 percent a year and borrowed the pennies of 1600 widows and orphans to get the money to play the stock market. His stock purchases were not profitable and he spent six years in jail.[24]

As economic booms progress, greed mounts and the excuses become thinner, more nearly gossamer bubbles. In 1720 and again in 1847 (two occasions when lists were compiled), such swindles were numerous, although they have been embroidered with hoaxes perpetrated by and on later historians.[25] In 1720, for example, there was one proposal for carrying on an undertaking of great advantage that would be revealed only in due time. The perpetrator charged two guineas a share and made off with £2000, keeping his secret intact by failing to attend a meeting with the investors.[26] Another scheme was for the 'nitvender', or selling of nothing.[27] In the late 1990s stock market boom, some firms were able to raise money from the public before they had any business plans.

A project of modest current interest offers a premature example of women's liberation:

> A proposal by several ladies and others to make, print and stain calicos in England and also fine linen as fine as any Holland to be made of British flax... They were resolved as one man [sic] to admit no man but will themselves subscribe to a joint-stock to carry on said trade.[28]

In later periods, both historians and novelists noted that stock promotions had little connection with reality. 'Many companies were founded

without undertaking operations, railroads with any right-of-way or traffic.'[29] 'Construction companies grew like mushrooms. Many of these companies speculated in building sites rather than in construction.'[30] 'Limehouse and Rotherhithe bridge ... It was not at all necessary for them that the bridge should ever be built; that, probably, was out of the question ... But if a committee of the House of Commons could be got to say that it ought to be built, they might safely calculate on selling out at a large profit.'[31]

Financial distress leads to fraud in the effort to dump the losses on others before they cascade into ever larger losses. If the market goes decisively the wrong way, for example, bucket-shop operators abscond. When new cash subscriptions failed to meet profits paid out to greedy insiders, Blunt borrowed the cash of the South Sea Company for his own use;[32] a preview of the story of the Rigas family and Adelphia. In 1861 Bleichröder characterized Bethel Henry Strousberg as 'clever but his manner of undertaking new ventures in order to mend old holes is dangerous, and if he should encounter a [sudden] obstacle his whole structure may collapse and under its ruins bury millions of gullible shareholders'.[33] Bleichröder was right. Another German financier, the Hamburg banker Gustav Goddefroy, lost heavily in railroad and mining shares in 1873 and then bled his overseas trading company to support his position in the stock market.[34]

These incurable optimists, who know they are going to win the first time, but lose, frequently try again, often doubling their bets and increasing their risks by transactions that either are of dubious morality or clearly illegal. In the late 1920s, when US banks were still allowed to underwrite securities (before the Glass-Steagall Act of 1932), Albert Wiggins of the Chase Bank and Charles Mitchell of National City continued to sell Chilean and Peruvian bonds at the old prices even after they had learned by cable from those governments that they had stopped paying interest.[35] Horace understood the position, if Sprague quotes and translates him accurately: 'Make money; make it honestly if you can; at all events, make money.'[36] Equally cynical is Jonathan Swift over the South Sea Bubble:

> Get money, money still
> And then let virtu follow, if she will.[37]

On this topic, Balzac has the last word: 'The most virtuous merchants tell you with the most candid air this word of the most unrestrained immorality: "One gets out of a bad affair as one can".'[38]

Noble gamblers

The literature abounds in condemnation of noble gamblers and insiders, who might have been thought to regard financial obligations as debts of honor but are better at promising than at paying their subscriptions.[39] The Austrian nobility

was worse than the Junkers, who at least ostensibly disdained money. Eduard Lasker maintained that 'when the dilettantes enter, they make it even worse than the professional swindlers'.[40] Daigremont in Zola's *L'Argent* sends Saccard to the Marquis de Bohain to help launch his Banque Universelle: 'If he wins, he pockets it; if he loses, he does not pay. That is known, people are resigned to it.'[41] Again in novels and in real life, nobles seek seats on boards of directors. Wirth enumerates Austrian princes, Landgrafs, counts, barons, Freiherren, and other nobles on the boards of railroads, banks, and other industrial firms 'for which they have no capacity'.[42] To control the accounts of the Banque Universelle, Saccard appoints a *sieur* Rousseau and a *sieur* Lavignière, the first completely subservient to the second, who is tall, blond, very polite, approving always, devoured by ambition to come on the board.[43] *L'Argent* is a *roman à clef* based on Eugène Bontoux and the Union Générale, whose subscribers included the Pretender, royalists, notables, and country squires.[44] In Britain *The Economist* in October 1848 included the nobility and aristocracy at the head of a list of dishonor:

> Present prostration and dejection is [sic] but a necessary retribution for the folly, the avarice, the insufferable arrogance, the headlong, desperate and unprincipled gambling and jobbing which disgraced nobility and aristocracy, polluted senators and senate houses, and traders of all kinds.[45]

Rosenberg claims that while the Austrian and French aristocracy led the other estates in pursuit of the golden calf, Berlin bureaucrats successfully opposed a similar movement in Prussia, noting an abortive attempt by Mevissen to get some counts on the board of a 50 million thaler bank. Junkers speculated in spirits and land products, he admits, but shied away from urban developments.[46] Perhaps this was so in 1857. In the following decade, the perception of money as evil had weakened. Railroad finance, both inside Germany and in Strousberg's maneuvers in Romania, was tinged with scandal, reaching up to the peaks of the aristocracy and virtually into the Prussian court itself.[47]

Venal journalism

Speculation generally was helped by the press. Some members of the press were for sale, some were critical, and some were both. Daniel Defoe excoriated stockbrokers in November 1719 when the South Sea stock sold at 120 yet turned around to defend them at the peak of 1000 in August 1720.[48] He expressed his 'just contempt' for people who claimed he wrote for the Royal African Company, stating he had sold his stock; but a modern critic concluded that he either continued to hold the stock or was hired by the company to attack individual traders who competed with the monopoly.[49] The press was still in an underdeveloped stage in France in the early nineteenth century; in

1837 a journalist wrote: 'Give me 30,000 francs of advertising and I will take responsibility for placing all the shares of the worst possible company that it is possible to imagine.'[50] Laffitte financed newspapers.[51] Charles Savary of the Banque de Lyon et de la Loire had 500 journalists singing the praises of his operations, using releases, largely paid for, as if they were stories created by the journal's staff.[52] Journals often sought favor with banks, the stock exchange, and the public by whipping up the speculative fever.[53] Bleichröder was cautious in avoiding speculation and outright misrepresentation, but he owned general and financial newspapers and used journalists to hype his financial interests. In 1890–91 he financed a trip to Mexico for one Paul Lindau, who wrote thirty-four articles and a book on the country without mentioning his connection with Bleichröder who then was selling Mexican bonds on the Berlin market.[54] A critical press developed slowly during the nineteenth century on the Continent.

In the 1890s in the United States, on the other hand, a financier closer to the line was pursued by the press and lived in fear of it; at least this is the inference from Theodore Dreiser's fictionalized biography of Charles Tyson Yerkes, the Chicago streetcar tycoon who operated along, and sometimes over, the knife-edge that separates legitimate from illegitimate transactions. Yerkes needed favorable publicity to sell securities, which he sought and won through his gift of an observatory to the newly formed University of Chicago. The well-publicized gift helped restore his reputation and enabled him to sell his streetcar bonds in Europe.[55] But later the press drove him from Chicago.[56]

Tipping off friends before hawking a stock to the world in an investor column may be an old trick, but it is more readily detected today, along with all buying and selling of shares based on insider information. Suspicious trades in a stock and in options to buy and sell the stock before the news that affects its price is made public are now analyzed by computer programs. The technique led to the arrests of investment columnist R. Foster Winans of the *Wall Street Journal*, who fed his tips in advance to a friend, and of former Undersecretary of Defense Thayer, who told a friend of forthcoming events in a company of which he was a director. (Thayer's friend speculated on the basis of the information; Winans capitalized on his experience by writing *Trading Secrets: Seduction and Scandal at the Wall Street Journal*, St Martin's Press, 1986.) Similarly, a young Merrill Lynch broker in New London, Connecticut bought information from the local printers of *Business Week* before the magazine appeared on the newsstands. Once the ease of tracing trading on insider information became known, some miscreants with advance information would call a broker in, say, Zurich, instead of their local broker. The Zurich bank would place a massive order in London or New York before brokerage houses there became wary and would delay execution for a couple of days to see if a large order was based on breaking news.

Most recently the internet has become the vehicle for manipulating stock prices. Jonathan Lebeck, aged seventeen, posted 'news' about particular thinly traded stocks (which he owned) on internet chat rooms. The stock prices would increase and Jonathan would sell his shares. The Securities and Exchange Commission made Jonathan pay a penalty of $500,000.

MSNBC is a business news TV channel. Many of the commentators on the programs promote the stocks that they own.

Dubious practices

There are many forms of financial felony. In addition to stealing, misrepresentation, and lying, other dubious practices include diversion of funds from the stated use to another, paying dividends out of capital or with borrowed funds, dealing in company stock on inside knowledge, selling securities without full disclosure of new knowledge, using company funds for non-competitive purchases from or loans to insider interests, taking orders but not executing them, and altering the company's books.

George Hudson, who may have been the greatest figure in British railroad history, practiced nearly all of these at the same time in the 1846 railway mania. At one time he was chairman of four railways, and he mistakenly believed he was above the law that applied to his less powerful competitors. His accounts were muddled, and he may not have understood that he had appropriated shares or funds belonging to the York and North Midland railway. As a private individual he made contracts with various companies of which he was an officer, in direct violation of the Companies Clauses Consolidation Act. He raised the dividend of the Eastern Counties railway from 2 to 6 percent just before preparing the financial statements and then altered the accounts to justify the payments. Dividends of the York and North Midland were paid out of capital. He defended his actions against similar accusations in the case of the Yorkshire, Newcastle and Berwick by noting that he had personally advanced funds to the railway to extend its network. The risk was his, and he was entitled to the advantages that ultimately accrued from the expansion of the rail system. He embarked, on his own authority, on transactions that he deemed advantageous to the company, but that were nevertheless of doubtful legality. Still, he had a career of great brilliance that greatly benefited the British railway network.[57]

A less interesting and imposing character in the United States in the 1850s was Robert Schuyler, who was president of the New York and New Haven, the New York and Harlem, and for a time the Illinois Central railroads. Called the 'genteel swindler' by one author, he absconded to Europe in 1854 with almost $2 million obtained from fraudulently selling New York and New Haven stock and keeping the proceeds. Van Vleck suggests that the crisis of 1857 in the

United States was precipitated by British withdrawals of funds following the publication of the news of Schuyler's defalcation. Schuyler had resigned from the presidency of the Illinois Central in 1853, but his fraud perpetrated on the New York and New Haven led to a massive sale of Illinois Central stock and bonds. The stock fell drastically, and the bonds declined to 62 from 100 by August 1855. British investors had been looking for this opportunity and they bought large amounts of these bonds; by February 1856 the bonds were back to 90. European investors owned more than 40,000 shares of the stock and 85 percent of the $12 million in bonds.[58] Schuyler was connected with a panic in September 1854 but not with the one in 1857.[59]

The 1920s in the United States has been called 'the greatest era of crooked high finance the world has ever known' – but that was before the 1990s.[60] Notorious swindlers of the period include Harold Russell Snyder, who stole to extricate himself from losses incurred from the stock market crash (a precursor of the Rigas family, but on a smaller scale, even after adjusting for inflation). Arthur H. Montgomery paid the sincerest form of flattery to Ponzi by organizing a foreign-exchange investment scheme that promised a 400 percent return in sixty days. Charles V. Bob sought and obtained favorable publicity through a $100,000 gift to the Byrd Antarctic Polar Expedition and won the right to call the admiral 'Dick', which presumably helped promote his aviation stocks, which enjoyed a boomlet after Lindbergh's 1927 flight to Paris.[61]

The 1930s produced many memorable examples of fraud and alleged fraud, led by the bankruptcies of the Bank of the United States, of Ivar Kreuger, and of Samuel Insull's Middle West Utilities holding company. In his riveting account of the New York stock market from 1929 to 1933, Barrie Wigmore observed that Insull's reputation was irreparably damaged when he fled the United States to escape trial before what he considered would be inflamed juries for crimes of which he was later acquitted. Wigmore asserted that Insull was a brilliant manager of operating utilities who went on a buying binge, acquiring badly run companies for multiples of their net worth. Insull's purchases burdened his holding company with a mountain of debt, and the interest payments on the debt wiped out the equity of those who owned the common stock when the utilities got into trouble in the 1930s depression.[62] Wigmore treads warily in assessing the marketing practices of New York banks and their security affiliates. Albert Wiggins of the Chase Bank had a reputation as 'the most popular man on Wall Street' but his reputation was shattered when a Senate investigation exposed a picture of self-dealing at the expense of his bank, its subsidiaries, and its clients. Charles E. Mitchell of the National City Bank and its security affiliate, the National City Company, also marketed securities intensely, even though he had the inside information that the profits of the companies concerned were declining rapidly.[63]

Corruption in the 1990s and 2000s

The amount of corruption seems greater in the 1980s and the 1990s than in the 1920s. One explanation is that there has been a decline in adherence to moral norms. A second is that the risk-reward trade-off has been skewed; stock options have provided a much greater pay-off from financial success. Finance has been democratized. A third is the rise of certified public accountants, which can be traced to the legislation that led to the Securities and Exchange Commission of the 1930s. The initial role of the accounting firms was to protect the public from the financial chicanery of corporate managers who might have had a tendency to overstate the value of inventories and of accounts receivable. The accounting firms were paid to verify or certify the data on financial performance presented by the firms' financial managers. The accountants may have been the guardians of the public interest but the fees for their services were paid by the financial managers. Some of the firms being audited leaned on their accountants; the accounting firms were then forced to choose between agreeing to the firm's demands or walking away from the account.

Securitization was a central feature of the bubble in US real estate between 2002 and 2007; the investment banks issued new bonds, which were claims on the interest income and principal of mortgages, credit card debt, and student loans that they had placed in a trust. The new bonds were much more liquid than the individual securities that were placed in the trust. The investment banks earned fees from securitization; their fee income soared. They then issued asset backed securities (ABSs), which were claims to the interest income and principal of the segments of the ABSs that had been deposited in the trust. The investment banks earned a second round of fees from creating these new instruments, which had been based on 'slicing and dicing' the MBSs into three, four, or five different tranches, which differed in terms of their risk characteristics – one set of trusts contained the tranches that had the first claim on the interest income of the MBSs and hence they were the least risky, while another set of trusts constrained the tranches that had the last claim on the interest income of the MBSs, and they were the most risky.

The investment banks relied on the credit rating agencies – Moody's, Standard and Poor's – to rank the securities that the investment banks were producing. (The credit rating agencies are like the CPA firms, established to provide 'objective' information. The CPA firms were paid by the firms that they were auditing and the credit rating agencies were paid by the investment banks that issued bonds.) But these agencies were relying on the 'models' that the investment banks had provided for them. The investment banks had indicated to the credit rating agencies that if they could not get rankings that would help

them sell these securities, they would take their business to one of the other firms that provided ratings.

Was this illegal? Who knows? Was it corrupt? You bet. Is anyone from a credit rating agency likely to go to jail? Are you kidding?

In the end there was some crude justice, since the investment banks were done in by the declines in the market value of their holdings of asset-backed securities that they could not sell.

The temptation of banks

There are no firm data that permit comparisons of financial chicanery and fraudulent behavior across several centuries. The development of journalism as a profession may mean that any untoward activities are much more likely to be exposed today (although the activities themselves may not have changed). Exposure is the risk side of the equation; the reward is that the gains in wealth can be much greater. Martin Mayer's *The Bankers* does not dwell on safeguards against defalcation in the same way that James S. Gibbons did in 1859: 'There is perhaps no record of a bank fraud extant of which the perpetrator was not honest yesterday.'[64] Gibbons added, with emphasis, 'It will occur to the reader that there is one peculiar feature running through the whole system; and that is *the apprehension of fraud*.'[65]

What Gibbons said in 1859 is still true. The tendency is to believe that the banks and the bankers are 'paragons of integrity' and perhaps some of them are. A large number of banks lent extensively to Long-Term Capital Management in the 1990s, at the time believed to be one of the most innovative of the 'hedge funds'. (The term 'hedge funds' is a Madison Avenue home run, since the impression conveyed is that the firm has arranged its portfolio to reduce risk while in fact these firms rely extensively on borrowed funds to increase the return to their shareholders and investors.) LTCM had been remarkably profitable – at least until its collapse. The banks were eager to lend to LTCM because they hoped to mimic its trades and profit accordingly. In the spring and summer of 1998, LTCM encountered financial difficulties.

The awkward handling of derivatives and unclaimed deposits by Bankers Trust and the money laundering for Russia by the Bank of New York suggest that standards are not much higher today than they were, say, in the 1920s.

In the boom years of the second half of the 1990s, everyone – well, nearly everyone – was getting rich. The major investment banks had very high incomes from the fees associated with underwriting new issues of stocks and bonds, especially those of firms associated with information technologies and bio-genetics. The traditional 'Chinese wall' between the investment banking activities of these firms and their asset management activities was supposed to be retained after the repeal of the Glass-Steagall Act, which had become law in

the early 1930s to force the separation of the traditional commercial banking activities of firms from their investment banking activities. The law had been adopted in response to the abuses of the 1920s. The firms 'promised' that they would maintain a Chinese wall and that the statements made by their security analysts would not be influenced by the desire of their investment bankers to sell more securities.

But consider the scorecard of Merrill Lynch. The firm was extensively involved in brokering deposits to the rogue thrift institutions in the 1980s. Henry Blodgett hustled information that he knew was misleading. Blodgett's superiors almost certainly knew that Blodgett was involved in a con. Merrill helped Enron falsify its income by paying an above-market price for the Enron barges in Nigeria.

Consider the scorecard of Citibank/Citigroup. The tale of Jack Grubman's hustles for the telecom firms has been noted; Citibank/Citigroup paid a settlement of $150 million. Citibank was obliged to close its private banking activities in Tokyo because of the failure of bank officials to deal fairly with customers by repeatedly buying inappropriate securities for them after ignoring the warning from the Japanese authorities to stop the practice. Several of the top managers of Citibank 'resigned' – a Park Avenue euphemism for being fired. Citibank's traders in London dumped a lot of German government bonds on the market and caused the prices to dive; they then repurchased these bonds at much lower prices. Several of the managers of Citibank's mutual funds took rebates based on the revenues of the bank rather than of the funds.

What happens when a swindler is found out? Charles Blunt, the brother of John Blunt and himself an insider in the South Sea Company, in early September 1720 cut his throat 'upon some discontent', as contemporary newspapers put it; Charles Bouchard, the retired manager of LeClerc, a small Geneva bank that lost money in unauthorized real estate investments, was found dead in Lac Léman, an apparent suicide, in May 1977. Psychiatry holds that suicide in these circumstances comes from an intolerable loss of self-esteem, stemming from the realization of the irrationality of past behavior. The picture of stockbrokers jumping from Wall Street windows in October 1929 as they faced bankruptcy is now believed to be a myth.[66] An upswing in suicides is also part of the legend of the Austrian *krach* of 1873.[67] Nonetheless, the response does occur: 'Thus died by his own hand [having taken poison on Hampton Heath], at the early age of forty-two, John Sadleir, one of the greatest, if not the greatest, and at the same time the most successful swindler that this [Britain] or any other country has produced.'[68] (Greatness and success seem curious as characterizations in the circumstances.) Denfert-Rocherau of 1888, and Ivar Kreuger, the 'match-king' of the 1920s, both committed suicide;[69] Clifford Baxter, one of the senior officials of Enron, took his own life. But suicide may be more usual in fiction. Mr Merdle cut his throat in a public bath with a tortoiseshell

penknife in Dickens's *Little Dorrit*, and Augustus Melmotte in Trollope's *The Way We Live Now* took prussic acid at his club.

Flight is a less final form of exit than suicide. The prize case is Robert Knight, who doctored the books of the South Sea Company and then escaped to the Continent, there to make another fortune in Paris after breaking out of an Antwerp jail.[70] Robert Vesco fled to Costa Rica and then Cuba with an embezzled fortune. Charles Savary, who swindled the Banque de Lyon et de la Loire, died in Canada. Eugène Bontoux returned to France after five years of self-imposed exile to take advantage of a loophole in French law that held that prison terms not begun within five years of sentencing had to be dropped.[71] The analogous case roughly a century earlier was that of Arend Joseph, whose failure in January 1763 initiated the financial distress that culminated in the bankruptcy of the brothers DeNeufville on 25 July, touching off the panic of the same year. Arend Joseph left Amsterdam with 600,000 guilders, in a coach-and-six, for the free city of Kruilenburg in Holland, where he was immune from further process. He left a debt of one million guilders in Amsterdam.[72]

Comparisons can be made across the cycles in terms of the number of individuals who were indicted and the number who did jail time. Consider several episodes, from the 1920s, the 1980s, and the 1990s. In the 1920s two individuals went to jail for hustling the purchases of securities in a market that already had turned bearish. Eight or ten of the participants in the junk bond transactions of the 1980s, including Michael Milken, Ivan Boesky, Dennis Levine, and Charles Keating, were jailed for crimes that included insider trading, 'parking' of securities, and collusion to defraud; among those who served the longest sentences were heads of the thrifts that had been large buyers of junk bonds. The number of those who will have served jail sentences for their transgressions in the 1990s is still expanding and includes five individuals who had received pay-checks from Enron and two who had been involved with MCIWorldCom. Most of the Enron officers who have been indicted have gone to jail. Several of the individuals involved with HealthSouth have gone to jail. Two executives associated with Rite-Aid went to jail. Several members of the Rigas family went to jail. Sam Waksal and his good friend Martha Stewart have gone to jail. Very few Wall Street bankers have been jailed. Frank Quattrone, one of the star investment bankers for Credit Suisse First Boston, was found guilty in a second trial of obstruction of justice by destroying e-mails but his appeal succeeded. One Arthur Andersen partner involved with the Enron account did jail time.

Hundreds of Andersen partners and former partners implicitly paid large fines when the firm was forced to close and the value of partnerships and former partnerships collapsed. (Thousands of Enron employees lost their

pensions and much of their financial wealth as well as their jobs when the firm went bankrupt.)

The Milken family probably had $2 billion in the bank when Michael Milken was released from prison. It would be impossible to figure out how much of the fortune was earned legally from innovation and how much was earned from illegal behavior. But assume that half of the family fortune could be tracked to transactions that were illegal. Consider how Milken might answer the charge, 'You were in the jail for 1000 days, you graduated with $1 billion so you were paid $100,000 for each day in the jail.'

Economists are not qualified to discuss the appropriate punishment for the white-collar crime of swindling. At the time of the South Sea Bubble, Molesworth, then a member of the House of Commons, suggested that parliament should declare the directors of the South Sea Company guilty of parricide and subject them to the ancient Roman punishment for that transgression – to be sewn into sacks, each with a monkey and a snake, and drowned.[73] The suggestion is echoed in Dreiser's novel *The Titan*. 'Here the punishment consists of strangling first, then being sewn into a sack, without company, and thrown into the Bosphorus, a punishment reserved for cheating girl friends.'[74] In Christina Stead's *House of All Nations*, written a quarter of a century later, a character suggests that old sultans used to punish a faithless wife by tying her into a sack with two wildcats and sinking the sack in the Bosphorus.[75] These suggested punishments seem excessively harsh. Still, those who commit white-collar crimes seem to get off lightly, and many keep most of their ill-earned fortunes. The fines paid by the Wall Street firms are a tax on the wealth of their shareholders and not a real burden on the malfeasants except in as much as they are also shareholders.

Whether swindlers are punished or live out their days in indulgent luxury is a more appropriate topic for corporate governance and business ethics than financial history. The revelation of swindles, frauds, and defalcation, and the arrests and punishment of those who violate trust are important signals that economic euphoria has reached too high a pitch and that there will be significant social consequences.

8
International Contagion 1618–1930

Allocating the blame for the crisis

That the review of the historical record suggests that crises occur in waves leads to two questions – one concerns when and where each of the shocks begins and the other relates to the mechanism that links the crises in different countries.

The tendency in every country is to believe that the crisis is imported. US President Herbert Hoover insisted that Europe was responsible for the 1930s depression because of its cartels and the 'European statesmen [who] did not have the courage to face these issues'.[1] There was global overproduction of wheat, rubber, coffee, sugar, silver, zinc, and cotton. Hoover accepted some US responsibility for stock market speculation. Friedman and Schwartz asserted that the crisis originated in the United States, although the gold-exchange standard rendered the international financial system vulnerable. The initial climactic event – the stock market crash – was American, and the developments that led to a decline in the stock of money in late 1930 were predominantly domestic.[2]

The Europeans believe that the recession in 2008 started in the United States. So do many of those in Iceland.

In 1837 President Jackson placed the blame for that year's crisis on both Britain and the United States:

It would seem impossible for sincere inquiries after the truth to resist the conviction that the causes of the revulsion in both countries have been substantially the same. Two nations, the most commercial in the world, enjoying but recently the highest degree of apparent prosperity, are suddenly, without any great national calamity, arrested in that career, and plunged into embarrassment and distress. In both countries have we witnessed the same redundancy of paper money, and other facilities of credit, the same

spirit of speculation, the same partial success, the same difficulties and reverses, and, at length, the same overwhelming catastrophe.[3]

Although observers in the 1850s called the panics of 1836–37 the 'American panics' because they originated in and were confined to the banks that traded with the United States,[4] one modern economist concluded that it was 'futile to try to draw any hard-and-fast line assigning to either country causal primacy in the cycle as a whole or in its individual phases'.[5]

Friedman and Schwartz relied on the data on the pattern of gold movements to conclude that the United States was responsible for the recession of 1920–21.[6] Another observer disagreed:

> How was [the early postwar fall in economic activity] brought about?...I think the answer must be: It was a deliberate policy inaugurated by the two economically dominating countries, the U.K. and the U.S.A. It is impossible to give priority to any of them. The earliest official statement of the policy was undoubtedly made by England. On the other hand, causally U.S.A.'s policy must have the greater weight.[7]

A few crises are purely national – the gold agio crisis in the United States in 1869, the City of Glasgow Bank in 1878, the Union Générale in France in 1882. In contrast the Canadian financial crises that occurred in 1879, 1887, and 1908 seemed related to major financial currents that bound Western Europe, Scandinavia, and the United States together.[8]

Some countries may not be affected by international crises that impact their neighbors and for obvious reasons. France was not affected in 1873 because it had undergone severe deflation in 1871 and 1872 as it transferred reparations to Prussia. The United States was insulated from the European potato disease and the tumultuous European wheat situation in 1847 because US railroads had not yet developed to the point where US food markets were closely linked to those in Europe.

For the most part however financial crises ricochet from one country to another. Juglar,[9] Mitchell,[10] and Morgenstern[11] noted that financial crises tend to be international, and either affect a number of countries at the same time or alternatively spread from the centers where they originate to other countries.

One of the several links among countries is that arbitrage connects national commodity and financial markets; the implication of the Law of One Price is that the difference in the prices of identical or similar goods in various countries cannot exceed the costs of transport and trade barriers. The links between commodity markets in different countries may involve only a modest amount of trade. When the price of cotton soared in the 1830s in one country, the price increased in all other countries; similarly the decline in

the price of cotton after 1864 was worldwide. A decline in the price of a given commodity – especially a widely traded product like wheat or cotton – may produce bankruptcies and bank failures far from the source of the original change in demand or supply.

Similarly the financial markets in the various countries are also linked, since the prices of internationally traded securities available in different markets must be virtually identical after a conversion of prices in different currencies into the same currency. The prices of internationally traded securities that are listed on the stock exchanges in several different countries increase and decrease together. The prices of domestic securities often move in a synchronous fashion with internationally traded securities as a result of psychological linkages or through impacts on interest rates transmitted through short-term capital movements.

When changes in stock prices are small, the correlations between the stock price movements in different national markets are low. As the changes in stock prices become larger, the correlations become stronger.

In 1929 all stock markets crashed simultaneously and again in October 1987 virtually all of the stock markets declined at the same time. Even though national financial markets generally were believed to be more fully integrated in the 1980s and the 1990s than in earlier periods, share prices in the 1920s in different countries were as strongly correlated as in later decades. Because of the strong correlation of the stock price movements in different countries, many of the investors who sought to reduce risk by diversification through ownership of stocks in different national markets obtained smaller reductions in risk than they had anticipated.

The correlations among changes in stock prices in different countries are somewhat asymmetric in that changes in US stock prices have a much more powerful impact on stock prices in various foreign countries than changes in stock prices in any one of these countries have on US stock prices. Thus US stock prices increased throughout the 1990s while stock prices in Tokyo were on a downward trend – but when US stock prices declined in 2001 stock prices in Tokyo, London, and Frankfurt declined. The Mexican financial crisis of 1994–95 impacted both Brazil and Argentina; US investors became more cautious about buying both Latin American bonds and stocks. The devaluation of the Thai baht in early July 1997 triggered the 'contagion effect' and induced depreciations of the currencies in nearby Asian countries in the following six months and then spread to Russia and Brazil and eventually Argentina.

Transmission mechanisms

Booms and panics are transmitted from one country to another in several different ways, including arbitrage in commodities or securities and movements

of money in various forms (specie, bank deposits, bills of exchange), cooperation among monetary authorities, and investor psychology.[12]

The security and asset markets in various countries are linked by movements of money. The inflation in the United States in the late 1960s and the early 1970s led to larger money flows from the United States to Germany and Japan and the inflation rates in these countries increased as their monetary bases and their money supplies increased.

Cross-border money movements may respond to real causes, including wars and revolutions, technical innovations and the opening of new markets and of new sources of raw materials, and to changes in the relationship among the growth rates in different countries, as well as to changes in monetary policy and in fiscal policy. The privatization of government-owned firms in a country often induces inflows of money as foreign investors buy shares in these firms.

Consider the connections between the appreciation of a currency and deflation in that country's goods market (or the connections between the depreciation of a currency and inflation in that country's goods market); the appreciation of the currency leads to declines in the prices of internationally traded goods and to bankruptcies and the de-capitalization of banks and other financial firms. The appreciation of the Japanese yen in the second half of the 1980s and the 1990s put downward pressure on the prices of internationally traded goods in Japan. The depreciation of the Argentine, Uruguayan, Australian, and New Zealand currencies in the early 1930s contributed to the decline in wheat prices in the United States which in turn led to bankruptcies among farmers and to the failures of banks in Missouri, Indiana, Illinois, Iowa, Arkansas, and North Carolina.[13]

Booms and busts are connected internationally in several different ways. An economic boom in one country almost always attracts money from abroad. Similarly, a boom in one country may reduce the flow of money to others; thus in 1872 Berlin and Vienna stopped lending to New York, while the US stock market boom in 1928 led to a sharp reduction in US purchases of German, Australian, and Latin American bonds. The collapse of the market in syndicated bank loans to Mexico, Brazil, and Argentina in 1982 led to a sharp decline in the values of their currencies.

Stock prices and real estate prices increased sharply in three of the Nordic countries in the late 1980s; with a three-fold increase in Norway and five-fold increases in Sweden and Finland.[14]

The *Kipper- und Wipperzeit*

This financial crisis occurred with metallic money; credit from banks or other lenders was not involved. Princes, abbots, bishops, even the Holy Roman Emperor debased the subsidiary coinage used in daily transactions (but not

gold and silver coin of large denominations) by raising the denomination of existing coins, substituting base for good metals, and reducing the metallic content of coins to extract more seignorage to prepare for the Thirty Years' War that broke out in 1618. Debasement was limited at first to their own territories. Some entrepreneurial spirit then found that it was more profitable to take bad coins across the borders into neighboring principalities to exchange for good coins with ignorant common people; the good coins were then brought home and debased. The territorial unit on which the original injury had been inflicted would debase its own coin in defense and turn to other neighbors to make good its losses and build its war chest. More and more mints were established to extract more seignorage.

Debasement accelerated until the subsidiary coins became practically worthless, and children played with them in the street, much as recounted in Leo Tolstoy's short story, 'Ivan the Fool'.

Some local sources set forth the view that the first invasion of debased money came from Italy and from there to southern Germany through the Bishop of Chur on Lake Constance. The same source at Ulm, however, claimed that the counterfeiting of the Upper Rhine Circle that included Strasbourg was especially outrageous. Beginning on a small scale around 1600, debasement slowly picked up speed after 1618 and spread to Germany and Austria and to what became Hungary and Czechoslovakia and into Poland and, according to some sources, to the Near and Far East by way of Lvov in Russia.[15]

The South Sea and Mississippi bubbles

Åkerman called the crisis of 1720 the first international crisis, because the speculation from 1717 to 1720 in France and in Britain affected the cities of the Netherlands, northern Italy, and Hamburg.[16] The South Sea and Mississippi bubbles were connected in several ways. As early as 1717 British investors began to follow trading in the shares of John Law's banks and companies on the Rue de Quincampoix in Paris. In May 1719 the British ambassador in Paris received letters from friends and relatives in Scotland begging him to buy stock for them in the Compagnie des Indes. Thirty thousand foreigners, including British nobility, traveled to Paris to subscribe in person. In May, Ambassador Stair urged his government to do something to compete with John Law and slow the flow of money from London to Paris. As Law's system peaked in December 1719, some speculators, including the Duke of Chandos, sold South Sea stock and bought Mississippi stock.[17]

While British speculators were buying Mississippi stock in Paris, many continentals were buying South Sea stock in London. Sir Theodore Janssen had a long list of subscribers from Geneva, Paris, Amsterdam, and The Hague. One of the French investors was the banker Martin, who as already noted was recorded

by Charles McKay as subscribing £500 with the remark: 'When the rest of the world are mad, we must imitate them in some measure.' When the early birds liquidated in July, the Canton of Berne, which had speculated with £200,000 of public funds, sold for a profit of £2 million.[18]

Amsterdam profited from its position between Paris and London. The Dutch sold their stock in Mississippi Compagnie des Indes at the right moment and lost little in the crash. In April 1720, a bit prematurely perhaps, David Leeuw liquidated his South Sea stock and bought Bank of England and East India Company stock. By the end of that month, the Dutch banker Crellius observed coolly that Exchange Alley resembled 'nothing so much as if all the Lunatics had escaped out of the Madhouse at once'.[19] In June and July there were twelve-hour relays by ship between Britain and Amsterdam, and on 16 July eighty Jews, Presbyterians, and Anabaptists, speculators from Exchange Alley, went to Holland and Hamburg to trade insurance stocks.[20] In the autumn of 1720 London and the Continent were sharing the financial disaster. Samuel Bernard, a French banker, was sent to London to sell South Sea stock against gold, to be brought back to France in revulsion against Law's system. Dutch banks 'shortened sail, recalling advances, refusing further credit, selling stocks held as collateral'.[21] The price of the British pound in terms of the guilder in Amsterdam, which had risen from 35.4 guilders to the pound to 36.1 when the first increase in South Sea stock took place in April and 'France, Holland and to some extent Denmark, Spain and Portugal' were buying, declined to 33.9 on 1 September as 'foreigners lost their taste for English securities'. At the height of the panic it recovered to 35.2.[22]

1763 to 1819

The crisis of 1763 involved mainly Holland, Hamburg, Prussia, and Scandinavia, with repercussions on and help from London. France was not involved; the Seven Years' War had been directed against France. George Chalmers, a perceptive contemporary observer, claimed that speculation in land in the United States was a factor in the crisis although the statement is not supported by others.[23] Amsterdam had been the entrepôt center for the payment of money to British allies, and the Dutch had been expanding credit by buying British government stock and *Wisselruiti* (chains of accommodation bills) that led to a giddy credit edifice on a small base (the proverbial 'house of cards') with bills drawn on merchant houses in Stockholm, Hamburg, Bremen, Leipzig, Altona, Lubeck, Copenhagen, and St Petersburg. Bills of exchange drawn with the security of goods shipped also circulated in Amsterdam in addition to the accommodation paper. When prices of commodities fell after the war – especially sugar as imports from the French West Indies were resumed – the bills could not be paid.[24] Hamburg warned Amsterdam houses that they would

suspend payment unless support was furnished to the DeNeufvilles. In one account, the letter arrived too late.[25] Another stated that a plan to save the firm failed because its reputation was too bad.[26] In the long run, the DeNeufvilles would have been able to pay 70 percent of their obligations, but their creditors settled for 60 percent before that became known. In the end, the Hamburg creditors had to wait thirty-six years to collect even that much.[27] The *coup de grâce* occurred when King Frederick II of Prussia who had debased the silver coins in 1759 to help fight the war recalled the old coins and had new ones minted in Amsterdam on the basis of credits from the Dutch bankers.[28] Withdrawing the old coins before issuing new ones led to deflation because the money supply declined.

London came to the rescue of Amsterdam and financed a considerable portion of Dutch trade and investment with Scandinavia and Russia. Very much against his will, King Frederick had to assist Berlin merchants that were caught in the crisis as their bills were protested.[29] Swedish houses complained early in the fall of 1762 that bills they drew were protested and not paid in Amsterdam while remittances sent to cover the bills were retained. Whether Amsterdam tried to save itself by selling British securities is debatable. Wilson claimed that in this way Amsterdam exported the crisis to London. Carter insisted she cannot find evidence of sales in the transfer books.[30]

The 1772 crisis spread from Scotland and London to Amsterdam and thence to Stockholm and St Petersburg. Large outflows of specie from Paris to London fed the canal and country bank mania during the Reign of Terror in 1792 that peaked with the guillotining of Louis XVI in January 1793; the direction of flow of precious metals reversed itself in 1797 when monetary order had been more or less restored under the Consulate after the Assignats.

The British crisis of 1810 was localized: British exporters first overdid sales to Brazil and then were cut off from their Baltic outlets by the blockade. There were echoes of this crisis in Hamburg and in New York.

The international aspects of the crises of 1816 and 1819 reflected that the prospect of the end of war in 1814 led to large British sales of manufactures on the Continent. Smart called this an exporting frenzy that soon broke like the South Sea and Mississippi bubbles. When prices collapsed, the goods were sent to North America which led to the US tariff of 1816. The result was a deep depression without a panic or even a crisis.[31] In 1818 and 1819 there were panics on both sides of the Atlantic that were connected in a non-obvious way. The 1819 crisis in Britain followed the collapse of commodity speculation in 1818, the discredit and distress 'originating clearly in great previous overtrading'.[32] The year 1819 was marked by the resumption of specie payments and by the Peterloo massacre, when protesting Manchester workers and their families were charged by cavalry who killed at least eight individuals; Smart called it a 'disastrous year'.[33] In America, the Second Bank of the United States precipitated a

panic by having its branches call on state banks to redeem large balances and notes that it held. The purpose was to assemble $4 million in specie to repay the borrowing undertaken in Europe in 1803 to pay for the Louisiana Purchase.[34] But the Second Bank itself was a bubble, having been re-established in 1817 after the dissolution of the First Bank in 1811. The bank was run by greedy and corrupt directors who accepted promissory notes in payment of stock, registered stock in different names to get around the law limiting concentration of ownership, voted loans on the security of bank stock, permitted other loans without collateral, and allowed accounts to be overdrawn. Hammond observed that the sober pace of eighteenth-century business had given way to a democratic passion to get rich quick, and that men imbued with this passion and unscrupulousness had gained control of the Second Bank.[35]

1825 to 1896

The 1825 crisis involved Britain and South America, although there was a distinct spillover to Paris that stretched out until panic struck there in January 1828. With the panic in London in December 1825, continental sales halted, impacting banks in Paris, Lyon, Leipzig, and Vienna and obliging Italy and other markets that depended on these centers for finance to reduce their purchases. Distress from burdensome inventories in the textile-producing area of Alsace was general; firms were low on money and circulated between 9 million francs and 16 million francs of their own IOUs as a substitute for money. When this edifice was toppled in December 1827 by Parisian banks' refusal to renew the Alsatian loan, the London crisis that resulted from overtrading in South American stocks had arrived on the Continent.[36]

Åkerman calls the crises of 1825 and 1836 'Anglo-American' in contrast to the one in 1847 that was Anglo-French.[37] The first two were Anglo-American in different ways: the 1825 crisis was Anglo-South American, the 1836 crisis Anglo-United States. In addition, the situation in 1836 was more complex than the one in 1825.

As noted earlier, President Jackson considered that responsibility for the crisis of 1836 to 1839 should be divided equally between Britain and the United States, while Matthews thought it futile to assign causal primacy. Monetary expansion in the two countries was vastly different. Wildcat banking aided by silver imports had started in the United States while new joint-stock banks had been established in Britain following new legislation in 1826 and 1833. British speculation was in cotton, cotton textiles, and railroads; American speculation was in cotton and land, especially land that could grow cotton. Moreover, Anglo-American banks in London financed British exports to the United States.

The crisis was not a purely Anglo-American affair, although it is often discussed in these terms with emphasis on its impact on the evolution of the Bank

of England's discount policy.[38] Hawtrey states that it started in Britain in 1836 and 1837, spread to the United States, and then in May 1838, when Britain was quietly recuperating, erupted in Belgium, France, and Germany to spread back to Britain and to the United States in 1839.[39] The crisis in the United States also affected France and Germany directly through the decline in the volume of imports, price declines and financial connections. Lyon felt the loss of outlets for silk immediately. American purchases were important to the success of the fairs in Frankfurt and in Leipzig. American commission houses in Paris, which financed their purchases largely in London, and the American banker Samuel Welles, who also relied on London for finance, were threatened with failure as early as the spring of 1837.[40] The Maison Hottinguer, a French bank, helped Nicholas Biddle of the Bank of the United States underwrite the corner in cotton, which strangled cotton spinners in Manchester, Rouen, and Alsace in the summer and fall of 1838 before the collapse of the corner in November of that year that followed from an Anglo-French boycott.[41] Moreover the Bank of France provided assistance to the Bank of England. By the 1830s, the financial world had complex transatlantic relations in trade, commodity prices, and capital flows among Britain, the United States and France.

In January 1847 distress developed in London in response to railroad calls and the crisis came late in the summer. Åkerman stated that the distress was Anglo-French, but it had echoes in British-Indian trade, in Amsterdam and the Low Countries, and also in Germany and even New York. Some sense of the spread of the crisis is evident from the number of bankruptcies collected by Evans but not the value of the assets of the failed banks and firms. The data are more complete for Britain than for other countries since Evans reported only the 'principal foreign failures'. Despite its serious deficiencies a table showing the monthly failures provides a useful impression of how the shock wave of the crisis spread. The British crisis is seen to have almost died away, except for London, when revolution in France and Germany produced the reactions of March and April 1848 that are probably under-recorded in Evans's data.[42]

The failure of the A. Schaaffhausen Bank of Cologne on 29 March 1848 played a role in the development of German banking. The Prussian government allowed the bank to be saved through its conversion into a joint-stock company, despite its standing policy of opposition to credit expansion; this precedent paved the way for the substantial expansion of German banks in the 1850s with important consequences for German economic growth.[43] Since Cologne had been a Hanseatic city, the bank was probably tied into the merchant banking network of London-Antwerp-Hamburg-Bremen-Le Havre-Marseilles featured so prominently by Evans. A local source claimed that Cologne was at the crossroads of trade between Holland, Brabant, France, and eastern and upper Germany and that the city suffered many bankruptcies as a consequence of the British crisis of 1825. The source admitted that apart from

Table 8.1 Reported failures in the crisis of 1847–48, by cities (number of failures)

City	1847					1848								Oct–Dec
	Aug	Sept	Oct	Nov	Dec	Jan	Feb	Mar	Apr	May	June	July	Aug	
London	11	19	21	25	7	3	7	3	1	8	2	1	1	1
Liverpool	5	4	28	10	4		3						1	
Manchester		6	11	8	1				1					
Glasgow	2	4	6	9	7	6				1				
Other UK	2	4	16	7	7	2		1	1					1
Calcutta				1		11	5		1	2				1
Other British Empire											1	2	1	4
Paris		1				2	1	14	2					
Le Havre				1			1	5	2					
Marseilles		1			1	1		2	13					
Other France			2			1		1		1	1			
Amsterdam				3	1	1		14	4		1			
Other Low Countries	1		1	4				4		1	1			
Hamburg	1			2		1			7	4	3	1		
Frankfurt					3	1			1					
Berlin								3	4	1				
Other Germany		2			1		1		6					
Italy		3		7	1									
Other Europe		1		3	2	1	1		1	1				
New York		1		3	1				1	5				4
Other US														7
Elsewhere		1			1					2	1			2

Source: Derived from names of firms and banks listed by D. Morier Evans, *The Commercial Crisis, 1847–48* (1849; reprint edn, New York: Augustus M. Kelley, 1969), pp. 69, 74, 91–2, 103–4, 105–6, 112–13, 118–20, 123, 127.

some financing of leather imports from South America, most banking finance was local and undertaken for heavy industry. Johann Wolter and Abraham Schaaffhausen got their start as leather merchants, purchasing South American hides from Spain, first through Amsterdam and then directly. Abraham, the son, was a merchant, commission agent, forwarder, and banker with international connections. The trouble in 1848, however, came largely from financing real estate investments in Cologne. Almost one-quarter of the portfolio of the bank consisted of land and loans to a single builder that together amounted to 1.6 million thalers compared to the bank's capital of 1.5 million. As social

unrest increased and depositors sought cash, the bank first took on a Dutch partner and then received help from the Prussian Bank's branch in Cologne, the same bank's branch in Münster, the Prussian Seehandlung (another state financial agency), and the Prussian lottery. Permission for A. Schaaffhausen to convert into *A. Schaaffhausen'scher Bankverein* may have been related to the fact that joint stock banks were forbidden to invest in building sites and all other forms of speculation.[44]

The boom leading up to the panic of 1857 was worldwide. Gold discoveries in California (1849) and Australia (1851) led to export spurts to those countries and enlarged the credit bases in Europe and the United States. It would have done so to a greater extent had India not been exporting far more than it was importing and beginning to receive, along with the United States, money from Britain which had been discouraged from investing on the Continent by the revolutions of 1848. The balance of payments surplus was financed by a payment in silver. Both Europe and the United States had railroad and banking booms. Expansion also came from joint-stock banks in Britain and Germany and from the *Crédit Mobilier, Crédit Foncier,* and *Crédit Agricole* in France which made large loans to trade and industry. Scandinavia in particular had been stimulated by the boom in trade generated by the repeal of the British Corn Laws, timber duties, and Navigation Acts.[45] Bad harvests and the Crimean War, which cut off Russian exports, raised the price of grain for farmers worldwide. These were, in fact, golden years for British farmers, despite the repeal of the Corn Laws in 1846. After the war, grain prices sank as Russian supplies came back on the market, and railroad building declined. The dominoes started their collapse in Ohio – or, rather, the New York branch of an Ohio bank – and fell in New York, Ohio, Pennsylvania, Maryland, Rhode Island, and Virginia, and then in Liverpool, London, Paris, Hamburg, Oslo, and Stockholm. Evans's data on bankruptcies for 1857 are sketchier than those for 1847 and the path of failures cannot be traced. The failure of the New York branch of the Ohio Life and Trust Company occurred around the same time as British withdrawals of funds from the United States in response to increases in interest rates in London.

The concentrated nature of the crisis from the Ohio Life revelation on 24 August to suspension of the Bank Act in London on 12 November through Hamburg's loan from Austria (the *Silberzug*) on 10 December is striking. Clapham observed that the crisis appeared at almost the same moment in the United States, Britain, and Central Europe and was felt in South America, South Africa, and the Far East.[46] Rosenberg calls it 'the first worldwide crisis'. The chamber of commerce of Elberfeld asserted: 'The world is a unit; industry and trade have made it so.'[47]

The crisis of 1866 is the tail end of one that began in 1864. Åkerman said that it parallels the 1857 crisis, insofar as it followed the Civil War as 1857 followed the Crimean War, and as the collapse of cotton in 1866 paralleled the collapse

of wheat a decade earlier.[48] The inclusion of 1864 eliminates a general view that the crisis was strictly British.[49] The timing of the panic on Black Friday, 11 May 1866, was intimately tied to the Prussian-Austrian war, largely through stock market collapses attributable to rumors of war and then the advent of the war, and then to the *corso forzoso* of 11 May 1866, when the Italian government suspended convertibility of the lira into gold and in return for this privilege borrowed 250 million lire from the national bank.[50] Like the Overend, Gurney collapse, the *corso forzoso* had been triggered by an internal run on notes against gold, stimulated by money withdrawals toward Paris that in turn had suffered from sales of foreign securities. The London market was shaky in mid-April because of rumors of war. The Berlin bourse panicked on 2 May with mobilization and again on 12 May when war broke out. The Prussian Bank raised its discount rate to 9 percent on 11 May. The panic in London that same day was part of a general rush for liquidity against a vulnerable company at a time of acute financial distress. Alfred André, a Parisian banker with major interests in Egypt, spent 'an exhausting week' in London looking after the interests of his firm at the time of the Overend, Gurney crisis. He returned to Paris on 17 May, having concluded that the finance companies were ruined and that business was paralyzed in Italy, Prussia, Austria, and Russia, with France standing up pretty well, but only for a moment.[51]

There is no obvious connection between the US gold crisis of September 1869 and the Austrian crisis of the same month. Both currencies were floating. Both countries had had investment booms following their wars, although the devastation from the conflict had been much greater in the United States. Wirth prefaced his brief discussion of the 'great crash of 1869', which preceded the real great crash of 1873, with remarks about German and Austrian investments in the United States, the invasion of European markets by US goods, and the extension of shipping and banking connections across the Atlantic.[52] Since Wirth did not mention the US gold crisis, however, it seems unlikely that he was suggesting a connection. The accounts of the 1869 gold crisis in the United States ignore Austria.[53] A possible link is through the price of wheat, which Jay Gould and Jim Fisk were trying to increase when they bid up the gold premium (for example, the discount on the greenback dollar). The difficulties following the September 1869 'crash' were concentrated in Hungary, which was a wheat-growing country.[54] Gould states that the United States could sell wheat to Britain in competition with the low-priced labor and water transportation from the Mediterranean with the gold agio at 45, but the United States could not sell wheat with gold below 40.[55] The decline in the gold agio in the United States in September should have assisted Hungarian economic prospects.

The 1873 story begins with the Franco-Prussian indemnity that was one-tenth paid in gold in 1871 and led to substantial speculation in Germany which then spilled into Austria.[56] Jay Cooke, a latecomer to railroad finance

and seeking capital for railroads in Europe, overextended himself with the Northern Pacific; he tried to borrow in Frankfurt but could not compete with the German and Austrian building booms.[57] Other shocks included the opening of the Suez Canal in 1869, the mistake of the German authorities in paying out new coin before withdrawing the old silver coins, the Chicago fire of 9 October 1871,[58] and especially the excitement generated by German unification under the leadership of Prussia's Bismarck. German acquisition of £90 million from the indemnity endangered stability in Britain because of the threat of conversion into gold. France deflated to pay the indemnity and was not affected by the inflation that occurred elsewhere in Europe.

A major question is the connection between the collapse in May 1873 in Austria and Germany after some months of distress and that in the United States in September. One link is through the changes in German investments in American railroads; initially German investors speculated in the railroads and western lands and then there was an abrupt halt to further investment. McCartney stated that 1873 is generally accepted as the first significant international crisis. The crisis erupted in Austria and Germany in May; spread to Italy, Holland, and Belgium and then to the United States in September; and then later involved Britain, France, and Russia. A second panic hit Vienna on 1 November but was shortlived.[59] Morgenstern noted 'clear evidence of transmission throughout the year, extended to Amsterdam and Zurich' in his table of international stock market panics.[60] In the fall of 1875, Baron Carl Meyer von Rothschild wrote to Gerson von Bleichröder and commented on the low state of stock prices everywhere, noting that 'the whole world has become a city'.[61]

A series of less intimately related failures and panics followed, including the City of Glasgow in 1878, Union Générale in 1882, and the New York stock market in 1884; the European-wide stock market panics in 1887 over the threat of war between Russia and Turkey; the copper corner in 1888 in Paris with the failure of the Comptoir d'Escompte; and the Baring crisis of 1890, the Panama scandal of 1892, and the New York panic of 1893. Their propagation was studied in detail by Morgenstern[62] and Pressnell who focused especially on the Baring crisis and attached particular importance to its impact on the Bank of England's gold reserves.[63] The 1890 Baring crisis produced financial stringency rather than panic in New York as British investors sold good US stocks to carry bad Latin American loans.[64] One view is that the financial crisis in New York in October 1890 precipitated the Baring Brothers collapse in November by producing failures in London, which made it more difficult for Baring to continue. The Baring crisis, induced by difficulties in Argentina, brought a sharp decline in British lending worldwide and contributed strongly to economic crises in South Africa, Australia, and the United States for the next several years.[65]

1907

The US crisis of 1907 began several years earlier in Italy, which had participated in the upswing of the first several years of the century.[66] Speculation fed by credit had been rife. There were fictitious ventures and a steel trust that used funds ostensibly borrowed for real investment to speculate in its own securities; high dividends were paid with the cash from borrowings to stimulate investor interest. Distress set in in May 1905 with the collapse of many new companies. A second relapse occurred on the Genoa stock market in October 1906. By April and May 1907 lending from Paris and London had slowed and the distress became more acute. The Società Bancaria Italiana had started in 1898 with capital of 4 million lire, which was raised to 5 million in 1899, 9 million in 1900, 20 million in 1904, 30 million in 1905, and 50 million in March 1906. As the capital was increased, old and often troubled banks and new personnel were acquired.[67] The head office of the bank in Milan did not know the risks that had been acquired by its branch in Genoa.[68] In particular, the bank was deeply involved in advances on securities (*riporti*). Governor Stringher of the Banca d'Italia was worried because of the poor quality of its loans and the large amount it had borrowed from the central bank by December 1906. When Paris and London cut off new credits to Italy and to the United States in the spring of 1907, the upstart, marginal bank was doomed. Direct connections between Turin-Milan-Genoa and New York were limited. But Italian centers were connected to Paris; New York was connected mainly to London; and Paris and London to each other. Bonelli asserted that when Paris sold its British securities, and Paris and London both stopped lending, the colonial countries suddenly found themselves deprived of capital and were obliged to halt ongoing investment projects with consequent downward pressure on demand and output and employment and prices. The analogy between Italy in 1907 and a colonial territory is striking. Bonelli asserted that the Paris cut-off of loans to Italy would have had much more serious consequences if it not been for emigrant remittances, largely from the United States.[69] There was a direct connection across the Atlantic, largely New York to Naples. Bonelli's account focused on the narrow direct connections and contrasted with that of a contemporary observer, a New York banker named Frank Vanderlip, in a paper called 'The Panic as a World Phenomenon'. Vanderlip asserted that the basic causes of the panic were the Boer War, the Russo-Japanese War, and the San Francisco earthquake. But after such a grandiose beginning, he discussed overtrading by newly formed trust companies and the need for an expansive currency.[70]

The international ramifications of 1929

President Hoover stated that part of the real cause of the Great Depression was expansion of production outside of Europe during World War I, which

contributed to global oversupply when European production recovered after the war. In addition, there were the financial complications of reparations and war debts; an overvalued British pound and an undervalued French franc, and the recycling of German reparations after the Dawes Plan by American purchases of the bonds of German corporations and public bodies. Some blame attaches to the reduction of US interest rates in the summer of 1927 to assist Britain in maintaining the parity for the pound when US domestic purposes would have been better served by higher interest rates. When US stock prices increased in March 1928 and especially after June, US purchases of foreign bonds came to a halt. For a time, Germany, the Latin American countries, and Australia borrowed at short-term. Germany responded to the reduced inflow of money by deflating its economy so it would have the cash to make reparations payments. The payments positions of Argentina, Australia, Uruguay, and Brazil turned sharply adverse. Unable to fund their accumulations of short-term indebtedness or to borrow more, their currencies depreciated shortly after the stock market crash of October 1929, as the prices of wheat, coffee, rubber, sugar, silk, and cotton fell sharply.

An open-market program undertaken by the Federal Reserve Bank of New York on its own initiative, over the protest of the Federal Reserve Board in Washington, alleviated the credit squeeze early in 1929. There was an increase in international lending in the first half of 1930; the volume of international lending in the April-June quarter was larger than in any other quarter in the 1920s and the 1930s. However, the lower level of prices and the loss of confidence in Germany, especially after the National Socialist gains in the September 1930 elections, meant that financial distress continued. Banks in Central Europe, largely Austria and Germany, tried to improve their positions by bidding up the prices of their own stocks. Two private banks, the Banque Adam and the Banque Oustric, failed in Paris, the latter unleashed a scandal that implicated three government officials and led to the fall of the government. The deflationary Laval government came to power early in 1931. And then the rolling deflation started: the failure of the Credit Anstalt in Vienna in May, the failure of the Danatbank in Germany in July, the German standstill agreement of July, and withdrawals of money from London in August, that culminated in the decision of the British in September 1931 to break the link between the pound and gold. The gold bloc of France, Belgium, the Netherlands, and Switzerland started buying gold with US dollars and the withdrawal of gold from the United States reduced the reserves of US banks. Japan went off gold in December 1931. Deflation in the United States came from the depreciation of the British pound and the currencies of the sterling area countries that remained pegged to the pound and from the reduction of bank reserves. In February 1932 the Glass-Steagall Act made it possible to reflate through open-market operations, but it was too late. Bank failures continued to spread in a

positive feedback debt deflation process of declining goods prices, bankruptcies, and bank failures. The US economy reached the bottom with the general Bank Holiday that began in March 1933 and the depreciation of the dollar in the spring of that year when the US gold parity of $20.67 was abandoned.

This history does not lead to the conclusion that the 1930s depression originated in the United States.

9

Bubble Contagion: Mexico City to Tokyo to Bangkok to New York, London, and Reykjavik

Four waves of credit bubbles in 30 years plus a bubble in US stocks in the late 1990s are unique in financial history. Either this succession of waves in a relatively short period was a coincidence, or there was a systematic relationship among several of them. Obviously the first wave – which involved the rapid growth in bank loans to governments and government-owned firms in Mexico and ten other developing countries – was *sui generis*. Was there a connection between the implosion of this wave and the credit bubble that was centered on Japan in the last half of the 1980s? Similarly was there a connection between the implosion of the asset prices in Tokyo at the beginning of the 1990s and the third wave of bubbles, which involved Thailand and its neighbors in Southeast Asia in the mid-1990s as well as Mexico, Russia, Brazil, and Argentina? Was there a link between the Asian Financial Crisis that began in mid-1997 and the bubble in US stocks in the late 1990s? Finally, was the wave of bubbles in real estate in the United States, Britain, Ireland, Spain, and Iceland between 2002 and 2007 – and in the debt of the governments of Greece and Portugal and Spain in 2008 and 2009 – related to these earlier events?

Asset price bubbles in major industrial countries are infrequent; the previous US stock price bubble had been in the late 1920s. Japan had never previously had an asset price bubble and neither had the other Asian countries. A bubble in three, four or more countries at the same time is extraordinary and suggests a common cause – a Minsky-type displacement or shock, and probably an external one because so many countries were similarly impacted at the same time. Four waves of bubbles in 30 years suggest that the reversal in the direction of cross-border money flows that follows the implosion of one bubble may contribute to the next wave.

Most credit bubbles led to asset price bubbles; every real estate bubble has resulted from the rapid increase in the supply of credit. The credit bubble of

the 1970s financed the deficits of the governments and of government-owned firms in Mexico and other developing countries when their GDP growth rates accelerated because of the sharp increase in commodity prices. The bubble in stocks and property in Japan in the second half of the 1980s resulted from the rapid increase in the supply of credit for real estate purchases. Similarly the bubbles in real estate and stock prices in Thailand, Malaysia, and the other Asian countries as well as in Mexico and other Latin American countries in the first half of the 1990s were responses to money inflows and domestic credit expansion. The increases in the prices of residential real estate in the United States, Britain, and other countries after 2002 resulted from a massive increase in the supply of credit that partly followed from the securitization of mortgage loans. The bubbles in the bonds of the governments of Greece and some of its neighbors in the south of Europe followed from a significant increase in money flows to these countries.

One of the three essential components for the development of a credit bubble is a large pool of money that can be accessed to provide loans. A second component is a shock that leads to sharp increases in the anticipated rates of return – or to a significant reduction in risk – on loans to a particular group of borrowers; the shock may involve a reduction in regulation that makes it easier for borrowers to access credit in foreign markets. A third component is a group of investors who are willing to extend more credit to this group of borrowers.

The 1970s surge in bank loans to Mexico et al.

The exceptional feature of the 1970s was the surge in the US and world inflation rates during peacetime; the two previous episodes of sharply rising prices in the twentieth century were during and after the First and Second World Wars. Payments imbalances surged in the late 1960s and early 1970s, which led to a massive increase in money supplies. The 1970s inflation followed from the breakdown in the Bretton Woods arrangement of parities for national currencies, which eliminated the national 'anchors' for monetary policies.

One dramatic impact of the increase in the US inflation rate that began in the late 1960s was that interest rates on both short-term and long-term US dollar securities increased. The interest rates that banks in the United States could pay on their deposits bumped into the ceilings set by the Federal Reserve on US dollar deposits. The result was a large shift of US dollar-denominated deposits to offshore banks (including the branches of US banks) in London and other financial centers that were not subject to these ceilings.

The surge in demand associated with rapid growth in money supplies led to sharp increases in the prices of primary products. There were were two massive oil-price shocks and the payments surpluses of some of the oil-exporting countries surged, and their demand for US dollar securities increased rapidly.

The increases in the GDP growth rates in the commodity-producing countries meant that their governments were more attractive borrowers.

British, Canadian, Japanese, and other foreign banks sourced the money for dollar-denominated loans to Mexico and other developing countries from the banks in these offshore centers. Previously these borrowers had relied on US banks and on institutions like the World Bank for most of their external financing; the number of lenders willing and able to increase their loans to these borrowers increased dramatically. These non-US banks were able to increase their assets much more rapidly than they could in their domestic economies. The US banks responded to the challenge of foreign banks entering 'their turf' by cutting price – reducing interest rates – to minimize the decline in their market share.

The borrowers in these developing countries were eager for the money, and the lenders believed that the credit risk was modest; a wise and respected New York banker said 'countries don't go bankrupt'. As long as the money available to the borrowers on new loans was several times larger than the interest payments on the old loans, these borrowers encountered no burden in servicing their external indebtedness. The chatter was that the major international banks were involved in 'petro-dollar recycling', which would have enabled the oil-importing countries to finance much larger trade deficits. The surpluses of the oil-exporting countries were recycled, but the increase in the indebtedness of oil-importing countries was much larger than the increase in the holdings of international reserve assets of the oil exporters. The increase in the flow of money to these developing countries enabled them to finance larger trade and current account deficits, which meant that the industrial countries had significantly larger current account surpluses.

The external indebtedness of Mexico and other developing countries increased by 20 percent a year for a decade, while the interest rate on these loans averaged 8 percent a year, although these rates trended upward in the 1970s as the world inflation rate increased. The cash flows from new loans were significantly greater than the interest payments on the outstanding loans, which meant that the currencies of these countries appreciated in real terms, and their trade deficits increased. As long as debt was increasing much more rapidly than income, there was a bubble in the pattern of cash flows, since it was inevitable that at some stage, the lenders would become more cautious, indebtedness then would increase less rapidly, and the currencies of these countries would depreciate.

The bubble in the credit flows to the developing countries was punctured by the sharply contractive monetary policy that the Federal Reserve adopted in the autumn of 1979. The ability of these countries to finance trade and current account deficits declined sharply – which meant that the trade surpluses of the industrial countries declined. Moreover there was a sharp flow of money to the United States from other industrial countries, which meant that their

currencies depreciated and their trade surpluses surged as the US trade deficit increased.

The European currencies and the Japanese yen depreciated until 1985 and the US trade deficit increased until 1987. The mirror of the increase in the US trade deficit until 1987 was that the trade surpluses of many countries including Japan increased.

The 1980s asset price bubble in Japan

The rapid growth in the supplies of money and credit in Japan in the second half of the 1980s resulted from the reluctance of the Japanese authorities to have the yen appreciate because the country's trade surplus would decline, with negative impacts on profits and employment in the export industries. The Bank of Japan bought US dollars to resist the upward pressure on the yen, which led to a rapid increase in the reserves of the Japanese banks, which then were able to increase their loans at a rapid rate. Regulations on bank loans for real investment purchases and construction were relaxed in the belief that increased spending on real estate would contribute to economic growth and lead to increases in imports and dampen the tendency toward a stronger yen. Regulations that limited the foreign investments by Japanese firms and banks were relaxed in the belief that the larger money outflow also would dampen upward pressure on the currency.

The 1980s real estate bubble in Japan was so massive that by the end of the decade the chatter in Tokyo was that the market value of the land under the Imperial Palace was greater than the market value of all of the real estate in California. The land area in California is several billion times larger than the grounds of the Imperial Palace, which meant that the price per hectare was a million times higher than the price in California. Not that there had been an auction or even a pseudo-auction for the Imperial Palace grounds. The analyst who made this comparison estimated the value of the palace grounds by multiplying the hundred or so hectares by the recent price per hectare paid for a small plot of land in the nearby Ginza entertainment neighborhood, which was much the most valuable land in Tokyo. The value of Californian real estate was obtained from Federal Reserve data on US household wealth.

All of the financial values in Tokyo were sky-high at the end of the 1980s. The market value of Japanese stocks was twice the market value of US stocks, even though Japanese GDP was less than half of US GDP. The comparison between Japanese and US firms in terms of the ratios of the market value of stocks to profitability was even more skewed. The market value of Japanese real estate was twice the market value of US real estate, even though the land area in Japan is 5 percent that in the United States and 80 percent of Japan is mountainous. The market value of land per capita in Japan was more than four

times that in the United States while per capita income in Japan was 60 or 70 percent that of the US.

The Japanese banks were at the top of the hit parades of the world's banks as measured by assets and by deposits (but not by profits); usually seven of the ten largest banks on this list were Japanese. Similarly the capital of Nomura, Japan's largest investment bank, was larger than the capital of the five largest US investment banks.

As we mentioned in the first chapter, the Mitsui Real Estate Company paid $625 million for the Exxon building on Sixth Avenue in New York City even though the asking price was $510 million because the company wanted to get into the *Guinness Book of World Records*. Other Japanese firms were also acquiring trophy properties and buildings in the United States. Mitsubishi Real Estate and a group related to Sumitomo Bank bought the Pebble Beach Golf Course in Northern California. Sony bought Columbia Records and then Columbia Pictures, and Matsushita, its dominant rival in the electronics industry, acquired MGM Universal.

By the 1980s Japan was the second leading industrial power, more economically powerful than Germany. Toyota, Nissan, and Honda were leaders in the global automobile industry. Sony, Matsushita, and Sharp and a seemingly endless list of firms dominated the global electronics industry. Nikon and Canon 'owned' the world's photo-optics industry.

The mandarins in the Ministry of Finance maintained low interest-rate ceilings on both bank deposits and bank lending rates; the interest rates on deposits were below the inflation rate so households had to save a high proportion of their incomes or else their wealth would have declined. The demand for loans from business firms at these low interest rates was much greater than the supply; government officials provided 'window guidance' to the banks identifying the firms that were to be given preference. The real rates of return on both real estate and stocks were positive and high.

Japanese banks owned large amounts of real estate and stocks; as the prices of these assets increased, their capital increased. They were then able to increase their loans, and some of these loans went to borrowers who used the money to buy real estate. In the early 1980s the banks established a new set of intermediaries, *'jusen'*, to make housing loans, which traditionally Japanese banks had been reluctant to make. The *jusen* would get the funds for housing loans by borrowing from the banks – in effect seven of the large banks established these specialized lenders. About the same time the Ministry of Finance established public-sector lending institutions that would also make housing loans. The banks then decided to make housing loans on their own.

In the first half of the 1980s Japan began liberalize its financial regulations, in part because of pressure from the Americans to 'open up' the markets in Tokyo so US firms would have access to clients and customers and trading

opportunities in Japan comparable to those that Japanese firms had in New York. Interest rate ceilings on deposits and on loans were raised. Window guidance became much less extensive. Moreover the restrictions on the foreign investments of Japanese firms were relaxed and Japanese banks were permitted to increase their foreign branches and subsidiaries.

Financial liberalization meant that the banks could increase their loans to borrowers who wanted to buy real estate and to build new office buildings and apartment buildings and shopping centers. Real estate prices increased. Firms involved in the real estate business accounted for a significant proportion of the market value of all of the firms listed on the Tokyo Stock Exchange. These real estate holding companies were somewhat like mutual funds; when the prices of the properties they owned increased, investors rushed to buy more of their shares, and so the share prices of the real estate companies increased. Increases in real estate prices led to a construction boom as new skyscrapers were constructed. The Japanese banks owned large amounts of real estate and of stocks, and the contribution of the increase in the value of both real estate and stocks to the capital of the banks was much greater than their operating profits. As their capital increased, the banks were able to increase their loans.

Japan had developed the financial equivalent of a 'perpetual motion machine'. The increases in real estate prices led to increases in stock prices; the increases in both real estate prices and stock prices led to increases in bank capital. As bank capital increased, banks were in a position to increase their loans, and because of financial liberalization, they were much better positioned to increase their loans to groups that had been restricted in their ability to borrow in earlier periods. Because most bank loans were collateralized by real estate, bank loan losses were trivially small as long as real estate prices increased. The profits of the firms that invested in real estate were increasing, and so many of these firms borrowed more in their search for larger profits.

The real estate loans were collateralized by property; traditionally banks would lend up to 70 percent of the appraised value of these properties. Some of the loans made by the *jusen* went to '*yakuza*', organized criminal groups, who used their connections to secure exceptionally high property appraisals. Since property prices were increasing by 30 percent a year, a modest 'error' made by an appraiser in putting too high a value on the property would soon be corrected.

Real estate prices increased much more rapidly than rents, with the consequence that the rental rate of return declined significantly below the interest rate on the borrowed funds. Investors who had bought properties in the last several years of the 1980s had a negative cash flow – the rental income on their properties after the payment of the operating costs was below the interest payments due to the lenders – but because property prices were increasing so fast,

they could raise cash to make the interest payments by increasing the amounts borrowed against properties that had been purchased in earlier years.

The rapid increases in real estate prices and stock prices led to a surge in household wealth – Japanese households owned currency and bank deposits, real estate, and stocks. While many of the stocks in Japanese firms were owned by other firms in the same *'Keiretsu'* (successors to the *Zaibatsu*), one-third of the stocks were owned by individuals. In effect each firm was a combination of an operating company and a mutual fund.

As the market value of Japanese stocks increased, investors resident in the United States and Western Europe bought more Japanese stocks. Global stock index funds wanted to own more Japanese shares. The rates of return to non-Japanese investors on their purchases of shares in Japanese firms were high since these investors benefited from the combination of the increase in the price of the stocks and the increase in the foreign exchange value of the Japanese yen.

The bubble in Japan reached its crescendo at the end of 1989. Real estate prices seemed so high that the quip by the much quoted baseball star Yogi Berra that 'It's so expensive that no one can afford to live there' seemed applicable. Banks developed one-hundred-year, three-generation mortgages. The incoming governor of the Bank of Japan was concerned that such high prices for homes would erode social harmony. A new central bank regulation instructed Japanese banks to limit the rate of growth of their real estate loans so that it would be no greater than the rate of growth of their total loans.

Once the rate of growth of bank loans slowed, some recent buyers of real estate developed a cash bind; their rental income was smaller than the interest payments on their mortgages, but they could no longer obtain the cash needed to pay the interest on their outstanding loans from new bank loans. Some of these investors then became distress sellers of their properties. The combination of the sharp reduction in the rate of growth of credit for real estate and these distress sales caused real estate prices to decline; the cliché that the price of land always rises was belied.

Stock prices began to decline at the beginning of 1990; stock prices declined by 30 percent in 1990 and 30 percent again in 1991. The stock price trend in Japan was downward although there were four significant rallies. At the beginning of 2010, stock prices in Japan were in the same ballpark that they were thirty years earlier.

Now the perpetual motion machine began to work in reverse. Property sales led to declines in property prices. The decline in real estate prices and stock prices meant that bank capital was declining; banks were now much more constrained in making loans. Because the value of Japanese stocks was declining while that of US stocks was increasing, the global stock equity funds sold Japanese stocks and bought US stocks.

One of the stylized facts in monetary economics is that the implosion of an asset price bubble is deflationary, the flip-side of the economic boom that occurred during the expansion phase of the cycle. Investment spending in Japan declined in part because the cost of capital had surged and the anticipated growth of profits had been revised downward, and in part because the splurge in investment spending during the expansion phase had resulted in significant excess capacity. Household spending increased much more slowly as millions of families increased their saving from their earned incomes to compensate for the decline in their wealth that followed from the fall in stock prices and real estate prices.

The recession in Japan in 1991 meant that import growth slowed markedly while exports surged; some Japanese firms greatly increased their efforts to sell abroad because the domestic market for their products was growing slowly relative to the growth in their supply capabilities. The result of the slowdown in the growth of imports and the surge in the growth of exports was that Japan's trade surplus increased. This increase was larger than the increase in the capital flow from Japan and led to the appreciation of the yen, which became a handicap to export-oriented Japanese firms. Many increased their investments in China, Malaysia, and Thailand to take advantage of lower labor costs. The increase in investment spending for firms in the export industries stimulated income growth.

The East Asian economic miracle and the bubble in emerging-market equities

Four innovations contributed to the surge in property prices and real estate prices in Mexico, Thailand, and other emerging-market countries in the early 1990s. The funding of the overhang of bank loans into Brady bonds, which effectively removed these countries from their bankruptcy – and they were re-christened as emerging markets. The discovery of 'emerging market equities as a new asset class', index-based mutual funds and pension funds began to acquire these stocks. The expectation was that these countries would industrialize at a rapid rate, and corporate profits would surge. Privatization of government-owned firms in resource extraction, communications, manufacturing, and other industries led to a surge in money inflows. The appreciation of the Japanese yen following the implosion of its bubble led to the 'hollowing out' of the Japanese economy and led firms to seek lower-cost production sites in Thailand, Malaysia, Indonesia, and their neighbors.

In 1992 the World Bank published *The East Asian Miracle*, an expressively descriptive title for the economic performance of the countries in the arc from Thailand to South Korea; the increases in their GDPs were in some ways comparable to the gains that Japan had made in the 1950s and the 1960s. The

Korean peninsula had been fractured in the war in the early 1950s; in the mid-1960s South Korea began a remarkable period of economic growth. Singapore had been a fortified swamp in the 1950s but by the 1990s had achieved a first-world standard of living. The change in political leadership in China from Mao Zedong to Deng Xiaoping in 1978 led to a dramatic change from a self-contained, isolated country to one that was open and eager for international trade and international investment; the annual rate of growth averaged nearly 10 percent for more than twenty years and the increase was even more dramatic in the provinces that bordered the seacoast and in the major cities such as Beijing, Shanghai, and Shenzen. Hong Kong had evolved from an outpost for peeking into China in the 1950s, the 1960s, and the 1970s into an entrepôt center for prepping Chinese goods for world markets.

Thus the expansion of the asset price bubbles in the Asian capitals followed from the implosion of the asset price bubble in Tokyo and the surge in the flow of money from Japan at the beginning of the 1990s. Most of the money was Japanese-owned; some had been owned by foreigners who sold Japanese stocks as their prices declined. The flow of money from Tokyo to Thailand and Indonesia and the other Asian countries led to increases in the foreign exchange value of their currencies if they were floating and to increases in the international reserve assets and the money supplies if these currencies were pegged. The rates of growth of domestic income in these countries increased in response to the increases in investment spending and in consumption spending that followed from the increases in the prices of real estate and stocks. The residents in these countries who sold their securities and assets to the Japanese used most of their receipts to buy other domestic securities and real estate.

Mexico was being prepped to join the North American Free Trade Agreement with the United States and Canada. The Salinas government adopted three sets of measures. Several hundred government-owned firms were privatized. The central bank adopted a severely contractive monetary policy to reduce the inflation rate, which had been above one hundred percent; as a result, the real interest rates on Mexican securities were extraordinarily high. Finally, many restraints on competitive business practices were relaxed or removed. The prospect was that Mexico would become a low-cost source of supply for the US and the Canadian markets. There was a massive surge in money flows to Mexico, partly in response to the high real interest rates and partly in response to the opportunity to buy some of the newly privatized firms.

Many of the once-developing countries were re-labeled emerging-market countries in the early 1990s after the overhang of past-due bank loans were funded into long-term bonds under the Brady plan. Some bright investment banker came up with the very profitable idea that emerging market equities were a new 'asset class'. Every pension fund and every mutual fund that was following an index fund approach toward developing a global portfolio

believed it had to acquire stocks that were available in these countries; the sales spiel was that the rates of return would be higher than the rates of return on equities available in the industrial countries because their rates of economic growth were higher. And moreover the pattern of price changes was not correlated with those of equities of the firms headquartered in industrial countries, so their inclusion would reduce the risk of the global portfolios.

In Thailand, Malaysia, and Indonesia in the first half of the 1990s stock prices increased by between 300 and 500 percent and manufacturing activity surged. Stock prices doubled in most of the East Asian countries in 1993 and continued to increase in 1994. Real estate prices soared. The economies boomed. Trade deficits grew. Since the asset price bubbles across these countries were pervasive despite marked differences in economic structures, per capita incomes, exchange rate arrangements, and whether they were international creditor countries like Singapore, Taiwan, and Hong Kong or international debtors like Thailand and Malaysia, there is a strong presumption that the bubbles had a common origin and that it was external.

China, Thailand, and the other East Asian countries were on the receiving end of outsourcing by American, Japanese, and European firms that wanted cheaper sources of supply for established domestic markets. Rapid economic growth was both the result and the cause of the inflow of foreign money, especially from Japan. Japanese investment initially took the form of the construction of manufacturing plants to take advantage of lower labor costs; high valued-added components would be produced in Japan and shipped to the affiliated plants for assembly. From there a large part of the production would be exported, some to the United States, some to Japan and some to third countries. The direct foreign investment by Japanese firms pulled supplier firms and the banks from Japan. The buzzword was export-led growth, which was almost always based on a low value for the country's currency. Many of these exports were produced by major firms headquartered in the United States, Japan, and Taiwan. Firms headquartered in South Korea began to invest in China and Indonesia because their wage rates were so much lower.

The surge in money flows to Mexico in the early 1990s led to the real appreciation of the peso, and the Mexican trade deficit increased to 6 percent of the country's GDP. External debt was increasing much more rapidly than GDP. Several political incidents associated with the presidential election in 1994 led to a sharp decline in money flows to Mexico; the Bank of Mexico supported the peso in the currency market by spending its own reserves, and eventually its ability to continue to support the peso was exhausted, and the peso depreciated sharply.

For a while money flows to other countries in Latin America diminished. Then in the winter of 1996 the consumer finance companies in Thailand – many of which had been established by the large Thai banks to circumvent

the regulations that limited their ability to make consumer loans – began to experience large losses on their loans. Foreign lenders to Thailand became increasingly concerned about the value of their loans to Thai borrowers, and the flow of money declined. The Bank of Thailand's ability to support the baht at its existing value was quickly exhausted, and in early July 1997 the baht depreciated sharply.

The depreciation of the baht triggered the contagion effect and within six months the values of each of the currencies on the Asian arc, with the exception of the Chinese yuan and the Hong Kong dollar, had lost 30 percent or more of their values. Stock prices declined by 30 to 60 percent, partly because foreign investors were seeking to cash out, partly because the domestic firms were no longer profitable. Real estate prices declined sharply. Most banks, with the exception of those in Singapore and Hong Kong, failed. The closing of many banks in Indonesia triggered racial strife, and an immense run on the currency which lost more than 70 percent of its value.

When the crises occurred, the play script was a reprise of similar events in Japan in the previous decade. The chatter about the East Asian miracle disappeared and new buzzwords arose – crony capitalism, spontaneous privatization, and destabilizing speculation.

The sharp depreciation of the currencies led to significant losses for those firms that had borrowed US dollars or Japanese yen or some other foreign currency. The banks that had lent to these firms also incurred losses; the banks in most of the Asian countries toppled into bankruptcy because of their own revaluation losses and the losses of the firms to which they had made loans.

The depreciation also resulted in a very quick reversal of the Asian countries' trade balances, from large deficits to large surpluses. The counterpart of the large swings in the trade and current account balances of the Asian countries was the mirror-image increase in the US trade deficit.

Irrational exuberance and US stock prices in the dot.com boom

Between 1982 and 1999 US stock prices increased by a factor of thirteen, the most remarkable run of annual increases in stock prices in the two hundred years of the American republic. In the very long run, US stock prices have declined every third year; in the last two decades of the last century, stock prices fell in only one year, and then only by 5 percent. The market value of US stocks increased from 60 percent of US GDP in 1982 – an exceptionally low ratio – to 300 percent of GDP in 1999, an extraordinarily high ratio.

The increase in US real estate prices during this period was modest for the country as a whole, although there were significant regional increases in areas that were experiencing large increases in per capita income or in the number of

employed individuals, including Silicon Valley and the broader San Francisco Bay area, Washington, DC, and Boston and New York.

The US economy boomed in the 1990s. The inflation rate declined from above 6 percent at the beginning of the 1990s to less than 2 percent at the end of the 1990s, the unemployment rate declined from 8 percent to less than 4 percent, the rate of economic growth increased from about 2.5 percent to 3.5 percent, and there was a remarkable increase in US productivity. The US Treasury's annual fiscal balance changed by more than 5 percent – from a deficit of nearly $300 billion at the beginning of the 1990s to a surplus of nearly $200 billion at the end of the decade.

One of the 'negatives' in terms of US economic performance was that the annual US trade deficit surged to $500 billion. Another was that the household saving rate declined to a new low.

The remarkable aspect of the boom was the focus on the 'new economy' and especially the role of information technology, the computer, dot.coms, and the firms that provided both the hardware and the software or exploited these developments to serve traditional needs. These technological developments led to sharp declines in the cost of sending and storing information. Ebay provided a nationwide auction market for tens of thousands of different products. Amazon developed the technology for the sale of books and electronic products. Peapod allowed individuals to shop for most of their groceries at home. Millions of accounts were established at the discount broker Charles Schwab and at its competitors. Firms were established that enabled investors to trade stocks using the computer at extremely low transaction costs. 'Day traders' emerged: individuals who quit their regular jobs to trade stocks either from their computers at home or from desks in specially designed shops. Priceline enabled airlines and hotels to sell seats and rooms at sharply discounted prices.

Entrepreneurs were able to get the cash to develop these ideas from venture capitalists (VCs) who provided seed money. The VCs developed a portfolio of investments in different firms in the hope that within three to five years they would be able to sell these shares – and make their profit – when the firms made their first public offerings of stock. The VCs' rates of return would depend on whether the firms in which they had invested were successful in their technological challenges and on the selling prices of their shares and the lengths of the holding periods.

As stock prices increased, the high rates of return earned by the VCs attracted lots of money, and the capital available to the VCs as a group surged by a factor of five. No investors wanted to be left behind. The money was there, so a large number of ideas were funded; money was chasing ideas and concepts. Three or four years on, the new firms would have an Initial Public Offering (IPO). The IPO would follow the traditional road-show, the occasion on which the

investment banks would parade the company to asset managers around the country seeking to induce them to buy its shares.

At the end of the road-show, the investment bankers would estimate the amount of the shares that they might sell at the IPO, and set both the price and the quantity. In 99.46 percent of the cases, the share price at the end of the first day of trading was significantly higher than the IPO price, and those fortunate enough to be able to buy at the initial offering price would make a significant capital gain.

One impact of this price pop was that more and more investors clamored to buy at the IPO price. The second impact was that the demonstration of the strength of the demand meant that more and more people wanted to get a piece of the action – the entrepreneurs were attracted to the immense wealth they might earn with a successful innovation, the VCs were attracted to the large profits they could gain by identifying the entrepreneurs that were likely to be successful, and the investment bankers wanted the fees from bringing a large number of firms to the public. The investors wanted the large capital gains associated with the pop between the IPO price and the price of the same shares at the end of the first day of trading, the first week of trading, and the first month of trading.

The size of the price pop was like dynamite or nitroglycerine or maybe even a nuclear explosion. The investment bankers appeared to set the price for the IPOs so as to maximize the price pop on the first day of trading – and not maximize the cash received by the shareholders who were selling. From this point of view, a lower price for the IPO might be preferable to a higher price, for the demand for the stock would increase – at least for a while – as the pop increased. The entrepreneurs sold only a small part of their total shares at the IPO; they calculated that the larger the pop, the greater their wealth. They were more interested in the apparent value of the shares they owned at the end of the first day's trading than they were with the amount of cash they might obtain from the IPO.

On some IPO days the number of shares traded was three or four times the number of shares that had been sold at the IPO. Since many of the buyers at the IPO had been told to hold their shares, the float was much smaller than the number of shares sold, and so these shares in the float might have been traded five or six times in the course of the day.

The United States seemed to have its own perpetual motion machine, one designed to enrich the fortunes of hundreds of thousands of families. The larger the price pop on the first day of trading, the greater the number of investors that were attracted to IPOs. The stronger the demand for IPOs, the larger the number of venture capitalists that were willing to back the entrepreneurs. The more capital that the entrepreneurs were willing to put in play, the larger the number of entrepreneurs that would seek their fortunes by breaking away from established firms.

In December 1996 Chairman Greenspan of the Federal Reserve Board first used the term 'irrational exuberance'; the Dow Jones was at 6400 and change and the NASDAQ was at 1300. Greenspan was cautious and careful with data; it seems highly unlikely that he would have commented on stock prices unless he believed they were then over-valued by a minimum of 15 or 20 percent. At the end of December 1999 the Dow Jones was at 11700 and the NASDAQ was at 5400, and the market value of NASDAQ stocks was 80 percent of the market value of stocks traded on the New York Stock Exchange.

In the late 1990s the prices of stocks representing the new economy – the dot. coms, e-commerce, fiber optics, servers, chips, software, IT, telecoms – which were traded on the NASDAQ had increased much more rapidly than the prices of old-economy stocks, those of firms such as GE and GM, AT&T and Time-Life, that were traded on the New York Stock Exchange. But there was more than a spillover effect, for the enthusiasm about the future that was characterizing the new economy stocks was infectious and also led to increases in the prices of the old-economy stocks.

It seemed as though developments in information technology were driving finance. Computers were becoming much more powerful and less expensive. The costs of transmitting and storing information and data were declining rapidly. Moore's Law came into play, and the cost of a unit of computing power declined by 30 percent a year. The World Wide Web was developing fast, and markets in separate centers were becoming linked. Computers were replacing humans in trading stocks. Individuals could order their airline tickets over the web. Fiber optics were linking the US east and west coasts and the prices of long-distance phone calls were declining to the level of local calls. Servers were big, and so was storage capacity. Thousands – tens of thousands – of new firms had been established to move information or data or to store information or data, and the profits earned by the venture capitalists that had funded them were so high that more money flooded in from pension funds, university and charitable endowments, and wealthy families. The demand for newly issued shares at the IPO prices was much larger than the supply, which led the investment bankers to engage in the process of 'spinning' – of allocating a goodly number of shares to the heads of the major firms that would bring them investment banking business. The investment banks grew rich; they had a lot of product to sell and a public convinced that share ownership would bring them large profits.

There is no easy answer to the question of when rational exuberance morphed into irrational exuberance. The idea that there might be an asset price bubble in US stocks occurred to different investors at different dates. The first date for the onset of the bubble in US stocks is the spring of 1995, eighteen or twenty months before Greenspan's 'irrational exuberance' comment. Stock prices had increased at an annual rate of 34 percent in 1995, and at 25 percent a year in the first eleven months of 1996.

The surge in stock prices in 1995 and 1996 can be attributed to two different aspects of the Mexican financial crisis of 1994; one was direct and one was indirect. The direct effect was the sharp depreciation of the Mexican peso which resulted in a sudden shift in the Mexican trade balance from a deficit of $20 billion in 1994 to a surplus of $7 billion in the next year; the mirror image was that the US trade deficit increased by about $25 billion, since the United States was much the largest trading partner for Mexico. The counterpart of this change in the Mexican trade balance was that there was a money flow to the United States (a modest preview of the events that would occur in 1997 following the Asian Financial Crisis and the massive turnaround in the trade balances of the Asian countries). In effect the flow of money from Mexico to the United States led to an increase in the prices of US securities. The second aspect was that the Federal Reserve eased its monetary policy and reversed its tightening policy of 1994.

An alternative starting date for the onset of the bubble is the summer of 1998, following the Asian Financial Crisis, the financial debacle in Moscow and the collapse of Long-Term Capital Management. The sharp depreciation of the Asian currencies led to an increase in the US trade deficit of more than $150 billion. Moreover, the Federal Reserve again eased policy, partly because of concern with the fragility of the monetary arrangements following the crisis in Long-Term Capital Management, until then the most professional and sophisticated of the many US hedge funds.

In the twelve months after the end of June 1998 the market value of the stocks traded on the New York Stock Exchange increased by 40 percent, from $9005 billion to $12,671 billion. The market value of NASDAQ stocks increased by 90 percent.

The surge in the flow of money to the United States had the same impact that the increase in money flows had on other currencies; the US dollar appreciated and the US trade deficit increased. The money from abroad went into the asset markets. Asset prices, especially stock prices, continued to increase, and brought about an increase in domestic investment and a dramatic decline in the domestic saving rate (which is identical with an increase in domestic consumption).

Those who moved funds to the United States then bought US dollar securities, which led to increases in their prices; US financial wealth surged. The Americans who sold some of the securities they owned to foreign investors then had to decide what to do with the money they received from the sale. They used most of the money to buy more securities from other Americans, but they also increased their purchases of US goods as their wealth objectives were achieved. The decline in the US saving rate and the increase in the US trade deficit were inevitable outcomes of the increase in the flow of money to the United States.

The data suggest that between 95 and 97 percent of the increase in household wealth that followed from the flow of savings from other countries to the United States was used to buy other US securities and only 3 to 5 percent was used to buy consumption goods. Yet to the extent each of the sellers of the securities spent some of the money receipts on consumption goods, the domestic saving rate declined.

The spending on consumption goods is like a 'leakage'. The smaller the amount spent on consumption goods, the larger the amount spent to buy securities and real assets and hence the larger the increase in their prices.

During 1999 the Federal Reserve, the banks and the country at large became obsessed with the Y2K problem – a neurosis that the economy would break down because some computers would not be able to change the date when the year changed to 2000. Precautionary behavior by the Federal Reserve led to an expansion of bank liquidity. Once again banks increased their loans in response to the increase in liquidity.

The increase in stock prices attracted European investors, and the euro depreciated. Because of the decline in US import prices and the surge in the trade deficit, inflationary pressures declined in the United States. Consequently, the Federal Reserve felt no need to adopt more contractive monetary policies.

With the advent of the new millennium, the Fed began to withdraw liquidity. Stock prices began to decline. The aggregate decline in the stock market between 2000 and 2003 was 40 percent and the decline in the market value of stocks traded on the NASDAQ was 80 percent. The decline in stock prices meant that the costs of capital to the firms increased, and the decline in household wealth led to less rapid increases in household spending. Once again the implosion of a stock price bubble was followed by a recession, which was dated from the beginning of 2002. Because the bubble in asset prices had not been based on increases in credit, the subsequent decline in stock prices had no significant impact on the banks.

The global bubble in real estate

The remarkable feature of the first years of the third millennium is that there significant increases in real estate prices in the United States, Britain, Ireland, Spain, Iceland, South Africa, Australia, and New Zealand, which began in about 2002. The price increases in the United States were in sixteen states that account for about half of the country's GDP. Most of these states were in the South and along both the Atlantic and Pacific coasts, and were among those that were growing most rapidly – but not all of the high growth states experienced significant increases in real estate prices. Spain and Ireland were members of the European Monetary Union, and hence did not have their own currencies.

Each of these countries had a current account deficit; each experienced an increase in money inflows and in most cases, their currencies appreciated. One exception was that the US dollar depreciated, and hence the increase in money flows to the United States dampened the appreciation of the euro that was underway after the implosion of the stock price bubble of the late 1990s.

The ratio of the increase in household wealth to GDP was much larger in Iceland than in any of the other countries that experienced sharp increases in asset prices. Moreover, the ratio of the increase in the flow of money to GDP was much higher for Iceland than for any other country. The unique feature of the bubble in Iceland was that the increase in stock prices was three to four times larger than the increase in real estate prices. The flow of money to Iceland during several years was more than 20 percent of the country's GDP. The Icelandic krona appreciated by more than 30 percent. The balance of payments accounting identity is that the changes in the value of a country's capital account balance and changes in the value of its current account balance were necessarily equal as long as currencies were floating. Thus the autonomous increase in the foreign demand for Icelandic securities led to increases in the value of the Icelandic krona and to increases in asset prices in Iceland.

After the implosion of the real estate bubble in the United States, and prices of homes and commercial real estate began to decline, and banks and other financial institutions began to collapse, the press chatter turned to the cause or causes of the bubble. The list is long and varied and includes excessively high leverage, the repeal of the Glass-Steagall Act, the corruption of the credit rating agencies, compensation based on increases in revenues rather than on increases in profits after adjustment for loan losses, 'derivatives', the opacity of the over-the-counter market in derivatives, the inflow of money from China, the proclivity of Chairman Greenspan to keep interest rates too low for too long, and so on.

This list is peculiarly US-centric, and ignores the bubbles in real estate markets in Britain, Ireland, Spain and Iceland, as well as in several other countries; the list also ignores the three previous waves of bubbles.

Sloshing money and asset bubbles

Just as water flows downhill, money flows across national borders toward higher anticipated rates of return. Both increases in stock prices in a country and increases in the value of its currency attract money from abroad. The surge in money flows to the United States following the Asian Financial Crisis illustrates how the change in the pattern of cross-border money flows is the link from one wave of an asset price bubbles to the next.

One factor that explains why there were four waves of bubbles in a thirty-year period is that there had been a surge in the pool of global money as a

result of the large payments imbalances that developed in the late 1960s and continued into the 1970s and the next several decades. Moreover the floating currency arrangement meant that asset prices in a country would increase in response to an increase in money inflows. In addition the efforts of banks headquartered in different countries to grow their loans in foreign markets contributed to the increase in the supplies of credit available to borrowers in various countries. Each wave of bubbles begins when the lenders become much more willing to extend credit to a group of borrowers, perhaps because their incomes or anticipated incomes increased, or because the regulatory environment had become less restrictive and the lenders were no longer prohibited from extending credit to these borrowers. Alternatively, the change in the regulations enabled borrowers to reduce their anticipated costs of credit by tapping into foreign markets.

The second feature of the period since the early 1970s is that the differences in interest rates and in anticipated rates of return on similar securities denominated in different currencies have become much larger once countries were no longer committed to parities for their currencies. The divergences in national inflation rates were larger – although the departure from the Bretton Woods system of adjustable parities occurred because of differences between the United States, and Germany and several other European countries on the maximum acceptable inflation rate. Cross-border money flows occur when the differences in the anticipated rates of return on similar securities available in different countries does not correspond with the anticipated rates of change in the currency values. In the long run, these cross-border flows reduce the differences in anticipated returns in various countries. However, because there may be an almost immediate feedback from the movement of money across a border to an increase in the economic performance in the destination country, the movement of money may initiate a bubble – and then the extension of credit can become self-fulfilling, at least in the short run, because higher rates of growth led to higher rates of return.

The onset of a bubble can be illustrated by the case of Iceland in 2002. Rates of return on Icelandic kronor securities were high and attractive to foreign investors at a time when interest rates were relatively low in most industrial countries. The increase in the flow of money to Iceland led to both the appreciation of the kronor and an increase in the prices of kronor securities, the feedback from the money flow to the value of the currency was almost immediate because the market for the kronor was thin. Hence the money flow that led to an appreciation of the currency and an increase in the price of Icelandic securities was small relative to the global pool of money and also small relative to the capital of the investors that acquired Icelandic securities.

The currencies of most of the countries that experienced an asset-price bubble appreciated in response to an increase in money inflows; the major exception was that the euro and the Swiss franc and other currencies appreciated

in 2002 and the next several years even as money flows to the United States increased. (The surge in stock prices in Tokyo in the second half of the 1980s attracted foreign money, but the Japanese purchases of foreign securities were larger than the money inflows and the yen appreciated as the Japanese trade surplus increased.) The increase in asset prices in each country as its currency appreciated was an integral part of the adjustment process to the money inflows to ensure that the change in the country's trade deficit would correspond with the increase in money inflows.

The increases in the external indebtedness of the countries involved in the first, third, and fourth waves were more rapid than the increases in their GDPs; similarly the rates of increase in money inflows were much higher than the interest rates on the indebtedness. Hence it was inevitable that these currencies would depreciate when the money inflows would slow. Similarly it was inevitable that the growth of credit for the purchase of real estate in Japan would slow, and that some of the recent buyers of real estate would encounter cash flow problems because their interest payments would be larger than their rental incomes. The increase in the domestic indebtedness of those who bought real estate in Japan in the 1980s was much more rapid than the increase in their incomes, and it was inevitable that the lenders would become more cautious in the extension of credit.

The increases in the supplies of credit often have been so rapid that the less creditworthy borrowers become attractive to the lenders. Often new or recent entrants to the credit market may cut prices to increase market share, and then the larger, more established lenders may respond by reducing price – that is, the interest rate markups over the costs of funds.

The first wave of credit bubbles involved sharp increases in the loans from the international banks to governments and government-owned firms in Mexico and other developing countries at a time when the world inflation rate was accelerating. The demand inflation led to rapid increases in commodity prices and in the anticipated growth rates in the primary producing countries. Moreover the pool of global savings surged as a result of large payments imbalances, the rapid growth in offshore dollar deposits, and the large payments surpluses of Saudi Arabia, Kuwait, and several other oil-exporting countries.

The link between the implosion of the first wave of credit bubbles and the bubble in Japan was the extensive intervention by the Bank of Japan to limit the appreciation of the yen in the second half of the 1980s. The Japanese trade surplus had surged in the 1970s and the first half of the 1980s. The appreciation of the yen in the second half of the 1980s would have led to a decline in profit rates and employment in the tradable goods sector; the Bank of Japan bought dollars to dampen the appreciation of the yen. Moreover the restrictions that had limited bank credit for real estate, which had been in place for

more than thirty years, were relaxed and the rapid increase in these loans led to a surge in real estate prices, a boom, and euphoria.

The appreciation of the yen that followed from the implosion of the bubble in Japan led to an increase in the anticipated returns on investments in Thailand and other emerging-market countries that would supplement and replace factories in Japan as sources of supply. The yen appreciated because the dramatic slowdown in the Japanese economy led to a decline in imports and a surge in exports as productive capacity was diverted from the domestic market to the foreign markets. Japanese firms began to invest extensively in China, Thailand, and their neighbors in response to the 'hollowing out' of the Japanese economy. About the same time, four regulatory changes contributed to the increase in the demand for securities available in these countries. The discovery of 'emerging market equities as a new asset class' followed from the remarkable increase in stock prices in Japan in the 1980s. Moreover the Brady bond initiative of 1988 funded the overhang of bank loans to the developing countries into long-term bonds, and these countries became 'bankable'. The privatization of government-owned firms, both in the industrial countries and in the emerging-market countries, led to a flow of money to Mexico, Brazil, Argentina, and other countries to purchase these firms and shares in these firms. One factor was Mexico-specific and involved the policy initiatives to stabilize the economy to prepare it for entry into the North American Free Trade Agreement; an extremely tight monetary policy adopted to reduce the monthly inflation rate led to high real interest rates on short-term peso securities that attracted lots of money from the Unites States.

Banks headquartered in Thailand, Malaysia, South Korea, and Indonesia increased the amounts borrowed in the offshore money market as their creditworthiness increased; the interest rates in the offshore market were lower than those in their domestic market, although the offshore borrowing meant that the banks were incurring a currency risk, either directly or at one remove if they on-lent to domestic borrowers in the foreign currency.

When the bubbles in Thailand and Malaysia popped, their currencies depreciated sharply in the second half of 1997; there was a surge in the money flows to the United States as they repaid loans. The Federal Reserve reduced interest rates three times within several months to dampen the flow of money from the Asian countries – which contributed to the acceleration of the bubble in US stocks.

The implosion of the bubble in US stocks led to a reduction in the flow of money to the United States and many foreign currencies – the euro, the Swiss franc, the Japanese yen – appreciated, which led to a reduction in the US trade deficit and contributed to an increase in the profit rate in US manufacturing. At about the same time, an autonomous event – an increase in the Chinese trade surplus – led to increases in the US trade deficit and the trade deficits of

other industrial countries. The money flows from China went to the countries with the most attractive security markets.

The pattern is that the implosion of a bubble leads to change in the cross-border money flows. The implosion of the credit bubble in Mexico and other developing countries in the early 1980s would have led to an appreciation of the currencies of the industrial countries; the desire of the Japanese to dampen the strengthening of the yen led to an increase in the money supply and the credit available for purchase of real estate. The increase in the flow of money from Japan in the early 1990s contributed to the increase in asset prices in various emerging market countries; similarly the flow of money from these countries contributed to the bubble in the United States.

Whereas the implosion of the bubble in Japan led to an appreciation of the yen and the increase in anticipated profit rates in manufacturing in Thailand and Malaysia, the implosion of these other three waves led to a depreciation of the currencies because money flows to these countries diminished.

One feature of the period is that there is a large – an increasingly large – volume of money that can move from one country to others – more precisely, from the dollar to some other currency – at relatively low cost. Another feature is a succession of shocks, including policy shocks like the shift from a pegged currency arrangement to a floating currency arrangement or the relaxation of regulations that limited bank loans for real estate in Japan or the relaxation on borrowing abroad by banks in Southeast Asia. The implosion of several of the bubbles induced a change in the pattern of cross-border money flows and contributed to the increase in asset prices in the countries that experienced an increase in these inflows.

10
Policy Responses: Benign Neglect, Exhortation, and Bank Holidays

If many financial crises have a stylized form, should there be a standard policy response? Assume plethora, speculation, panic; what then? Should the government authorities intervene to cope with a crisis and if so when? Should they seek to forestall increases in real estate prices and stock prices as the bubble expands so the subsequent crash will be less severe? Should they prick the bubble once it is evident that asset prices are so high that it is extremely unlikely that increases in rents and in corporate earnings could be sufficiently rapid and large to 'ratify' these lofty prices? When asset prices begin to fall, should the authorities adopt measures to dampen the decline and ameliorate the consequences?

Virtually every country has established a central bank to prevent or minimize shortages of liquidity, especially during a financial crisis. Many countries have a deposit insurance arrangement to reduce the likelihood of runs on their banks and to forestall what otherwise could be a self-fulfilling prophecy that a shortage of liquidity would trigger a solvency crisis. Even when there is no formal insurance for bank deposits, individuals in many countries believe that their governments will ensure that they will not incur losses if the banks should fail – in effect, the banks will be nationalized and the government will become their owners.

Chairman Greenspan of the Federal Reserve initially took the view that the Fed should focus monetary policy on the achievement of a stable price level – or at least a low inflation rate – and the macro-employment objectives. Then as the US financial system imploded in the last few months of 2008, he seemed more open to the idea that the Fed should have paid more attention to increases in real estate prices.

This chapter centers on the management of financial crises – it first deals with regulatory measures that might be adopted to reduce the susceptibility of financial arrangements to manias and then with the approach to panics. The next chapter focuses on the domestic lender of last resort and the following one

on the lender of last resort in an international context. This chapter considers the Austrian view that the best remedy for panic is to 'leave it alone' – to let it run its course, and to allow the economy to adjust to the decline in household wealth, the de-capitalization of the banks, and the slowdown in household and business spending that would follow from the declines in the prices of real estate, stocks, and commodities.

The moral hazard argument is the primary rationale for non-interference by the government; the view is that the more interventionist the authorities are in the current crisis, the more intense the next mania will be. The argument is that intervention skews the risk and reward trade-off in the minds of many investors – both the shareholders in banks and perhaps some of the creditors of the banks including those who own the bonds or other fixed price claims on the banks – by reducing both the likelihood and scope of future losses. A secondary rationale for this view is that if the adjustments to the imbalances created by the implosion of the bubble are delayed, the economic recovery will be slowed.

The moral hazard argument usually does not distinguish shareholders from bondholders. Shareholders in Lehman Brothers and Fannie Mae and Freddie Mac lost virtually all their money when these firms failed, as did the shareholders of the Icelandic banks. The shareholders of Northern Rock, the Royal Bank of Scotland, and Lloyds TSB lost most of their money – probably 90 to 95 percent. Similarly the shareholders of AIG and Bear Stearns lost more than 90 percent of their money. Moreover, a substantial part of the personal wealth of the employees of both Lehman and Bear was in the shares of these firms and in options to buy more shares, so some of these individuals lost 60 or 70 percent or more of their personal wealth.

At times the concern about moral hazard is expanded beyond the shareholders to include the creditors of the financial firms, including both the bondholders and those who have short-term claims on the banks; unlike the shareholders, these creditors have no significant upside gain if the banks' strategies prove immensely profitable. The returns of these creditors are skewed – if the banks fail, these creditors will incur substantial losses and if the banks are successful, the creditors do not share in these gains. Creditors may be lulled into complacency about potential losses because of the view that the banks will not be allowed to fail – but the failure of the banks is not unambiguous. If the shareholders lose 90 or 95 percent of their money before the government provides financial assistance, has the bank failed? Most of the shareholders are unlikely to agree to the statement that they have been bailed out. These bondholders may receive somewhat higher interest rates because the managers of the banks believe that they are 'too big to fail' and undertake somewhat riskier investments – and they will be able to pay higher interest rates – at least for a while. The 'too big to fail' doctrine may have the same impact as

deposit insurance – unless the authorities reduce the value of the claims of the creditors below their face value if the banks are closed and re-organized with government assistance. Thereafter the owners of the deposits that are much larger than the amounts covered by deposit insurance may be ready to withdraw their money whenever there is modest skepticism about the solvency of the bank.

The choice for each country is whether to design a system that would allow large institutions to fail, with losses to creditors, or instead to provide some form of effective insurance for all of those with claims on the banks to reduce the likelihood of runs.

Financial reform: bank regulation and supervision

Can financial crises be forestalled by strict regulation and supervision? Some observers advocate this approach. Others recommend deregulation. Most of the rules for sound banking are already incorporated in the regulations or are implicit in banking tradition. Many of the rules are ignored by both the banks and the regulators. Banks are supposed to 'mark to market', that is, to value their loans and investments each day (or each week or month) at the prices that would be realized if they were sold in the market rather than at their historic costs. Reserves should be established against 'problem loans' and write-offs against 'doubtful' ones. As a bank's loan losses increase and its capital declines, the bank might be required to raise more capital or be closed under the traditional rules. As an illustration of the unusual character of banks following these rules, the press was full of the news in the spring of 1987 when Citicorp wrote down the value of its Third World loans and the FSLIC allowed 500 insolvent banks to remain open in the hope that they would become sufficiently profitable to rebuild their capital. As part of the regulation process, the Federal Reserve began to collaborate with other central banks in the Group of Ten countries to strengthen bank structures worldwide by applying the same risk-based capital requirements to banks headquartered in different countries.

Emphasis on capital requirements as a percentage of assets or bank liabilities led some banks to develop 'off-balance sheet' operations, which generate fees and commissions and interest income, but the assets and liabilities are contingent and are shown only as a footnote on the balance sheet. The banks established special investment vehicles (SIVs) that circumvent formal capital requirements; hence they increased their effective leverage. The demand for the IOUs of the SIVs depended on the reputational capital of the banks. These off-balance sheet transactions include interest rate and currency swaps, futures contracts, options, underwriting risks, 'repos' (sales of securities with a guarantee to repurchase them at a later date), and note-issuing facilities. Each of

these positions can be valued as an option and included among assets or liabilities when calculating the appropriate required capital.[1]

A strong case can be made for stricter regulations and supervision of banks to forestall lending in euphoric periods that may end in financial crisis. Responsibility for bank examination in the United States is divided among the Comptroller of the Currency, the Federal Reserve Banks, and the state banking commissions. In one view there is competition not in deregulation but in re-regulation.[2] 'Divided responsibility,' said a famous German banker-politician, 'is no responsibility.'[3] The astute examination personnel needed when asset prices tumble are unwilling to submit to the boredom of long periods of calm between crises. The mismanagement of banks is hard to detect before a crisis. In booms, entropy in regulation and supervision permits the build-up of danger spots that only burst after asset prices decline. The question then is whether to liquidate, stall, guarantee, bailout, takeover, or rely on other means of last-resort lending.

The almost universal response to failure or near failure of banks is that there is need for more regulation or for more effective regulation. The problems of Bear Stearns, Lehman Brothers, and other US investment banks have been attributed to the decision of the US Securities and Exchange Commission to relax the restraints on the leverage of the investment banks. At times the collapse of some of the banks is attributed to the repeal of the Glass-Steagall Act – which conveniently neglects the failures of hundreds of firms in the 1980s and the failures of banks in Britain, Iceland, and Ireland.

Regulation imposes costs. As a result the history of regulation is that new types of institutions are developed that are outside the scope of regulations – and yet they are otherwise similar to regulated firms. Money-market funds were developed as a way to pay interest on demand deposits. The offshore deposit market developed to avoid the costs that domestic banks incur in the form of reserve requirement and deposit insurance premiums; the offshore branches of US banks could pay higher interest rates than their domestic branches.

Manias are macro phenomena and result from the excessively rapid growth of credit. Regulation is a micro phenomenon. Regulation is not likely to affect whether a mania develops, although it will impact some institutions and it might impact the pace of the development of the mania.

Benign neglect

Many economists believe that the panic will work its own cure, and that 'the fire can be left to burn itself out'.[4] 'Cool if not very imaginative heads in the Bank [of England] parlour thought it in the nature of panics to exhaust themselves.'[5] Lord Overstone maintained that support of the financial system in crisis is not really necessary because the resources of the system are so great

that even in times of the utmost stringency those that offer a sufficiently high rate of interest could borrow a large amount of money.[6] In 1847 an increase in the discount rate to 10 and 12 percent in London stopped the flow of gold to the United States; a small sloop was sent to overtake a ship that had already sailed for New York and got it to turn around and unload £100,000 in gold.[7] Testifying before the 1865 French *Enquête* (inquiry) into monetary circulation, Baron James de Rothschild stated that increases in interest rates could be relied upon to reduce speculation in commodities and securities. He added: 'If speculators could find unlimited credit, one can't tell what crises would ensue.'[8]

The moral hazard dilemma is that policy measures undertaken to provide stability to the system during this crisis may encourage speculation during the next expansion by those who seek exceptionally high returns and who have become somewhat convinced that there is a strong likelihood that the government will adopt measures to prevent the economy from imploding – and so their losses on the downside will be limited. A 'free lunch' for the speculators today means that they are likely to be less prudent in the future. Hence the next several financial crises could be more severe. The moral hazard problem is a strong argument for non-intervention as a financial crisis develops, to reduce the likelihood and severity of future crises. Will the policymakers be able to devise approaches that penalize individual speculators while minimizing the adverse impacts of their imprudent behavior on the rest of the country? Even then the cost-benefit question is whether the benefit to the economy from not allowing the panic to run its course is worthwhile in terms of the undeserved reward to the speculators.

The losses incurred by shareholders of Bear Stearns, Fannie Mae and Freddie Mac, AIG, Citicorp, and Bank of America were immense. Moreover the leadership and directors of these institutions were replaced. The firms have continued in business, but for all practical purposes they failed. Individuals who invested on the assumption that these institutions were 'too big to fail' may have incurred large losses.

The view that a panic should be allowed to pursue its course has two elements. One element takes pleasure in the troubles – or enjoys *schadenfreude* – that the investors or speculators encounter as retribution for their excesses; this puritanical view welcomes hellfire as the just deserts for the excessively greedy. The other sees panic as a thunderstorm 'in a mephitic and unhealthy tropical atmosphere' that clears the air. 'It purified the commercial and financial elements, and tended to restore vitality and health, alike conducive to regular trade, sound progress and permanent prosperity.'[9] One powerful statement of this position was made by Herbert Hoover as he characterized – without approval – the view of Andrew Mellon:

> The 'leave-it-alone liquidationists' headed by Secretary of the Treasury Mellon felt that government must keep its hands off and let the slump

liquidate itself. Mr. Mellon had only one formula: 'Liquidate labor, liquidate stocks, liquidate the farmers, liquidate real estate.' He insisted that when the people get an inflationary brainstorm, the only way to get it out of their blood is to let it collapse. He held that even panic was not altogether a bad thing. He said: 'It will purge the rottenness out of the system. High costs of living and high living will come down. People will work harder, live a moral life. Values will be adjusted, and enterprising people will pickup the wrecks from less competent people.'[10]

The neo-Austrian economic historian Murray Rothbard added: 'While phrased somewhat luridly, this was the sound and proper course for the administration to follow.'[11] The conservative historian Paul Johnson commented: 'It was the only sensible advice Hoover received during his presidency.'[12]

The opposing view conceded that while it is desirable to purge the system of bubbles and manic investments there is the risk that if the government does not intervene to staunch the crisis than a deflationary panic would spread and wipe out sound investments by the non-speculators who would not be able to obtain the credit they need.

One feature of many liquidity crises is that interest rates seem extremely high for private borrowers, especially because they are expressed as annual percentage rates they are really premiums for liquidity for one, two, or at most a few days. The more fundamental question is whether money is available at these high interest rates or whether the credit market is cleared by non-price rationing. The evidence from many crises is that it is difficult and sometimes impossible to borrow in a panic and that the quoted interest rates are irrelevant because money is not available.

- After first Arend Joseph and then DeNeufville failed in 1763, and panic broke out on 22 July, a succinct, not very informative or convincing report commented: 'Panic even on securities and on goods, no money was to be had.'[13]
- The statement in 1825 was: 'A panic seized upon the public, such as had never been witnessed before: everybody begging for money – money – but money was hardly on any condition to be had. "It was not the character of the security," observes the *Times*, "that was considered, but the impossibility of producing money at all".'[14]
- The interrogation of Thomas Tooke before the Select Committee on the Commercial Crisis in 1847: Question 5421: 'For several days, if not some weeks, the Bank of England was the only establishment that was discounting?' Answer: 'Yes.' Question 5472: 'The Governor of the Bank of England said he could not sell £1 million of stock [English government bonds] in the week after October 14, if there had been no letter. Do you think

possible?' Answer: 'No, perfectly impossible, taking the word impossible to signify with the exception of such a reduction in price as could not be contemplated.'[15]

- The evidence of Mr Glyn in the same inquiry: Question: 'Are you aware that it was the opinion of the Bank broker that a very large sum might have been sold without materially affecting the prices of Consols [English perpetual bonds]?' Answer: 'I was not aware that the Bank broker had stated that. I should say, from what I saw at the time, that a sale of a million or two million, which were the figures talked of, would have been almost impossible without knocking down the funds to such a price as would have created a further panic.'

- Mr Browne, MP, did not think such sales could have been effected, unless at a great sacrifice, adding that 'if the panic had been equal to what we might suppose it might have been, under such circumstances, I doubt whether they could have been sold at all.'[16]

- Then in 1857 'At one stage during the crisis it was impossible to negotiate paper at all, the charge under the most favorable circumstances being 12 and 15 percent.'[17]

- A letter from Liverpool: 'Bills of exchange of the first Quality in themselves, and to which this and other Banks were willing to add their Endorsement, were absolutely inconvertible into Cash, and it is my Belief that many Houses, who were not merely solvent but able to pay 40s and 60s to the Pound, must have stopped had not the Government letter been issued.'[18]

- 'Commercial confidence in Hamburg is entirely at an end. Bills of only three or four of the first houses are negotiable at the highest rate of interest...A government bond advance of 15 million marks banco payable by banks failed to help. The panic was so great that government bonds could not be discounted, and on no security whatsoever would capitalists part with their money...When it was known on December 12 that assistance would help all, the panic ceased. Government bonds which had not been discountable at 15 percent on the first of the month were readily taken at 2 and 3 percent.'[19]

- Edward Clark in New York wrote to Jay Cooke in Philadelphia before Cooke left Clark's firm to start his own firm: 'Money is *not tight – it is not to be had at all*. There is no money, no confidence & value to anything. A week more of such times and the Bks [sic] will fail.'[20]

- Then in 1866 'The Bank court raised the discount rate to 9 percent and intimated that loans on Government securities were available at 10 percent. Before that announcement it was impossible to sell either Consols or Exchequer bills. Jobbers in other securities refused to deal.'[21]

- During the 1873 crisis in New York the 'National Trust Company of New York had eight hundred thousand dollars worth of government

securities in its vaults, but not a dollar could be borrowed on them; and it suspended.'[22]

- And finally, in 1883, 'The growing demand for money finally led to a money famine. Time loans were unobtainable, call loans were 72 percent in June, 72 percent on July 28th, 51 percent on August 4. First-class commercial paper was quoted 8 to 12 percent nominal, with a very small amount of money available.'[23]

The evidence is not unambiguous, since the sale of government bonds is somewhat qualified by such remarks as 'with the exception of such a reduction in price [i.e., increase in the rate of interest] as could not be contemplated'. Moreover, there is occasional information on the other side of the argument, especially in the United States under the national banking system, in which a lender of last resort was unavailable. In 1884:

> To add further to the discomfiture of dealers, money became exceedingly stringent, and at one time commanded as much as 4 percent for 24 hours' use. This caused a further sacrifice of stocks since few could afford to pay the high rate asked. The exorbitant charge was, of course, the direct result of the distrust prevailing, since there was no actual scarcity ... It was to ... the desire to realize and obtain cash that the large decline on Thursday and Friday of nearly 7 percent on United States Government bonds is to be attributed. There was no loss in confidence in these, nor was there in good railroad bonds and stocks.
>
> One result of the phenomenal and temporary rise in rates for money was to bring a vast amount of foreign capital to the market. Some of it was sent here to buy stocks at their depressed prices, and more to loan on stocks or on any other good securities at the high rates of interest. The effect of this was to completely turn the foreign exchanges which had been running so heavily against the US.[24]

This statement in its turn is not unambiguous, since the panic started before an acute liquidity shortage had developed.

The International Monetary Fund acted as a lender of last resort in the Asian crisis of 1997 – although long after the Thai baht, the Indonesian rupiah, and other currencies had depreciated sharply – and insisted that the government in each of the Asian countries balance its budget and that the central bank in these countries increase interest rates. A number of economists objected to these measures because they would lead to an increase in unemployment, especially among the poor, while the financial problems had been set in motion by comfortable officials and well-to-do bankers.

Some support for orthodoxy, however, came from the Japanese experience in the 1990s when the combination of an expansive monetary policy and the depreciation of the yen produced a 'liquidity trap'. Both Japanese interest rates and bank loans decreased after the declines in stock prices and real estate prices, and the inference was that there was a 'credit crunch'. Banks were reluctant to lend because their loan losses had eroded their capital and firms were reluctant to borrow because of the sluggish growth in the demand for their products.

The decline in short-term interest rates in Tokyo to 1 percent and below led to a surge in the 'carry-trade'; US hedge funds borrowed yen, which they sold to buy US dollars which were used to buy US dollar securities that yielded 3 or 4 percent. The 'carry-trade' transactions led to an increase in the flow of funds from Tokyo to New York and to a decline in the value for the yen that in turn led to an increase in the Japanese trade surplus and increases in Japanese output and employment. The increase in the Japanese trade surplus was a useful response to the liquidity trap and an effective supplement to expansive Japanese fiscal and monetary policies. The carry-trade transactions would remain profitable for the US hedge funds as long as any appreciation of the yen was smaller than the excess of the interest rates on US dollar securities over the interest rates they were paying on their yen loans.

Moral suasion and other exhortatory devices

The dominant argument against the *a priori* view that panics can be cured by being left alone is that they almost never *are* left alone. The authorities feel compelled to intervene to forestall a further decline in asset prices. In panic after panic, crash after crash, crisis after crisis, the authorities or some 'responsible citizens' try to halt the panic by one measure or another. The authorities may be unduly alarmed and the position might correct itself without serious harm. The authorities may be stupid and unable to learn. (The Chicago School assumes that the market participants are always more intelligent than the authorities, in large part because the authorities are motivated by short-term political objectives.) The uneven distribution of intelligence cannot be tested against crisis management because the authorities and leading figures in the marketplace both exert themselves to halt the spread of falling prices, bankruptcy, and bank failures. If there is a learning process at work – and the assumption of rationality requires one – the lesson has been that a lender of last resort is more desirable and less costly than relying exclusively on the competitive forces of the market.

One insight from the historical record is that there are many examples when the authorities initially were resolved not to intervene but eventually did so reluctantly. Lord Liverpool threatened to resign as Chancellor of the Exchequer

in December 1825 if an issue of Exchequer bills was provided to rescue the market after he had warned against excessive speculation six months earlier.[25] William Lidderdale, Governor of the Bank of England at the time of the Baring crisis, refused categorically to accept a 'letter of indemnity' to permit the Bank to exceed its lending limit.[26] On both occasions face was saved by finding some other approach to avert the panic. The initial strong stand not to intervene was reversed on many other occasions as the panic escalated. These included the intervention of Frederick II in the Berlin crisis of 1763,[27] the Bank of England's refusal to discount for the 'W banks',[28] and the US Treasury's decision in 1869.[29]

The view of the US authorities in September 2008 was that Lehman Brothers should be allowed to fail, even though the government had been centrally involved in JPMorganChase's purchase of Bear Stearns and the conservatorship of Fannie Mae and Freddie Mac. One explanation is that the then Secretary of the Treasury did not want to become known as 'Mr. Bailout'. A day after Lehman closed, the 'policy' was reversed in response to the massive run on AIG. In effect a new policy of 100 percent deposit insurance was implicitly provided to all large financial firms.

Stallings

In a run, each depositor rushes to get his or her money from the bank before the doors are closed because its money holdings have been exhausted. Banks are often reluctant to pay the depositors that want their money because the banks' money holdings are much smaller than their short-term deposit liabilities. During the Great Depression, banks took their time to pay off depositors, hoping, like Micawber, for something to turn up. The technique goes back to the eighteenth century.

Macleod's *Theory and Practice of Banking* describes how the Bank of England defended itself in September 1720 against a run brought on by its reversal of a promise to absorb the bonds of the South Sea Company at £400. The Bank organized its friends in the front of the line and paid them off slowly in sixpence coins. These friends brought the coins back to the Bank through another door. The money was deposited, again slowly counted, and then again paid out. The run was staved off until the feast of Michaelmas (29 September). When the holiday was over, so was the run, and the Bank remained open.[30]

A second story, which may well have the same origin and is likely to be more accurate, is that the Sword Blade Bank, a supporter of the South Sea Company, resisted attempts to redeem its paper with silver coins. When the run started on 19 September, the bank brought up wagonloads of silver that it paid out 'slowly in small change'. One depositor is reported to have received £8000 in shillings and sixpences before the bank closed its doors on Saturday 24

September.[31] The circumstances suggest one story; the dates, two. Since the Sword Blade Bank and the Bank of England were mortal enemies, it is unlikely they cooperated.

The lessons of 1720 were not lost on the Bank of England a quarter century later. The Young Pretender (Charles Edward, grandson of James II) landed in Scotland in July 1745, unfurled his banner in September, invaded England in November, arrived in Carlisle on 15 November, and reached Derby on 4 December. Panic broke out on Black Friday, 5 December 1745. British Consols fell to 45, the lowest price on record, and a run began on the Bank of England. The Bank resisted, partly by paying off its notes in sixpence coins. The time gained was used to induce London merchants to proclaim their loyalty and readiness to accept Bank of England notes. The second half of the prescription, collecting pledges of faith in notes, was used again in similar circumstances when the French landed at Fishguard in 1797. On that occasion, 1140 signatures of merchants and investors in government stock were collected in a single day.[32] The time gained in 1745 by both the slow payout and the petition of support enabled the government to organize the army that defeated the Young Pretender at Culloden in April 1746.

Complete shutdowns and bank holidays

One way to stop a panic is to close the market so trading stops. Trading on the New York Stock Exchange was halted in 1873, and in London and many other centers at the outbreak of war in 1914. In both cases the motivation was to stop a run by providing the market participants with more time to think through whether it was necessary or desirable to sell at depressed prices.

However, shutdowns may drive the trading underground and intensify the panic. Moreover, short-run and long-run goals are in conflict. Closing the stock market during one panic may exacerbate the next, as investors sell their stocks or withdraw their money from the call money market because they are fearful that trading will be halted. The New York Stock Exchange was closed in a panic in September 1873, but a financial editor suggested that fear that trading on the exchange might be halted in October 1929 hastened the withdrawal of call money by out-of-town banks and other market participants.[33] The closing of stock exchanges in Pittsburgh and New Orleans for two months in 1873 had fewer serious consequences, since they traded only the securities of the local firms that had brought on their own difficulties.[34]

The New York Stock Exchange and the other US stock exchanges were closed for a week after the bombing of the World Trade Center towers on 11 September 2001 because the communications and the technical support systems were inoperative. Many of the same securities could have been traded on

the regional stock exchanges in the United States although these exchanges would have been overwhelmed.

The declaration of a legal holiday by the government is another technique for closing the market, which was used during the panic of 1907 in Oklahoma, Nevada, Washington, Oregon, and California.[35] The device was the forerunner of the bank holidays that started at the local level in the fall of 1932 and were generalized throughout the country on 3 March 1933, the day that Franklin D. Roosevelt was inaugurated as president. (A bank holiday closes only the banks, while a legal holiday shuts down all businesses.)

Another device is to suspend the publication of bank statements, as in 1873, in the hope that 'what you don't know won't hurt you'. The technique was designed to hide the large losses of a few banks since the fear was that accurate news would further reduce depositor confidence.[36]

Some commodity and financial markets set daily limits on the maximum change in prices; when the limit is reached, trading is suspended for the rest of the day. Specialists in individual stocks have often taken a 'time out' whenever the imbalance between the buy orders and the sell orders has been exceptionally large. This 'circuit breaker' was recommended for US stock markets after the meltdown of Black Monday, 19 October 1987. The proposal for the New York Stock Exchange was to postpone trading for a stated interval – such as twenty minutes – in those stocks whose prices increased above or declined below the limit.

Time has been gained by moratoriums on payment of all debts or on particular types of obligations, such as bills of exchange that have less than two weeks to run. The most ubiquitous measure of this sort is that the bank examiners ignore the bad loans that banks own as long as they can – an implicit moratorium on marking the loans to their market value. Regulatory forbearance was used in the US savings and loan debacle in the 1980s. The International Monetary Fund and the World Bank continued to allow the indebtedness of many African countries to increase by the amount of the interest that was due; if these institutions had declared the loans in default, they would have had to recognize the losses on the loans. The banks that were the lenders to the Real Estate Investment Trusts (REITs), landlords of mortgaged shopping centers and the owners of mothballed Boeing 747s allowed the interest on their bank loans to compound because they wanted to delay the recognition of the loan losses to a more propitious moment when their capital would be larger. But the lenders need forbearance from the bank examiners.

Official moratoriums may be less effective than informal ones, however. A moratorium on the settlement of differences in payments due on the 1873 Vienna Stock Exchange lasted a week, from the stock market collapse to 15 May. A guarantee fund of 20 million gulden was put together by the Austrian

National Bank and the solid commercial banks; these imitations of earlier measures were of little assistance.[37] Another moratorium was noted in Paris after the July Monarchy when the municipal council decreed that all bills payable in Paris between 25 July and 15 August should be extended by ten days. This moratorium sterilized the commercial paper in banking portfolios and did nothing to discourage a run by holders of notes.[38]

Clearing-house certificates

The major device used in the United States to cope with bank runs prior to the establishment of the Federal Reserve System was the clearing-house certificate, which is a near-money substitute that was the liability of a group of large local banks. A bank subject to a run could pay the departing depositors with these certificates rather than with coin. The New York clearing-house was established in 1853 and the one in Philadelphia in 1858 after the panic of 1857. During the panic of 1857, New York banks failed to cooperate to halt the run. The Mercantile Agency of New York took the position that if four or five of the strongest banks had come to the assistance of the Ohio Life and Trust Company, so it could meet its obligations, the business and credit of the country would have been preserved.[39] By 1873 the New York banks were ready to accept payment on cleared checks in clearing-house certificates. The advantage of these certificates was that they reduced the incentive for any bank to bid deposits away from its competitors. Sprague insisted that this system had to be accompanied by an agreement to pool bank reserves; otherwise, a bank that was not subject to a net drain might be forced to suspend payments after it paid cash to its own depositors if it had not received cash in settlements from other banks.[40] In 1873 reserves were pooled.

One serious drawback of clearing-house certificates was that they were acceptable only in the local area – New York, Philadelphia, Baltimore. Thus these certificates helped maintain domestic payments such as payrolls and retail sales within a city but they dampened the effective flow of payments between cities. In the 1907 panic, 60 of the 160 clearing-houses in the United States used these certificates to facilitate local payments. Nevertheless Sprague claimed that the dislocations of the domestic exchanges were no less complete and disturbing than on previous occasions. The prices of New York funds in Boston, Philadelphia, Chicago, St Louis, Cincinnati, Kansas City, and New Orleans between 26 October and 15 December 1907 varied from a discount of 1.25 percent in Chicago on 2 November to a 7 percent premium in St Louis on 26 November, an increase from 1.5 percent the previous week.[41] In December 1907 Jacob H. Schiff wrote: 'The one lesson we should learn from recent experience is that the issuing of clearinghouse certificates in the different bank centers has also worked considerable harm.

It has broken down domestic exchange and paralyzed to a large extent the business of the country.'[42]

Other devices of the same general character were clearing-house checks and certified checks that were both close substitutes for money and increased the means of payment in circulation.

Non-bank groups can also organize to mitigate a panic. Consider, for example, the stock market consortium. On 24 October 1907, a bankers' pool, headed by J.P. Morgan, loaned $25 million at 10 percent in call money in an attempt to stem the collapse of the stock market.[43] Twenty-two years to the day later, on Black Thursday in 1929, Richard Whitney went from post to post on the floor of the New York Stock Exchange and placed bids to buy stocks on behalf of a syndicate headed by J.P. Morgan and Company – once again.[44]

Bank cooperation

Banks have also collaborated through rescue committees (as in Vienna in May 1873 and earlier), loan funds, funds for guarantees of liabilities, arranged mergers of weak banks and firms, and other devices whereby the strong banks support the weak and failing banks.[45] Three examples include the role of the Paris banks in the 1828 crisis in Alsace, various devices employed by Hamburg in meeting the difficulties of 1857, and the Baring Brothers loan guarantee of 1890.

The Alsatian crisis of 1828

Three firms in textiles failed in December 1827 at Mulhouse. The Paris banks then refused to accept any Alsatian paper, and the Bank of France set a limit of 6 million francs on the amount it would support, a figure that was 'scarcely the fortune of two Alsatian houses'. The Bank of France then decided against accepting any paper with Mulhouse or Basel endorsements and that decision precipitated a panic. On 19 January two more Mulhouse merchants failed. On 22 January in Paris there were rumors of the failure of two Schlumberger firms. The Paris banks sent Jacques Laffitte as an emissary to Mulhouse; he arrived on 26 January and offered to lend 1 million francs on the consignment of merchandise. Before he came, however, two textile merchants, Nicholas Koechlin and Jean Dollfuss, had left Mulhouse for Paris. To raise cash, these merchants had been selling inventories on the market at discounts of 30 to 40 percent from the traditional market prices for these goods. Nine houses failed from 26 January to 15 February. Lévy-Leboyer wrote that it could have been worse. At the last minute a syndicate of twenty-six Paris banks, presided over by J.-C. Davillier, extended a credit of 5 million francs to Koechlin and Dollfuss, who returned to Alsace on 3 February and distributed 1 million francs to those of

their colleagues who offered guarantees and kept 4 million for themselves. These measures restored confidence.[46] Those who qualified for neither the Koechlin-Dollfuss fund nor the Basel money failed.[47]

The Hamburg crisis of 1857

The background of the crisis of 1857 in Hamburg was that trade had expanded, particularly because the Crimean War had led to an expansion of credit. Hamburg was the 'all-English' city of Germany, but had close relations with the United States in sugar, tobacco, coffee, and cotton, and with Scandinavia. When the deflationary tidal wave swept across the Atlantic, Hamburg was inundated. The panic touched off by the failure of Ohio Life on 24 August arrived in Hamburg three months later (following price declines of 30 percent) with the suspension of Winterhoff and Piper, a firm that was engaged in the American trade.[48] Daily dispatches from the British consulate in Hamburg by date tell the story:

- **21 November:** Some of the leading merchant houses and two banks plan for relief.
- **23 November:** Two major houses engaged in the London trade fail, and the Discount Guarantee Association grows more cautious in endorsing Hamburg bills.[49] On one authority the Discount Guarantee Association is exhausted in three days.[50]
- **24 November:** A Discount Guarantee Association (Garantie-Diskontverein) is formed, initially with a capital of 10 million marks banco, later raised to 13 million (about £1 million), of which the sum of 1 million marks is to be paid in immediately.
- **28 November:** The chamber of commerce and leading merchants induce the Senate to call parliament (Burgerschaft) to arrange to issue government bonds in order to lend 50 to 66 2/3rds percent of the value of hypothecated goods, bonds, and shares to merchants in distress.
- **1 December:** With the suspension of Ullberg and Cremer, ten to twelve houses in the Sweden trade have gone down. The Discount Guarantee Association will not issue any more guarantees. Business is at a standstill.
- **2 December:** A suggestion is made to change the laws of bankruptcy to enable creditors to share in attachment of goods.
- **7 December:** A proposal is made to establish a state bank for discounting good bills to the amount of 30 million marks banco (about £2.4 million). Parliament rejects this, wanting instead to issue 30 million marks banco of paper currency as legal tender. The senate rejects this, insisting on clinging to the silver standard.

In the end, a compromise was reached for a State Loan Institute fund of 15 million marks that included 5 million marks banco of Hamburg government bonds and 10 million in silver to be borrowed abroad.[51] The story of the silver train (*Silberzug*) is recounted in Chapter 12 as an example of an international lender of last resort.

One observer totaled the sums available for rescue operations to 35 million marks banco, which included 15 million in the Discount Guarantee Association, 15 million in the State Loan Institute, and 5 million from the chamber of commerce. He compared this amount with 100 million marks banco of protested bills and noted that if merchants speculated with capital equal only to one-sixth the value of their goods, a 17 percent decline in prices of these goods would be sufficient to wipe out their capital position. To the suggestion that the senate was 300 years behind the times, he reported with approval the senate's answer: The merchants have been 300 years ahead of the times in issuing debt. State help in these cases, he insisted, merely means assistance for speculation and perpetuation of higher prices at the cost to the consumer.[52]

Guarantees of liabilities: the Baring crisis

The most famous guarantee of liabilities was that worked out by William Lidderdale when he was Governor of the Bank of England during the Baring crisis of 1890. Similar guarantees had occurred earlier in Britain. In December 1836 the private bank Esdailes, Grenfell, Thomas and Company, which served as London agents for seventy-two country banks, was in financial difficulties. The view was that this firm could not be allowed to fail because of its relationships with the country banks; moreover, its paper included all the best names in the City. The assets of the bank far exceeded its liabilities, and the London bankers offered guarantees. The Bank of England led the list with £150,000. Esdailes survived, but only for two years.[53]

The guarantee was worked out as an alternative to a letter of indemnity that permitted suspension of the Bank Act of 1844. The letter was offered by the Chancellor of the Exchequer, Lord Goschen, to Lidderdale, who refused on the ground that 'reliance on such letters was the cause of a great deal of bad banking in England'.

If Lidderdale refused to quiet the market by the usual means that had been employed in 1847, 1857, and 1866, he was not one to let the market take its medicine. In August 1890 he warned Baring Brothers that the firm would have to moderate its acceptances for its Argentine agent, S.B. Hales. Baring Brothers revealed its acute distress to Lidderdale on Saturday 8 November. Fearful of a panic when Barings' condition was made public, the Bank of England met with the Exchequer on Monday 10 November, turned down the

letter of indemnity, prepared for the problem by seeking assistance from foreign countries (a subject for Chapter 12), and formed a committee headed by Lord Rothschild to address the question of the large overhang of Argentinean securities.

As the days passed, the rumors circulated and Barings' bills were increasingly discounted at the Bank of England. By Wednesday Lidderdale had learned that although Baring was solvent it would still need £8 to £9 million. On Friday John Daniell, the leading man at Mullens and Co., used by the Bank of England in open-market operations, came to Lidderdale, crying 'Can't you do something, or say something, to relieve people's minds: they have made up their minds that something awful is up, and they are talking of the *very highest* names – the very highest!'[54]

On 14 November Lidderdale met with two cabinet ministers who represented the Exchequer, Lords Smith and Salisbury. They agreed that the government would increase its balance at the Bank immediately and that the government would share with the Bank in any losses suffered on Baring Brothers' paper discounted by the Bank between 2pm Friday and 2pm Saturday. On the basis of this agreement, Lidderdale met with eleven private banks to get them to contribute to a fund that would guarantee Barings' liabilities, and he got the State Bank of Russia to agree not to withdraw its £2.4 million deposit at Barings. The private banks as a group contributed £3,250,000, but this included £1 million from the Bank of England as well as £500,000 from each of three lenders, Glyn, Mills & Co., Currie and Co., and Rothschilds. Lidderdale then used these commitments to obtain the agreement of the five London joint-stock banks to join the guarantee fund for another £3,250,000. On the basis of these assurances, *The Times* of 15 November announced that Baring Brothers would fail but that there would be no loss. The work of the guarantee fund continued on Saturday because the directors of the joint-stock banks had to meet to approve their subscriptions, which was done by 11am. Then other banks and financial institutions raised the fund from £7.5 million in the morning to £10 million by 4pm; the guarantee fund eventually reached £17 million and was taken as a measure of the strength of the London financial system. Martin's Bank was in distress over its loans to Barings and to Murriettas, another bank involved in Argentina. Martin's Bank joined the guarantee fund for £100,000 on 18 November (Tuesday), too late to afford much help to Barings, but early enough to demonstrate to the world its own strength.[55] In summarizing the episode, Powell stated: 'The Bank is not a single combatant who must fight or retire, but the leader of the most colossal agglomeration of financial power which the world has so far witnessed.'[56]

On 25 November a new firm, Baring Brothers and Co., Ltd, was formed as a joint-stock company with a capital of £1 million. The form of the guarantee may be of interest.

Guarantee Fund
Bank of England, November, 1890

In consideration of advances which the Bank of England have agreed to make to Messrs. Baring Brothers and Co., to enable them to discharge at maturity their liabilities existing on the night of the 15th of November, 1890, or arising out of business initiated on or prior to the 15th of November, 1890.

We, the undersigned, hereby agree, each individual, firm, or company, for himself or themselves alone, and to the amount only set opposite to his or their names respectively, to make good to the Bank of England any loss which may appear whenever the Bank of England shall determine that the final liquidation of the liabilities of Messrs. Baring Brothers and Co. has been completed so far in the opinion of the Governors as practicable.

All the guarantors shall contribute rateably, and no one individual, firm, or company, shall be called on for his or their contribution without the like call being made on the others.

The maximum period over which the liquidation may extend is three years, commencing the 15th of November, 1890.[57]

The rescue of Long-Term Capital Management in September 1998 provides a contrast to the rescue of Barings. William McDonough, the president of the Federal Reserve Bank of New York, induced fourteen large banks and investment banks, including Merrill Lynch, Morgan Stanley Dean Witter, J.P. Morgan, Chase Manhattan Bank, and the Union Bank of Switzerland to provide $3.6 billion of capital to prevent the collapse of LTCM; in exchange they acquired 90 percent of LTCM's equity.[58] These firms were large creditors of LTCM so the 'bailout' involved a change in the legal nature of their claim on LTCM. The Federal Reserve was concerned that if LTCM failed the markets would be paralyzed by the need to unwind its massive position in futures and options contracts and other types of derivatives.

Deposit insurance

Since 1934, federal deposit insurance in the United States has prevented bank runs by providing an *ex ante* guarantee of deposits. Initially the deposits were guaranteed up to $10,000 but gradually the ceiling on deposits was raised in

stages and eventually reached $100,000 – and then $250,000 during the 2008 crisis. When large banks got into trouble, the FDIC deliberately removed all limits on the amounts of deposits covered by the guarantee to halt imminent runs and in practice it established that banks with significant deposits over $100,000 were 'too big to fail' (although the shareholders of these banks would probably lose all of their money). Similarly the owners of the subordinated debt of these banks might lose much or all of their money. The deposits of the foreign branches of these banks had implicitly become insured even though the banks had not paid deposit insurance premiums against their offshore deposits. Although deposit insurance was designed to prevent bank runs by taking care of the 'little man', this ceiling had in practice been raised to the sky.[59]

When the limit on deposit insurance was $100,000, 'deposit brokers' arranged for the placement of larger amounts of money in guaranteed deposits. John Doe could have an insured deposit of $100,000, his wife Jane could also have an insured deposit of $100,000, and together John and Jane could have a third insured deposit of the same amount. And the Does could follow the same strategy with the bank across the street.

The effect of this innovation was that it provided a guarantee to wealthy individuals and hence circumvented one purpose of the ceiling. Moreover it encouraged banks to make riskier loans since they were confident that they were protected against runs by the owners of the larger deposits – if these riskier loans proved profitable, the owners of the banks would benefit and if the loans went into default, the owners would not have to worry about bank runs (although the market value of their own shares might decline and even become worthless).

The implosion of the bubble in Japan in the 1990s caused the value of the loans of the banks headquartered in Tokyo and Osaka and various regional centers to decline sharply below the value of their liabilities. Nevertheless, there were no bank runs; the depositors were confident that they would be made whole by the government if any of the banks were closed.

Deposit insurance has limited both bank runs and contagion in the runs from one troubled bank to other banks in a neighborhood. What accounts for the reluctance to provide insurance at an earlier date?

In the long tradition of the United States, free banking, even wildcat banking, was the rule. Anyone could start a bank, and many did. Risks were large, banker turnover rapid. A guarantee of bankers' deposits would have constituted a license to speculate, if not embezzle, and would have removed the threat of withdrawal of deposits, which was the major check on the irresponsibility of bankers. Deposit guarantees were rejected as conducive to bad banking as late as 2 March 1933, when the Board of Governors of the Federal Reserve was not prepared to recommend such a guarantee, or any other measures, on the eve of the national bank holiday.[60]

The Federal Deposit Insurance Corporation had an excellent financial record until the 1970s; the deposit insurance premiums that it collected were much larger than the amount it paid out to make good the losses when banks failed. From the beginning in 1934 through 1970, only one bank with deposits of more than $50 million failed, and most failures were of banks with deposits of less than $5 million. The FDIC had in most cases arranged for takeovers of the failed banks so that individuals and firms with deposits above the maximum insured amounts did not incur losses when banks failed.

Beginning in the late 1970s the financial problems of the FDIC and those of the Federal Savings and Loan Insurance Corporation (FSLIC) mounted sharply as more and more banks and thrifts were closed. The FDIC rescued a large number of banks including two giants, the Continental Illinois Bank of Chicago and the First Republic Bank of Dallas; honoring the deposit insurance guarantee cost these agencies billions of dollars. The usual operating procedure was to close the bank or the thrift institution when its capital had been depleted, and then to carve a 'good bank' from the rescued institution while the remaining assets of the failed institution would be retained for eventual sale by another newly created government agency, the Resolution Trust Corporation (RTC). Both insurance agencies incurred such large losses in honoring their guarantees that their capital and reserves were exhausted; they then borrowed from the US Treasury. In the early 1990s, the estimates were that the losses to the US taxpayers would amount to $150 billion but the increase in the US growth rate meant that the RTC received more money than anticipated from the sale of collateral and bad loans so the losses totaled a bit more than $100 billion.[61] There was some question whether a portion of this cost to the taxpayers could be reduced by increasing the insurance premiums on bank deposits – a suggestion resoundingly opposed by sound banks.[62]

During the 2008 crisis the US government extended deposit insurance to the money market funds; the fear was that otherwise the owners of these funds would transfer their money to banks whose deposits were guaranteed by the FDIC. The money market funds then would be obliged to sell assets, and the rapid sale of assets would depress their prices. Some money market funds then would 'break the buck'; the net asset value per share would decline below $1.00 – which would trigger a massive, self-justifying run.

Exchequer bills

One ancient device short of lending money to a firm in trouble was to issue marketable securities to the firm against appropriate collateral. (Of course, when markets break down, even the most liquid securities may not be sold readily.) The securities have been private and public; both types were part of the complex package put together by Hamburg in 1857. In 1763 and 1799, in

an equally complex and jerry-built system of support, admiralty bills were an integral feature.[63] The widest development, however, concerned the Exchequer bills issued in Britain in 1793, 1799, and (without enthusiasm) 1811, but sternly rejected in 1825.

The Exchequer bill was widely thought to have been the idea of Sir John Sinclair, although it may have originated with the Bank of England. On 22 April 1793, City leaders met with the Prime Minister, William Pitt, to devise means to combat the crisis that arose from the failure of 100 of the 300 country banks and the calamitous decline in commodity prices. The next day, eleven members of this group met at the Mansion House to formulate a scheme for state assistance. According to Clapham, there was no clear guide to what ought to be done. Then the idea emerged of having the government issue £3 million in Exchequer bills, a total that was later raised by parliament to £5 million, to be issued to merchants on the collateral of goods that they would deposit in the customs houses. An additional feature of the plan was to issue £5 notes – the previous minimum was £10 – to economize on the use of gold and silver coin. The Exchequer bills were issued by special commissioners rather than the Bank of England; £70,000 of these bills was immediately sent to Manchester and an equal amount to Glasgow. The device worked like a charm, according to MacPherson. Three hundred and thirty-eight firms applied for £3 million of the total amount. A total of £2.2 million was granted to 228 firms, only two of which subsequently went bankrupt. Applications for more than £1.2 million were withdrawn after the panic abated.[64]

In 1799 the panic in Hamburg had an echo in Liverpool, and Exchequer bills again were used. Parliament provided £500,000 in Exchequer bills, used solely in Liverpool, against £2 million of goods stored in warehouses.[65]

In 1811 the question arose again. A Select Committee on the State of Commercial Credit was appointed. Among its members were Henry Thornton, Sir John Sinclair, Sir Thomas Baring, and Alexander Baring. The committee's report, completed in a week, recorded the distress of exporters to and importers from the West Indies and South America, as well as the piles of goods bound for the Baltic that had been cut off and stored in London warehouses, and recommended a new issue of £6 million of Exchequer bills. Support was moderate in the House of Commons because of the overtrading in South America; the opposition, while sympathetic to the distress, doubted the wisdom of bailing out the speculators. William Huskisson, who later made his mark as the president of the Board of Trade, claimed that the evil came from too easy credit:

Did gentlemen not see that the race of old English merchants, who never could persuade themselves to go beyond their capital, was superseded by a set of mad and extravagant speculators, who never stopped so long as

they could get credit, and that persons of notoriously small capital had now eclipsed those of the greatest consequence; so that speculations now took place even in the lowest articles of commerce ... If the relief given was used for further speculation, it would only aggravate the evil – and he feared that this might be done – in which case the present measure would go only to add six million to the circulation and to raise the prices of all our commodities.[66]

Smart gave the fullest account of the debate and noted that many criticized the measure, though few were bold enough to deny it. In the end the proposal passed, but few applications were made and only £2 million was advanced. 'Not many of those who were in embarrassed circumstances were able to furnish the desired security, and it is difficult to see what remedy there was in being enabled, by advances, to produce more goods when the radical evil was that there was no market for them.'[67]

Luck: a tailpiece

Lady Luck once worked effectively to assist a bank in distress. Wirth wrote that the Brothers Kauffmann in Hamburg was failing during the crisis of 1799 when one of the brothers sent his bride a Hamburg city lottery ticket, which carried a first prize of 100,000 marks banco. She bought the same number in another lottery in the Duchy of Mecklenburg, the prize for which was an estate worth 50,000 Prussian thalers, then equal to 100,000 marks banco. She won both, and the Brothers Kauffmann was fully rehabilitated.[68] The odds at Las Vegas may be higher for the solution to a global crisis.

11
The Domestic Lender of Last Resort

The hallmark in the development of 'the Art of Central Banking' over the last two hundred years has been the evolution of the concept of a lender of last resort. The expression comes from the French *dernier ressort*, and centers on the last legal jurisdiction to which a petitioner can take an appeal. The term has become thoroughly anglicized, and now the emphasis is on the responsibilities of the lender rather than the rights of the borrowers.

The idea is that the lender of last resort can and should forestall a run by depositors and other investors from real assets and illiquid financial assets into money by supplying the amount of money that is needed to satisfy the demand; the concept is of an 'elastic supply of money' that expands to meet the demand in panics. How much money? To whom? On what terms? When?

The dilemma for the lender of last resort – the moral hazard argument – is that if owners and managers of banks believe that their firms will be supported in moments of distress, they will be less cautious in the extension of loans during the next boom. The public good that credit may be available from the lender of last resort weakens the responsibility of private lenders to ensure that they make 'sound' loans. If, however, in a panic the rush from the securities and commodities into money cannot be halted, the fallacy of composition takes center stage. The sale of these assets by investors leads to declines in their prices with the consequence that a large number of otherwise previously solvent and well-capitalized firms may be dragged into bankruptcy.

The opposition to a lender of last resort has been made continuously. Napoleon's minister of the public treasury, François Nicholas Mollien, wrote strongly against the interventionist instincts of his mentor, who wanted to save the failing manufacturers damaged by the Continental System (blockade); he asserted that a start in this direction would only get the Treasury in deeper and deeper.[1] Louis Antoine Garnier-Pagès, French minister of finance in 1848, claimed later that it was useful to precipitate a crisis to render it less durable: 'Do nothing to save the *rente*, clean out stocks; sell merchandise.' The policy,

he asserted, contributed to the brilliance of the French recovery from 1850 to 1852.[2] Murray Rothbard asserted that 'any propping up of shaky positions postpones liquidation and aggravates unsound conditions'.[3] The most trenchant formulation is that of Herbert Spencer: 'The ultimate result of shielding man from the effects of folly is to people the world with fools.'[4]

Such a view is understandable in a Darwinian age.

Origin of the concept

The development of the lender of last resort evolved from the practice of the market. Ashton asserted that the Bank of England was already the lender of last resort in the eighteenth century,[5] although this statement does not entirely square with his statement that 'long before the rules for the treatment of crises were laid down by economists, it was recognized that the remedy [for a financial crisis] was for the monetary authority (the Bank of England or the British government) to make an emergency issue of some kind of paper which bankers, merchants and the general public would accept. When this was done the panic was allayed'.[6]

The indecision about whether the central bank or the government was the final monetary authority remains to this day and qualifies the statement that the Bank of England emerged as the lender of last resort in the 1700s. E.V. Morgan maintained that the Bank of England's realization of its responsibility was delayed by the government's action in issuing Exchequer bills in 1793, 1799, and 1811, and that the Bank assumed the role as lender of last resort only gradually during the first half of the nineteenth century 'in spite of the opposition of theorists'.[7] The same evolutionary process can be seen in the Bank of France. In 1833 the majority of the Conseil General overrode Hottinguer's idea for a policy on the English model as well as Odier's plea for an entirely new policy and concluded that the major function of the Bank of France was to defend the French franc. Capital outflows were not to be feared. Interest rates should not be held artificially low or speculation will be encouraged and crises intensified. When a crisis occurred, however, the Bank should provide abundant and cheap discounts to moderate its intensity and shorten its duration.[8]

The role of the lender of last resort was not respectable among theorists until Bagehot's *Lombard Street* appeared in 1873, although Sir Francis Baring called attention to the idea at the end of the eighteenth century[9] and Thornton's classic, *Paper Credit*, developed both the doctrine and the counter-arguments in his discussion of the financial problems of the English country banks.[10] Bagehot traced the origin of the doctrine to David Ricardo in his statement before the Parliamentary Select Committee on Banks of Issue in 1875: 'The orthodox doctrine laid down by Ricardo is that there is a period in a panic at which restrictions upon the issue of legal tender must be removed.'[11] Bagehot

had articulated the doctrine in his first published article, written in 1848, when he commented upon the suspension of the 1844 Bank Act in the panic of 1847:

> It is a great defect of a purely metallic circulation that the quantity of it cannot be readily suited to any sudden demand... Now as paper money can be supplied in unlimited quantities, however sudden the demand may be, it does not appear to us that there is any objection on principle to sudden issues of paper money to meet sudden and large extensions of demand... This power of issuing notes is one excessively liable to abuse... It should only be used in rare and exceptional cases.[12]

Some analysts continue to reject the doctrine and powerful minds have argued both sides of the issue. Should one worry about the present panic or the next boom, the condition or the principle? 'There are times when rules and precedents cannot be broken; others when they cannot be adhered to with safety.'[13] The dilemma is that breaking the rule creates a new precedent and a new rule. Lord Overstone, the distinguished Currency School theorist, strongly opposed expansion of the money supply in a crisis but reluctantly admitted that a panic may require 'that power, which all governments must necessarily possess, of exercising special interference in cases of unforeseen emergency and great state necessity'.[14] On one occasion he produced a ringing metaphoric defense: 'There is an old Eastern proverb which says, you may stop with a bodkin a fountain, which if suffered to flow, will sweep away whole cities in its course.'[15] Friedman and Schwartz similarly and metaphorically flirted with lender-of-last-resort doctrine:

> The detailed story of every banking crisis in our history shows how much depends on the presence of one or more outstanding individuals willing to assume responsibility and leadership... Economic collapse often has the characteristics of a cumulative process. Let it go beyond a certain point, and it will tend for a time to gain strength from its own development... Because no great strength would be required to hold back the rock that starts a land-slide, it does not follow that the landslide will not be of major proportions.[16]

The paradox is equivalent to the prisoner's dilemma. Central banks should lend freely to halt the panic, but leave the market to its own devices to reduce the likelihood of future panics. A dilemma: actuality inevitably dominates contingency, today wins over tomorrow.

The Bank Act of 1844 represented a victory for the Currency School, which stood for a fixed supply of money, over the Banking School, which thought it

useful for the money supply to grow as output and trade grew. Both schools were concerned with the long run rather than with the short run, and neither approved of increasing the money supply as a temporary expedient to meet a crisis. When the Bank Act was being considered, the idea that the government has the power to suspend its provisions in an emergency was rejected. After 1847 and again after 1857, when it proved necessary to suspend the Bank Act and issue more money as a last resort, parliament conducted inquiries to determine whether there was a need to change the legislation. Both inquiries concluded that it was not desirable to have built-in provision for suspending the act even though the suspension had been useful and necessary. To limit precedent-setting, the bill brokers who had been sudden borrowers 'wanting incalculable advances in 1857 were told not to expect the like again'.[17] The principle of having a rule but breaking it if one had to was so widely accepted that after the suspension in 1866 there was no demand for a new investigation.

In the 1850s Jellico and Chapman had proposed rules for adjusting the discount rate of the Bank of England to the state of its reserves by a mathematical formula written into the legislation. Wood criticized them as having no real grasp of Bank transactions and methods of procedure.[18] Robert Love, Chancellor of the Exchequer in June 1875, introduced a bill that would authorize a temporary increase in Bank of England notes in exchange for securities (under certain contingencies, including panic), a bank rate above 12 percent, and the foreign exchanges favorable. The bill was tabled and given a first reading on 12 June but it never received a second reading and was withdrawn in July.[19] Hard-and-fast rules were agreed not to be workable. *The Economist* and Bagehot thought it proper that the Bank of England rather than the banks themselves should hold the reserves necessary to get the country through a panic. Mr Hankey, a former governor of the Bank, called this 'the most mischievous doctrine ever broached in the monetary or banking world in this country; viz. that is the proper function of the Bank of England to keep money available at all times to supply the demands of bankers who have rendered their own assets unavailable'.[20] The public, however, sided with Bagehot and practice against Hankey and theory. If the expansion of credit in boom periods cannot be controlled, then measures should be adopted to halt the contraction of credit in crisis.

Who is the lender of last resort?

The lack of clear agreement in Britain about whether the Treasury should relieve panics through the issue of Exchequer bills or whether instead the Bank of England should discount freely at a penalty rate, even if it is necessary to suspend the limits imposed by the Bank Act of 1844, has already been noted. Uncertainty about the answers to these questions may be optimal, along with

the question of whether the government authorities will rescue firms in distress and whether they will arrive in time. Thus there was no explicit provision for a lender of last resort in Britain and no fixed rule as to which agency should fill this role. In 1825 the Exchequer was not the selected agency. The task was given resolutely to the Bank of England whose reluctant acceptance was 'the sulky answer of driven men'.[21] In 1890 guarantees were used rather than the Bank or the Exchequer. Gradually the responsibility devolved on the Bank, which led Alfred Marshall to write that 'its directorate came to be regarded at home and abroad as a committee of safety of English business generally'.[22]

The Bank of France had agreed by the 1830s that it had responsibilities in a crisis but it thought it had other responsibilities as well, such as to ensure monopoly of the bank note circulation which permitted it to let the regional banks fail in 1848 and then to convert them into subsidiaries (*comptoirs*). The provinces feared Paris; their concern was that in a crisis Paris would take care of its own needs first at the expense of the regions and hence they wanted the privilege of note issue. But after Le Havre needed assistance, after it acquired its own bank, which made too many illiquid industrial loans including some to shipyards and to importers of cotton, the Banque du Havre went to Paris in 1848 seeking help. 'The return was not glorious. The Bank of France had been impitiable.'[23] It refused to lend on mortgages, saying: 'The statutes forbid it, and you have refused to accept a *comptoir*.'[24]

The Bank of France wavered over this question even as it wanted to destroy the regional banks. From America, Chevalier observed that the Bank of France had discounted freely in 1810, 1818, and 1826 – with Jacques Laffitte as governor for the first two years – making great efforts to sustain commerce; but it lacked the same courage in the crisis of 1831–32.[25] In 1830, after the revolution, the task was left to local authorities. A regional bank, conducted with honesty but not prudence, threatened a provincial crisis. The local receiver-general undertook to discount its doubtful paper, apparently after consulting Paris, where, Thiers testified, after 'mature reflection the public interest was put above that of the Finance Minister, M. Louis', 'with happy results', that is, the avoidance of the collapse of the bank and a resultant disturbance.[26]

After the Bank of France achieved its monopoly of the note issue and the conversion of the banks in the regions into its branches, the Bank began to act as a lender of last resort. Its statutes required it to discount only three-name paper; the task became one of producing acceptable names. Sixty *comptoirs d'escompte* were established throughout France, as well as a number of *sous-comptoirs* organized by various branches of trade to hold stocks of goods and issue paper against them. With the names of the merchant, the *sous-comptoir*, and a *comptoir*, the Bank of France could discount the paper and relieve the liquidity crisis. Louis Raphael Bischoffsheim of Bischoffsheim & Goldschmidt mocked the requirement of three names: 'The number is not important. With

bad signatures one can collect 10 instead of three. I prefer one good to 20 bad.'[27] After the crisis was over, a number of the *comptoirs* were taken over by bankers, merchants, and industrialists and became regular banks. The most famous of this group, the Comptoir d'Escompte de Paris, took its place among the leading banks in the country.[28]

The Crédit Mobilier of the Pereire brothers was not saved in 1868; the Bank of France refused to discount its paper, which might be interpreted as the revenge of the establishment on an outsider as punishment for not conceding the Banque de Savoie note issue to the Bank of France when the Pereires took over the Savoy bank after the region had been ceded to France by Italy in 1860.[29] The alternative interpretation was that this was the entirely normal refusal of a lender of last resort to bail out an insolvent institution.[30] Cameron accuses the Bank of France of conducting guerrilla warfare against the Pereire brothers in the interest of a quarrel between the Rothschilds and the Pereires that went back to the 1830s.[31]

The Bank of France and Paris bankers did not come to the rescue of the Union Générale in 1882 but seven years later they rescued the Comptoir d'Escompte de Paris. Critics of the Bank of France ascribe the difference in outcomes to venality. A less emotional position asserts that a second large bank failure in seven years might have completely destroyed the French banking system and that on this account Rouvier, the minister of finance, took the necessary measures to have the Bank of France and the Paris banks advance 140 million to the Comptoir d'Escompte.[32] In the Union Générale operation, as was noted in an earlier chapter, the Paris banks withdrew from the speculative activity when it began to peak in August 1881 and advanced 18.1 million francs to the Union Générale after the crash the following January to permit its more orderly liquidation rather than to save the bank.[33] Led by the Rothschilds and Hottinguer, and including the Comptoir d'Escompte and the Société Générale (but not the Lyons rival of Bontoux, the Crédit Lyonnais), the consortium represented the establishment, in which it was not really necessary to distinguish the Bank of France from the leading private banks (*hautes banques*) and deposit banks.

In Prussia in 1763 the king was the lender of last resort. In 1848 various state agencies, including the Prussian Bank, the Seehandlung, and the Prussian lottery vainly tried to help the Cologne bank, A. Schaaffhausen, before it was allowed to reorganize as a joint-stock bank. In the absence of a central bank in 1763, 1799, and 1857 the Hamburg city government, the chamber of commerce, and the banks – any and all leading agencies took part in the rescue operation.

The experience of the United States is especially pertinent to the question of the identity of the lender of last resort. There was some ambiguity

as to whether the First Bank of the United States and then the Second Bank of the United States were lenders of last resort despite the designation of the Bank in each case as the chosen instrument. On various occasions, the US Treasury assisted the banks by accepting customs receipts in post-dated thirty-day notes (1792), by making special deposits of government funds in the banks that were in trouble (1801, 1818, and 1819), and by relaxing the requirement that a commercial bank pay the Bank of the United States in specie (1801).[34] After the failure to renew the charter of the Second Bank of the United States in 1833, the US Treasury was even busier, both before and after passage of the 1845 law that required the Treasury to keep its funds out of the banks. In times of crisis and in periods of stringency caused by crop movements, the US Treasury would pay interest or principal on its debt in advance, make deposits in banks, offer to accept securities other than government bonds as collateral for deposits of government funds. Banks began to look to the Secretary of the Treasury for help in emergencies and to relieve seasonal tightness. In the fall of 1872, Secretary of the Treasury George S. Boutwell served as a lender of last resort by reissuing retired greenbacks – which may have been illegal. His successor, William A. Richardson, did the same thing a year later.[35]

The US Treasury could absorb money in deposits and pay out surplus cash that it had previously acquired but apart from the greenback period it could not create money. Thus the Treasury was unsatisfactory as a lender of last resort, unless it had built up its holdings of cash from budget surpluses. In 1907, when its cash holdings were low, the Treasury issued new bonds – $50 million of Panama Canal bonds, which were eligible for collateral for national bank notes, and $100 million of 3 percent certificates of indebtedness – that it hoped would entice existing cash and specie from hoards. In the end, the crisis was averted by a flow of more than $100 million from Britain.[36] Moreover the devices used to cope with a crisis were *ad hoc*. An analysis of the crisis of 1857 suggested that the Federal government was incapable of intervening effectively and that the public, including the banks, was left without guidance to stem the crisis.[37] In fact intervention proved to be too much and too early.

The complex record of involvement by the US Treasury raises the question of whether the market should not have regulated itself and, if so, how. O.M.W. Sprague, the historian of crises under the National Banking System for the 1910 Aldrich Commission, believed that the banks should have taken responsibility to ensure that they had enough reserves to meet all needs.[38] But Sprague was vague on which banks should take this responsibility or why the duty fell on them in the absence of responsibility embodied in legislation. Noblesse oblige? Duty? Several statements by Sprague indicate why a limited number of New

York banks had the obligation to stabilize the system and behave differently from other banks:

> During the period before the crisis of 1873 some 15 of the 50 New York banks held practically all the bankers' deposits in the city, and 7 of them held between 70 and 80 percent of these deposits. These 7 banks were directly responsible for the satisfactory working of the credit machinery of the country. (p. 15)

> It must always be remembered that in the absence of any important central institution, such as exists in other commercial nations, the associated banks are the last resort in this country, in times of financial extremity, and upon their stability and sound conduct the national prosperity greatly depends. (From the New York Clearing House report of 11 November 1873; p. 95)

> The fundamental characteristic of our banking system was illustrated [in 1890], that for any extraordinary cash requirements the reserves of the country banks are an unused asset. Evidence was again given which should have brought home to city institutions the heavy responsibility which they have incurred in attracting the reserves of other banks. (p. 147)

> The New York banks did not normally maintain the large reserves which the responsibilities of their position demanded. (p. 153)

> ...there was the possibility that the contraction of loans by outside banks, trust companies and foreign lenders might come together, creating a situation...well nigh impossible if in normal times the important clearing house banks failed to exercise great caution and maintain large reserves. (p. 230)

> The failure of the banks holding the ultimate reserve of the country to live up to the responsibilities of their positions is evident in still another direction. While the exact moment of the outbreak of the crisis of 1907 could not be foreseen, the imminence of a period of trade reaction had for many months been so probable that precautionary measures might reasonably have been expected from these banks if not from banks and the public in general. (pp. 236–7)

> The outside banks feel no responsibility for the course of the market. They will naturally withdraw from it when affairs at home require more of their funds or when they come to distrust its future. It therefore becomes necessary for the local banks to be able at all times to shoulder at least a part of the loans that may be liquidated by outside banks, and also to supply the cash which they thus secure the power to draw away. (p. 239)

> It is certainly an element of weakness in our central money market that influential credit institutions should have to be dragooned into doing what is after all in their own interest as well as to the general advantage. (p. 255)

...feeling common among New York bankers that they cannot reasonably be expected to remit funds which are the proceeds of loans made in the New York money market by outside banks and liquidated in an emergency...It should be remembered, however, that responsibilities are incurred in return for the advantages which accrue to the New York banks from their peculiar position. London holds its commanding position because it is known that money lent there can be instantly recalled. Similarly, New York is not meeting the obligations of its position as our domestic money center, to say nothing of living up to future international responsibilities, so long as it is unable or unwilling to respond to any demand, however unreasonable, that can lawfully be made upon it for cash. (pp. 273–4)

Sprague believed that the market needed a stabilizer and that the banks could not depend on the US Treasury to help supply the seasonal needs for cash, but he also did not believe that the biggest and most profitable US banks should take on the responsibility. These banks should be aware of seasonal currency requirements, the prospect that the out-of-town banks may withdraw deposits, and the state of the international balance of payments. Not all New York banks should take on this responsibility but only those that charged interest on out-of-town deposits, or the largest, or those with intimate connections with the stock exchange, or the leading members of the New York Clearing House.

The leading bankers in New York drew a different conclusion; they believed that the difficulties arose from lack of elasticity of the money supply and thus they fell into the trap of the Banking School. This was the real-bills doctrine, the idea that the money supply should expand and contract on the basis of changes in the supply of trade bills, which represented goods moving in domestic and foreign trade. Thus increases in the money supply would not be inflationary, and would have the necessary elasticity as long as they were discounted at banks and rediscounted at the central bank. There could be no doubt about it: 'The laws of finance are as well known, and as sure in their operation as the laws of physics.'[39] The lesson that Frank Vanderlip, Myron T. Herrick, William Barret Ridgely, George E. Roberts, Isaac N. Seligman, and Jacob H. Schiff drew from the panic of 1907 was that there should be a central bank that could respond to increases in demand for currency by producing more currency.[40]

Some ambiguity as to location of ultimate responsibility for the supply of currency may be helpful because it leaves some uncertainty so that the bankers are more self-reliant – provided there is not so much uncertainty that the market is disoriented. In London it was vaguely understood that there should be no formal provision for a lender of last resort but that there should be one in time of crisis. The intuitive politicians in the British government and the merchant bankers who ran the Bank of England thought it best to give power to grant

relief neither wholly to the Bank nor wholly to the government, but to leave it uncertain.[41] If giving relief were formally within the power of either the Bank or the government, it would be difficult to resist the pressure from the public.[42]

No one would bear the responsibility in too large a group. If only one entity were responsible, pressure for action might become irresistible. The optimum may be a small number of actors, closely attuned to one another in an oligarchic relation, like-minded, who apply strong pressure to keep down the chiselers and free-riders and who are prepared ultimately to accept responsibility. To give a more up-to-date example, tension in 1975 and 1976 among New York City officials, the unions, bankers, the State of New York, and the Federal government as to who would be the lender of last resort for New York City may have ensured uncertainty at a high level, and to have encouraged Yonkers, Buffalo, Boston, Philadelphia, and others not to slacken in their efforts to right themselves; yet action to save New York was eventually taken.

During the early stages of the 2008 crisis the Federal Reserve provided credit to facilitate the acquisition of Bear Stearns by JPMorganChase. This facility may be a credit, but if the losses are large, then this commitment would be viewed as an investment. Thereafter the Chairman of the Federal Reserve, Ben Bernanke, suggested that the Bush Administration needed congressional authorization before it could make further investments in banks.

Bonelli's statements about the 1907 crisis in Italy offer a moving account of a muddle. The Società Bancaria Italiana was failing, and dragging down a host of small financial, mercantile, and industrial firms. A consortium of the larger banks put together a support fund. The Bank of Italy was involved early and deeply, and almost became too heavily committed. The Treasury finally came to the rescue, at the insistence of Bonaldo Stringher, governor of the Bank of Italy, and paid the interest on the national debt early and thus relieved the liquidity crisis. Bonelli saw that the episode inevitably involved both the Bank of Italy and the government and suggested that this indecision might develop when the economy works for more than ten years with no one in charge.[43] Part of the difficulty may reflect the lack of sufficient cohesion among Turin, Genoa, Milan, and Rome and the resulting uncertainty, buck-passing and indecision.

Some uncertainty was inevitable because as a House of Commons committee noted in 1846 'looking to the impossibility of foreseeing what the precise character of the circumstances might be' it was thus 'more expedient to leave to those with whom the responsibility of government might rest at the time, to adopt such measures as might appear to them best suited to the emergency'.[44] Consider Sir Robert Peel's statement on the Bank Bill on 4 June 1844:

> My Confidence is unshaken that we have taken all the Precautions which legislation can prudently take against the Recurrence of a pecuniary Crisis.

It may occur in spite of our Precautions; and if it be necessary to assume a grave Responsibility, I dare say Men will be found willing to assume such a Responsibility.[45]

George Harrison, then president of the Federal Reserve Bank of New York in the late 1920s, opened the discount window wide during the October 1929 stock market crash and then was criticized by the Board of Governors in Washington when the New York Fed bought $160 million in government bonds on the open market in October and another $210 million in November. The Board of Governors resented the New York bank because of the earlier high-handed dominance of the system by Benjamin Strong (who died in 1928), and had little compunction in reining in Harrison when he tried to emulate Strong's penchant for filling a power vacuum by strong leadership. Ambiguity as to whether there will be a lender of last resort, and who it will be, may be optimal in a close-knit society. The division in experience and outlook between Washington and New York handicapped more effective action in coping with the collapse of stock prices in 1929.

There was no hint of criticism or second-guessing when the Federal Reserve Board, under the new chairmanship of Alan Greenspan, set about expansive open-market operations immediately after the 19 October 1987 crash, and poured in high-powered money 'right and left', to use Bagehot's expression. The Fed under Greenspan provided liquidity to cope with the Asian Financial Crisis in 1997, the debacle in Russian finances and the collapse of Long-Term Capital Management in the summer of 1998, in anticipation of the Y2K crisis in the last few months of 1999, and in response to the sharp decline in stock prices in 2000 and 2001.

The Federal Reserve under the chairmanship of Ben Bernanke was slow to recognize the sharpness of the housing bubble and impacts of the implosion of real estate values on the banks and financial markets and the economy. Only after the panic triggered by the failure of Lehman Brothers in mid-September 2008 did the Fed open 'all of its windows'. The Fed was willing to accept many different types of collateral for loans at the discount window. The Fed, moreover, made money at the discount window available to non-banks. The Fed became a direct lender to business and the Fed's assets nearly tripled in a year.

To whom on what?

The rule laid down by Bagehot was that loans should be granted to all comers on the basis of sound collateral 'as largely as the public asks for them'.[46] But in his testimony before the 1875 inquiry, two years after the publication of *Lombard Street*, Bagehot resisted the suggestion that last-resort lending be turned over to a body of commissioners appointed by the government on the grounds that they might make loans to 'improper persons'. They would be

subject to political pressure while the Bank of England is 'a body withdrawn from the political world and not subject to political pressures'.[47]

Bagehot's suggestion that central banks are immune from political pressure seems naive. The dilemma about collateral is that its soundness depends on when and whether the panic is stopped; the longer the panic continues, the sharper the decline in the prices of securities and bills of exchange and commodities and hence the less sound the collateral. In this case, it becomes necessary to look at the character of the borrowers, something that J.P. Morgan was reported to consider exclusively. Here the dilemma relates to the wry comment that bankers lend money only to those that do not need it.

Central banks typically have rules.[48] When the rules cannot be easily broken – as in the Federal Reserve Act of 1913, which permitted only gold and negotiable bills of exchange but not government securities to be held as reserve against Federal Reserve notes and demand deposits – there is frequently trouble. There is also trouble when rules are too readily broken. The beauty of the Chancellor of the Exchequer's letter of indemnity was that it preserved the rule while violating it and did not create a precedent, at least not for a while. The Bank of France and the Reichsbank occasionally discounted only three-name paper. But discretion to reject paper because it was 'unsound' or the borrower because of his character gave the lender of last resort a life-or-death power that might not always be used with complete objectivity. The literature is filled with accusations of venality on the part of the directors of central banks. Protestant and Jewish directors of the Bank of France were alleged to have punished the Catholic (and worse-off) supporters of the Union Générale in 1882, while saving the insider Comptoir d'Escompte in 1888.[49] In the crisis of 1772, the Bank of England's issuance of new regulations about discounting and refusal to discount doubtful paper were interpreted as an attempt to break the Jewish houses in Amsterdam that had been most involved in the speculation. Then there was the Bank's decision to refuse the bills of Scottish banks, and finally to stop discounting altogether, which was probably 'a step taken quite deliberately to break up a group of Dutch speculators'.[50] Outsiders particularly suffered. The Bank of the United States was allowed to fail in New York in December 1930 by a syndicate of banks amid accusations that the Bank was being punished for its pushy ways.[51]

The rule of discounting for everyone with good paper evolved slowly in Britain. For a time 'the invariable practice' was respectable London names on paper with no more than two months to run; but this description of 1793 is accompanied by a statement that while a request from Manchester had been turned down (along with one from Chichester, where refusal helped to bring a bank down), £40,000 had been advanced to banks in Liverpool. Only in July 1816 did the Bank, breaking a rigid precedent, agree to accept 'country securities of undoubted respectability if the firm cannot get enough London names'.[52]

The Bank of England made advances on a wide range of different types of assets much beyond two-name paper. In 1816 the Bank broke its rule against lending on mortgages, undertaking a 'Transaction quite out of the ordinary course of Business' to relieve the distress of poor people in the Black Country. The Bank resolved to lend only in the old way 'on notes of respectable parties' but a few years later the Bank began a regular mortgage business on the ground that the volume of discounts and especially the income from discounts had collapsed – a private rather than a public purpose.[53] At one stage the Bank made loans on the security of a mortgage on a plantation in the West Indies (ultimately the Bank foreclosed on this loan[54]) and on unimproved land in England. The land was unencumbered by a mortgage but belonged to a duke, an indication that collateral and the character (or status) of the borrower were not unrelated. Loans were not made on land in Scotland and in Ireland.[55]

With the growth of railroads, the discounts of the Bank of England were made on the collateral of railroad debentures. In 1842, as the second railway mania got underway, the Bank voted to make occasional loans to firms in difficulty and to well-tried firms for development.[56] The Bank of France began lending to a railroad syndicate in 1852; in fact, it was accused of supporting, if not starting, the feverish speculation in railroads.[57] Bagehot thought the Bank of England mistaken for not lending on railroad debentures when it did so on Consols and Indian securities; Bagehot stated that a railway was less liable to unforeseen accidents than the Empire of India.[58] But Indian securities were guaranteed by the Colonial Office and in effect were British government obligations.

Exchequer bills were issued on the collateral of goods, as were Admiralty bills in Hamburg. Clapham observed that many of the Bank's advances in 1825 were not actually on goods but rather on personal security;[59] the Bank loaned freely and was not 'over-nice'.[60] In a few weeks in 1847 the Bank advanced £2.25 million in both usual and unusual ways, including the securities of the Company of Copper Miners, through which it involuntarily acquired a copper works.[61]

The rule is that there is no rule. One does not lend to insolvent banks, except to avoid the mischief that would occur if the Lord Mayor of London were to go bankrupt (1793),[62] or to maintain for a time a payroll in Newcastle, a town used to banking disasters.[63] The Bank of France had never discounted as much as 4 million francs for anyone but Jacques Laffitte when Samuel Welles, an American banker, applied in 1837; he proved to be an exception.[64] (The Laffitte transaction had also been exceptional, with a political motivation.) The Conseil General could not abandon such an important bank, so the bank received a line of credit of 15 million francs.[65] In the crisis of 1830 the Bank of France discounted royal and municipal bonds, customs receipts, woodcutting receipts, obligations of the city of Paris, and canal bonds repayable by lottery.[66]

Some of the decisions that the lender of last resort must make are easy, such as whether to discount Treasury bills. Some are difficult, such as whether to discount shaky collateral from shaky banks. The record is full of firms that were refused help, failed, and paid off 20 shillings to the pound, and of banks that were helped in one crisis and failed in the next. The 241-page appendix to Evans's book on the commercial crisis of 1857 was devoted to court records of bankruptcies in Britain between 1849 and 1858. The reading is doleful. G.T. Braine, who was refused accommodation by the Bank of England in 1848, paid 20 shillings to the pound and finished with a surplus that was twice that originally estimated. One also finds petitions in bankruptcy brought by the Bank, such as that against Cruickshank, Melville and Co., for the unpaid residue of a bill it drew on another bankrupt firm that had paid only 12s. 6d. to the pound.[67]

Even the judgment of history is not always helpful. The Bank of England first refused to help the three American 'W banks' (Wiggins, Wilsons and Wildes) in the fall of 1836 and then relented and advanced them credit in March 1837. Andréadès noted that the Bank took a bold step and had no occasion to regret its courage.[68] Clapham held, on the contrary, that the Bank lent most reluctantly and was not surprised when Wilsons and Wiggins failed at the end of May and Wildes thereafter and the consequence was 'a long, dreary tale of debt lasting 14 years'.[69] To Matthews, the Bank of England's aid to the three banks 'in the vain hope of avoiding their suspending was a matter of faulty judgment but the principle on which they operated was a sound one'.[70]

When and how much?

'Too little, and too late' is one of the saddest phrases in the lexicon of central banking. But how much is enough? When is the right time?

Bagehot's rule is to lend freely at a penalty rate. 'Freely' means only to solvent borrowers and with good collateral, subject to the inevitable exceptions. It means rejecting the expedients that various central banks are tempted to indulge in crises. Early in 1772 the Bank of England tried to apply the brakes to overtrading by selective limitation of discounts and was criticized.[71] In 1797 the Bank began to pro-rate discounts, and Foxwell thought that might have been undertaken again in 1809.[72] Another technique when a central bank feels it is getting overcommitted is to tighten eligibility requirements for collateral, shortening the maturity of eligible bills from 95 or 90 days to 65 or 60 days, or adding to the number of names required. In May 1783 the Bank of England had discounted so heavily for its own clients that it departed from its regular practice and refused to make advances on subscriptions to government bonds issued that year. Clapham commented that fortunately no public or private catastrophe of the sort that starts a panic occurred that summer since the Bank

had limited its capacity to meet one.[73] In this case the Bank was behaving like a private bank worried about its own safety rather than a public institution with the responsibility to provide for the stability of the system.

The lender of last resort might supply funds to the system through open market purchases rather than through the discount mechanism. How much should the central bank expand the money supply? Were the $160 million in October 1929 and the additional $210 million through November 1929 adequate? In the view of the Federal Reserve Bank of New York they were not. The New York Fed was operating under a directive from the Board of Governors that permitted it to buy $25 million of government bonds each week. It violated this rule in October by buying $160 million, and on 12 November recommended to the Board that the limit of $25 million a week be removed and that the Open Market Investment Committee be authorized to buy $200 million of bonds for the system. After considerable negotiation, the Board reluctantly approved this request on 27 November, and $155 million was purchased between 27 November and 1 January 1930. By this time, discounts were running off rapidly, interest rates had fallen sharply, and the need for a lender of last resort – to accommodate the liquidation of call loans in the market – was over.[74]

Some monetarists seem ambivalent on the role of the lender of last resort. Friedman and Schwartz quoted Bagehot approvingly on not starving a panic.[75] They asserted that the action taken by the Federal Reserve Bank of New York in buying $160 million in October 1929 was 'timely and effective' although they were gently skeptical about Harrison's claim that the open market purchases kept the stock exchange open.[76] Friedman was opposed to all discounting.[77] An ultra-monetarist view maintains that the open market operations of the period constituted a renewal of the credit inflation of the 1920s.[78] But most monetarists believe that there is no need to have a lender of last resort as long as the money supply increases at a constant rate. In retrospect the open market operations were woefully inadequate in the weeks from mid-October to the end of November 1929. They enabled the New York banks to take over the call loans of out-of-town banks but at the cost of reducing the amount of credit available for purchases of stocks, commodities, and real estate, which led to declines in their prices and unleashed the depression.[79]

The timing of the Federal Reserve Board under the chairmanship of Alan Greenspan in the Black Monday crash of October 1987 was impeccable, as was the help for the US capital market when Long-Term Capital Management collapsed in September 1998.

Timing presents a special problem. As the boom mounts to a crescendo, it must be slowed without precipitating a panic. After a crash has occurred, it is important to wait long enough for the insolvent firms to fail, but not so long as to let the crisis spread to the solvent firms that need liquidity – 'delaying the death of the strong swimmers', as Clapham put it.[80] In a speech during the

debate on the indemnity bill on 4 December 1857, Disraeli quoted the leader of an unnamed 'greatest discount house in Lombard Street' who said that 'had it not been for some private information which reached him, to the effect that in case of extreme pressure there would be an interference on the part of the Government, he should at that moment have given up the idea of struggling any further, and [that] it was only on that tacit understanding that he went on with his business.'[81] Questions could be raised about the equity of private information and of tacit understandings for insiders but not outsiders. Still, the remark underlines the importance of timing. Whether too soon and too much is worse than too little and too late is difficult to answer.

In 1857 the US Treasury came to the rescue of the market too early and helped it inflate still further. In 1873 the response was too slow, no steps were taken during the first part of the year.[82] Sprague refers to 'the unfortunate delay of the Clearing House', that is, the slowness of any authority to respond to the 1907 crisis, in which as in no other crisis since the Civil War, things were allowed to drag on too long.[83]

If the need for a lender of last resort is accepted after a speculative boom and it is believed that restrictive measures are not likely to slow the boom without precipitating a collapse, the lender of last resort faces dilemmas both of amount and timing. The dilemmas are more serious with open market operations than with a system of discounts. In the latter case, Bagehot specified the right amount: all the market will take – through solvent houses offering sound collateral – at a penalty rate. With open market operations the decision for the authorities is more difficult, but Bagehot was right not to starve the market. Given a seizure of credit in the system, more is safer than less since the excess can be mopped up later.

Timing is an art. That says nothing – and everything.

12
The International Lender of Last Resort

The primary argument for an international lender of last resort is the historical record of the transmission of deflationary pressure from one country to others, which often has been associated with changes in currency values, both with devaluations when currencies have been pegged and with depreciations when they have been floating. The devaluation of the Finnish markka in 1992 transmitted deflationary pressure to Sweden, one of its major competitors in the production and sale of timber and other products. The depreciation of the Thai baht, the Malaysian ringgit, and the South Korean won in the second half of 1997 transmitted a massive deflationary impulse to the United States; the exports of these countries surged relative to their imports which led to a counterpart increase in the US trade deficit. The increase in the Japanese trade surplus in the 1990s after its asset price bubble imploded had a deflationary impact on its major trading partners and especially on the United States, whose trade deficit increased sharply. The sharp increase in China's trade surplus after 2002 led to counterpart increases in the trade deficits of its trading partners and especially in the US trade deficit.

An international lender of last resort has a problem that has no domestic counterpart; as long as there are separate national central banks each with its own money, changes in currency values are inevitable. Some of these changes will be necessary, especially when a country has not been as successful as its major trading partners in achieving a low inflation rate; a reduction in the value of a country's currency may be a less costly way to achieve a satisfactory trade balance than relying on declines in costs and prices. Some of these changes in currency values may be necessary because of a major structural shock, including the loss of an export market because of technological changes, the depletion of raw materials, and the productive gains in countries that are in earlier stages of industrialization. Some of the changes in currency values may not have been necessary, but instead were adopted because a low value for the currency was viewed as the preferred way to stimulate domestic employment and economic growth.

Changes in currency values can be an unnecessarily costly form of adjustment to a shock; the 'overshooting' of a country's currency as its market price increases relative to its long-run equilibrium value has almost always had a deflationary impact on the country's tradable goods sector as imports increase relative to exports. Conversely the 'undershooting' of a country's currency as its market price declines relative to its long-run equilibrium value can trigger a surge in the inflation rate and in domestic interest rates. Moreover sharp real depreciations have often led to widespread failures of business firms and banks.

The episodes of massive currency undershooting have often been preceded by significant overshooting of the same currencies. The Mexican peso was significantly overvalued in the first half of 1994, and significantly undervalued in the first half of 1995. The sudden reversal in relative values has almost always occurred as a result of a sudden turnabout in the direction of cross-border movement of money. Institutional arrangements and policy innovations that dampen or reduce the scope of overshooting and undershooting would enhance economic welfare.

The primary responsibility of a domestic lender of last resort is to reduce the likelihood that a shortage of domestic liquidity will cascade into a solvency problem and cause bankruptcies that would not have occurred in the absence of distress and precautionary selling. The domestic lender of last resort walks a tightrope between avoiding saving financial institutions that are already bankrupt because of their risky investments or bad business decisions and saving their healthier competitors from insolvency that would occur if the price level declines and deflation follows. The primary responsibility of an international lender of last resort is to provide liquidity to ameliorate the necessary changes in currency values and obviate those changes that are not consistent with the economic fundamentals.

International credits that have been extended for at least four centuries have enabled countries to dampen swings in currency values both when governments have borrowed abroad, and when private banks in one country have relied on their counterparts in other countries for assistance to meet a sudden withdrawal of funds. In the 1920s the first concerted concern with a shortage of international liquidity – and more precisely of monetary gold – developed in response to the view that the increases in national price levels during and immediately after World War I would lead to both a decline in gold production and increases in both the official and the private demand for gold. The financial authorities in the major countries were concerned that the gold shortage would be more intense because the establishment of new central banks in what had been the Austro-Hungarian empire would lead to an increase in the demand for monetary gold. Several conferences sponsored by the League of Nations in the 1920s recommended that central banks hold a larger proportion of their

international reserve assets in the form of liquid securities denominated in the British pound and the US dollar.

The International Monetary Fund was established in the 1940s to provide a formal arrangement for the extension of credit among national governments and to assist countries in coping with currency crises; the motivation was the belief that much of the destructive competitive behavior among countries in the interwar period that contributed to World War II had resulted from a shortage of international credit. Countries allowed their currencies to depreciate to increase their exports and stimulate domestic employment. The fallacy of composition was relevant and the policies that might have been successful for an individual country were costly for all countries as a group. When a country became a member of the IMF, it agreed to limit the movement in the value of its currency around its parity; the country also agreed to seek the approval of the IMF before it changed its parity by more than a modest amount. The IMF was endowed with a pool of gold and currencies from the capital subscriptions of its members. A member country could borrow from this pool to help finance a payments deficit.

A devaluation of a country's currency should almost always lead to an improvement in its international competitive position and a reduction in its trade deficit. The scope of the improvement is time-dependent and increases in the long run as firms within the country increase their exports and export capacity and their ability to produce import-competing goods. Thus the country's currency could depreciate sharply in the short run before the supply and demand structures changed and the country's trade deficit declined to a new and sustainable value. While these adjustments occurred and before they have been completed, momentum-oriented traders might induce an even sharper depreciation of the country's currency.

Undershooting and overshooting are inevitable developments in response to changes in cross-border money flows. The scope of undershooting and overshooting often expands as currency traders follow the canard that 'the trend is my friend'. In recent years the cliché of the 'vicious and virtuous cycle' was applied as the descriptive of this currency market behavior. An earlier generation of economists used the term 'destabilizing speculation' for the same pattern of currency transactions.

At times these adjustments in currency values may occur in propitious circumstances, but at other times the circumstances are less than propitious. Thus in 1994 and 1995, the Mexican trade deficit was much too large and the adjustment to a sustainable value would require a lower value for the peso. Mexico needed a decline of 3 to 4 percent in the ratio of its current account deficit to its GDP. This necessary and inevitable decline would have dramatic impacts on Mexican saving, the fiscal position of the Mexican government, and Mexican GDP. When the shock occurred, there was a surge in the outflow

of money from Mexico, the peso lost half of its value and the change in the ratio of Mexico's current account balance to its GDP exceeded 10 percent. The shock to the Mexican economy was immense.

Similar statements can be made about the depreciation of the Argentinean peso in 2001. The South Korean won lost nearly one-half of its value in December 1997; prior to the crisis the country's current account deficit was 1 percent of its GDP and afterward the country developed a current account surplus of 3 percent. The scope of the overshooting of the Indonesian rupiah before the Asian crisis was large; the undershooting after the crisis was much larger.

While overshooting and undershooting are transitional phenomena associated with changes in the direction and scope of international money flows, they have extremely powerful real effects on the domestic economies and some of these effects are permanent. One reason analysts have underestimated the extent of the changes in currency values associated with the move to new equilibrium values is that they have tended to ignore the permanent effects of overshooting and, especially, of undershooting.

An international lender of last resort would help countries moderate the deviations of the market values of their currencies from long-run equilibrium values. One inference from financial history is that in the absence of an international lender of last resort the economic depression that follows a financial crisis can be long and drawn out, as in 1873 and 1890 and the early 1930s.

One problem in developing an international lender of last resort is formulating the legal framework and the set of rules that would govern its activities and especially the rules and procedures for the allocation of its credits. In the absence of a world currency, the international lender of last resort would lend the currency of one or several countries – most probably those of the countries with large holdings of international reserve assets or large payments surpluses. The historical record suggests that the lender of last resort would supply the currency of the country that is the leading financial center perhaps together with the currencies of a few other countries.

The debt-deflation cycle of the 1930s involved a feedback loop that encompassed business failures, bank failures, and declines in consumer price levels. When businesses failed, their inventories were sold, which depressed the price level and reduced the net worth of other firms in the same industries. The increase in business failures led to larger bank loan losses and the banks become less eager lenders; they refused to renew some maturing loans because some of these borrowers might be on the proverbial slippery slope. Even though nominal interest rates were low, real interest rates were high because of the declines in the consumer price level, and investment spending was depressed.

Japan was in the throes of a very modest debt-deflation cycle at the end of the 1990s as its price level declined by 1 percent a year which led to an exceptionally high level of business failures. Hong Kong was in a debt-deflation cycle for six years following the Asian Financial Crisis; the depreciation of the currencies of virtually all the other Asian countries meant that the international competitive positions of firms producing in Hong Kong were declining.

The environment for the international lender of last resort differs from that of the domestic lender in two basic ways – one is that liquidity crises are often associated with changes in currency values and the other is that the rule of law is more fragile in the international context. The international lender of last resort also walks a tightrope between providing liquidity to countries so that unnecessary changes in currency values are avoided and minimizing the likelihood that its provision of liquidity would enable countries to postpone the necessary changes in the values of their currencies. In 1999 and 2000, Argentina would have borrowed and borrowed some more to delay the adjustments that were necessary to reduce its trade deficit. (After the devaluation, Argentina snubbed its foreign creditors and wanted them to settle for 25 or 30 cents on the dollar.) Moreover, even when a change in the value of a country's currency is necessary, the extension of credit from the international lender may reduce the undershooting of its currency in the move to a new equilibrium value. Finally, there are times when the determination of whether there is a need for a change in the value of a country's currency is a judgment call; the country's payments deficit might be reversed by developments in the business cycle or by changes in commodity prices – or there might be a miracle or just good luck.

The analogy between the domestic lender of last resort and the international lender breaks – or at least bends sharply – because there is no good domestic counterpart to the changes in the values of national currencies. Normally, a domestic lender of last resort would not lend to an insolvent institution – except perhaps temporarily until losses or capital shortages have been made good by a deposit guarantee agency or some other governmental group. Many international financial crises have involved changes in currency values that have become seriously overvalued. The problem is to determine when a country should change the value of its currency because of a fundamental dis-equilibrium – a term popularized by the IMF when countries adhered to the adjustable parities arrangement. As long as countries retain the ability to change the values of their currencies, the international lender of last resort would have to determine when access to its funds would enable a country to delay the necessary change in the value of its currency and when such access would enable a country to avoid a change that is not necessary.

The number and severity of financial crises in the last thirty-five years have been significant. These crises often have followed from large changes

in currency values. An effective international lender of last resort could have moderated the tumultuousness in financial markets. Consider some of the major currency crises of this period:

- Mexico, Brazil, and Argentina in 1982
- The British pound, the Italian lira, and other European currencies in 1992
- The Mexican peso in 1994
- The Thai baht, the Malaysian ringgit, and other Asian currencies in 1997
- The Russian ruble in 1998
- The Brazilian real in 1999
- The Argentinean peso in 2001
- The Icelandic krona in 2008

The United States cobbled together a financial assistance package for Mexico in 1995. Thailand, South Korea, Indonesia, and the Philippines received large credits from the IMF after their currencies had depreciated sharply. Russia received a large credit from the IMF in the spring of 1998, even though the devaluation of the ruble seemed inevitable; similarly Argentina received a large credit from the IMF in the early autumn of 2000 even though a major miracle would have been necessary to maintain the parity of the peso with the US dollar. (In both cases some member countries leaned on the IMF to extend the credits.)

A historical view of international financial crises

Often analysts have distinguished an internal drain that involves an increase in the domestic demand for currency or gold from an external drain that involves a flow of gold or other reserve assets to foreign countries. An external drain could generally be reversed by an increase in interest rates; hence the quip that an increase in the Bank of England's discount rate of 1 percent would 'draw gold from the moon', although often there was a lag before the higher interest rates would lead to the desired improvement in the country's gold holdings. If investors viewed the increase in the discount rate as a sign of weakness and sold the currency, the external drain would increase, and the central bank would need to raise interest rates further. At times the increase in interest rates would not be adequate to correct the imbalances – perhaps because the lags were long and the time limited – and the central bank would need to borrow to avoid a sharp depreciation. A central bank might borrow from other central banks. Domestic firms that owned foreign assets might sell them and remit the proceeds.

What is the appropriate public policy to reduce the likelihood of international financial crises and to ameliorate their impacts? During the financial

crisis of 1846–48, when bond prices all over Europe were plummeting, some by as much as 75 percent, and the Rothschild houses in Paris, Vienna, and Frankfurt were threatened with bankruptcy, Nathan Rothschild in London, with the help of Auguste Belmont, the Rothschild representative in New York, helped privately. Belmont shipped silver from New York to London that was then shared among the Rothschild brothers. Central banks helped by keeping interest rates low, but Nathan, as the effective lender of last resort, saved the day with the assistance from New York.[1] In the 1930s, Paul and Felix Warburg of Kuhn, Loeb in New York served in the same capacity for their brother, Max Warburg, then in Hamburg, with a credit of more than $9 million.[2]

Last resort lending centers on central banks as the providers of cash. In 1694 the Bank of England borrowed 2 million guilders from the States General in Holland to assist the British Exchequer in remitting funds to the Continent to support British troops and allies during the Nine Years' War.[3]

In the years that followed (1695–97) Amsterdam assisted the Bank of England by taking over its protested bills of exchange on the Continent; the Dutch charged 10 percent for the service – a penalty rate in Bagehot terms, but it was, after all, business, not the provision of a public good.[4]

In the crisis of 1763 the Bank of England and London private bankers rescued their Dutch correspondents by granting, in distress, credits larger than those previously given in periods of prosperity. Five consignments of gold were shipped in August and two in September. In addition, the Bank of England and other banks delayed presenting bills for payment. Wilson comments that none of this was pure altruism. Instead, it represented a practical policy based on the knowledge that British prosperity was intimately associated with Dutch prosperity and that the intensification of the Dutch crisis would cut off a source of money for Britain.[5]

As the 1772 crisis reached a peak in January 1773, Anglo-Dutch trade was paralyzed. Amsterdam was helpless, and only the Bank of England, Wilson says, could rescue the city. On 10 January, a Sunday, the Bank opened its windows and allowed specie to be drawn against presentation of notes and government bonds. Loads of bullion were sent on the first packet boat, and one Dutch banker was said to have drawn £500,000. At the same time, the Bank refused to discount doubtful paper, which had the effect of breaking many Jewish-owned banks in Amsterdam.[6] In the same crisis, Catherine the Great of Russia helped her best customers, the British merchants, in the first of a number of crises when czarist Russia assisted Western Europe.[7]

In the crisis of 1825 there was a rumor that the Bank of France was trying to add to the Bank of England's difficulties. Clapham insisted that instead France participated in an early example of international financial cooperation by shipping gold to London in exchange for silver.[8] That the price of gold in terms of silver was higher in London than in Paris (15.2 to 1 compared with

14.625 to 1) favored the exchange,[9] but the arrival via the Rothschild banks of £400,000 (mostly in sovereigns) from Paris on Monday 19 December 1825 helped because the Bank of England coffers were virtually empty after the run on the previous Saturday.

Twice during the extended crisis of 1836 to 1839 the Bank of England sought help from the Bank of France and the city of Hamburg to maintain its liquidity. The first time the Bank of England drew bills on Paris for £400,000. In 1838 the Bank of England arranged for a line of credit and in 1839 drew on this line for £2 million, using Baring Brothers and ten Paris banks as intermediaries. A similar line of credit with Hamburg brought in £900,000 in gold, an arrangement possibly designed also to help Hamburg, which needed silver.[10] In 1838 the Bank of England, not the most usual of gold dealers, sent nearly 700,000 sovereigns to the United States. Clapham said the operation was without precedent, and damaging insofar as it encouraged American bankers to issue more securities in the British markets in 1838 and 1839, but he acknowledged that the Bank was wise to recognize the interlocking interests of Britain and the United States.[11]

The Bank of France borrowed 25 million francs from British capitalists in the second half of 1846, according to French sources;[12] a British source states that the sum was borrowed from the Bank of England in January 1847.[13] At that point, the emperor of Russia offered to buy 50 million francs of the French 3 percent *rente* to assist in financing the imports of wheat that France and Britain needed. Britain benefited since the French used half the money to repay the British advance.[14] Palmer, then governor of the Bank of England, testified before the Select Committee of Parliament that it was preferable to have an understanding with the principal banks in the United States, Hamburg, Amsterdam, and Paris than to ship gold.[15]

Central bank cooperation was not universally applauded. Viner wrote that the Bank of England in 1836 found itself obliged to appeal to France for help 'no doubt reluctantly' and added that the necessity was regarded in Britain as humiliating because it came at a time when relations between the two countries were not particularly cordial, 'especially as it was reported that the followers of M. Thiers were boasting of the generosity of Frenchmen…while recommending that under no circumstances should such liberality be repeated in future'.[16] Thomas Tooke thought the loan a 'discreditable expedient', a 'circumstance of almost national humiliation'.[17] The Bank of France was criticized in Paris by some who thought it irresponsible to profit from the British arrangement rather than aid those at home who sought its help.[18]

In the 1850s there was less international cooperation in crisis periods. Clapham wrote that the Bank of England contemplated joint action with the Bank of France in November 1857, but he did not indicate what the action was and said little beyond that 'nothing came of it'.[19] Perhaps the most interesting operation

was the December *Silberzug* in Hamburg. Hamburg was at the end of the line of the rolling crisis that swept from Ohio to New York to Liverpool and then to the Continent, particularly to Scandinavia. On 4 December the Hamburg Senate voted a 15 million mark banco fund composed of 5 million in Hamburg bonds and 10 million in silver that would be obtained by borrowing abroad. The next task was to borrow the money. Appeals for a loan were made to Rothschilds, Baring, and Hambros in London, to Fould in Paris, and to various political and financial bodies in Amsterdam, Copenhagen, Brussels, Berlin, Dresden, and Hanover. Each request was turned down. From Fould came this answer: 'Your message is not sufficiently clear.' From Berlin: 'Bruck and the Kaiser are not financially ambitious.' On 8 December when every bank in Hamburg except Heine's was threatened with bankruptcy, and when captains of ships were unwilling to unload their cargoes because they were concerned that they would not be paid, word came from Vienna that it would provide all the money that was needed. A train bearing the silver (the *Silberzug*) arrived shortly.[20]

The silver was removed from the train, and loans in silver were made to Merck, Godeffroy, and Berenburg, Gossler and Co., among the leading bankers, plus five smaller ones. The panic ended on 12 December when it became known that there was enough silver. Some firms, like Donner & Co., which had initially been allotted 700,000 marks banco, turned out not to need any when confidence was restored. Böhme, who gave the most detailed account of the crisis, said that for years the episode kept coming up whenever Hamburgers and non-Hamburgers talked about currencies.[21] The political aspects of the rescue operation were revealed in British diplomatic dispatches. The British consul in Hamburg noted that it was fortunate for Britain that Austria and not Prussia had brought the aid, since there would then be no pressure on Hamburg to join the Zollverein.[22] From Berlin on 29 December came a dispatch with a translation of Baron Manteuffel's statement to Hamburg that explained why Berlin had been unable to help. Lack of ambition gave way to a series of lame explanations that underlined that Berlin had 'missed an opportunity'.[23]

The distress in Scandinavia was relieved as the panic in Hamburg subsided. A positive item of international official aid, however, was a loan on 18 December by the Bank of England on promissory notes of the Norwegian government in support of overdue bills the bank held from Norwegian houses.[24]

At the outbreak of the Civil War in November 1861, panic in New York drew specie from Paris and London. French reluctance to raise the discount rate and the increasing departure of the French gold-silver ratio from the ratio in the market led France to exchange £2 million in silver for £2 million in gold with the Bank of England. The funds made available by this transaction failed to remedy the situation, so in 1861 the French bought gold in London at a price above the export point. The Bank of France then arranged through Rothschild and Baring to draw £2 million of bills on London.[25]

There was no international cooperation during the crisis of 1873 but there were two aspects that underline the sensitive political nature of central bank transactions. In the letter books for 1872–73 the Bank of England refers to and denies a 'ridiculous rumor' that it had thought of applying for a loan from the Bank of France. And in the second week of November the governor of the Bank of Prussia – a predecessor of the Reichsbank, which was not established until 1875 – wrote a letter to the Bank of England offering a loan in gold now or at any future time. (Clapham commented earlier that Germany was half drunk with victory and that Berlin had swelled up like Aesop's frog.) The Bank of England politely but curtly declined the offer: 'The Bank is not, nor has it been in want of such aid and need not avail itself of the arrangement you so kindly suggest.' Clapham adds that this suggestion from the nouveau riche would have seemed impertinent to the governor.[26]

In 1890 William Lidderdale, then the Governor of the Bank of England, prepared on two fronts for the crisis that would follow the revelation of Barings' weakened position. In addition to the domestic guarantee, he arranged for the Russian government not to draw its £2.4 million deposit from Barings, and for loans of £3 million from the Bank of France and £1.5 million from the State Bank of Russia, both in gold. Lidderdale told the governor of the Bank of France that the ordinary operations of bank rate would have brought the gold in time and that there was nothing discreditable in using exceptional measures to meet an unusually sudden storm. Nonetheless, Lidderdale and the City were uneasy about asking the French and the Russians for help. Clapham put it this way: 'Suppose for some political-financial reason they had been unwilling to oblige?'[27]

The sensitive character of such international assistance prior to World War I is revealed in the 1906–07 incidents, when much discussion was devoted not to whether the Bank of France helped the Bank of England – which it clearly did – but whether the Bank of England had asked for such help or, if it did not, whether the steps taken by the Bank of France were largely in its own interest. In Sayers' account of the Bank from 1890 to 1914, in a chapter titled 'Supposed Continental Support of the Bank', he concluded that the Bank of England did not ask for help. Sayers quotes *The Economist* in September 1906:

> Some talk of the Bank of France helping the Bank of England cope with the American demand for gold … but it would not for a minute be supposed that the Bank would really put itself in so humiliating a position merely in order to permit American speculators getting gold here on easy terms.

Again in the autumn of 1907, the Bank of France forwarded 80 million francs in US gold eagles to London. The Bank of France report for 1907 refers to both incidents in its decisions. French journals cite the alleged reason: to relieve the

Bank of England of the need to raise its discount rate, which implies that the Bank of England requested help. British sources emphasize that there was no announcement on the British side, as there was in 1890, and that the French did not want the Bank to increase its discount rate. Hartley Withers, financial editor of *The Times*, wrote later:

> The determination it [the Bank of England] showed finally compelled the Bank of France to take some share in the international burden, and to send three millions of its gold, not to America but to London, whence it knew it could rely on getting it back. It is commonly supposed the Bank of England asked it to carry out this operation, but this is quite untrue.[28]

Another view is that when the Bank of France entered into transactions in sterling bills in London in 1906, 1907, 1909, and 1910, it was involved in open market operations, which it did not undertake in Paris until 1938.[29]

London vs Paris as the world financial center

The general view among Anglo-Saxon students of economic history is that London was the world's financial center from the beginning of the nineteenth century until 1914, and that Paris, Berlin, Frankfurt, New York, and Milan were satellites.

According to a German observer, 'England had a monopoly of capital exports to 1850. Then France moved in, largely for *gloire*, undertaking capital exports in the service of national policies, expansionary commercial interests, and the opening of new markets.'[30] In the 1850s Paris had a central role in international monetary relationships. Van Vleck wrote about the panic of 1857: 'Just as France was the political nerve center of Europe during the first half of the nineteenth century, so during the years from 1850 to 1857, it was the center from which fluctuations in the economic cycle radiated.'[31]

The center for the economic cycle is not necessarily the pivot of the international financial system. Between 1820 and 1840 Paris assisted in clearing London's foreign payments in the Baltic, Russia, China, Latin America, and the United States and then between 1850 and 1870 became the 'first place in Europe for foreign exchange'.[32] If this view were correct, the situation was changed by the Franco-Prussian War. According to Bagehot:

> Since the Franco-German war, we may be said to keep the European reserves... All great communities have at times to pay large sums in cash, and of that cash a great store must be kept somewhere. Formerly there were two such stores in Europe; one was the Bank of France, and the other the Bank of England. But since the suspension of specie payments by the Bank of

France, its use as a reservoir of specie is at an end. No one can draw a cheque on it and be sure of getting gold or silver for that cheque. Accordingly the whole liability for such international payments in cash is thrown on the Bank of England... all exchange operations are centering more and more in London. Formerly for many purposes, Paris was a European settling-house, but now it has ceased to be so... Accordingly London has become the sole great settling-house in Europe, instead of being formerly one of two. And this pre-eminence London will probably maintain, for it is a natural pre-eminence... The pre-eminence of Paris partly arose from a distribution of political power, which is already disturbed; but that of London depends on the regular course of commerce, which is singularly stable and hard to change.[33]

Sprague echoed this view from the United States in 1910, when he explained that the Bank of England raised its discount rate in 1907 to secure payment from other countries of money they owed the United States rather than to check the flow of gold to New York. London was the central money market in the world and its interest rates were increased to avoid having to finance all the payments flowing to the United States.[34]

In contrast, Bonelli asserted that Paris was the center that regulated world liquidity when he discussed Italy's links to the 1907 crisis.[35] One of the most penetrating students of pre-World War I finance stated that 'Paris emerges in this study as the *strongest* [italics in original] financial center in the world before 1914, if the fact that its short-term rate was relatively the lowest is an indication of strength. This conclusion seems to contradict the generally held opinion that London was the world's money center.' Morgenstern attempted to reconcile these statements by distinguishing the abundance of funds in Paris and the 'machinery' in London for setting funds into motion.[36] The distinction seems forced. Each center had its own particular clients: Italy and Russia for Paris, the United States and the British colonies for London. Belgium, the Netherlands, and Germany were more closely linked to London than to Paris. Moreover London lent funds on a worldwide basis while Paris lent funds to a smaller number of countries.

The lender of last resort after World War I

There was no international lender of last resort in the 1920–21 crisis. Part of the slack was taken up by the depreciation of the various European currencies. In a crisis caused by balance-of-payments weakness and money outflows, the depreciation of a country's currency led to increases in the domestic prices of tradable goods (those goods and services that are exported and imported, and those capable of being exported and those that compete with imports). A

sharp inflationary shock may produce another kind of crisis – out of money into goods – leading to hyperinflation, as happened in Eastern and Central Europe in 1923 in the newly independent countries that had been part of the Austro-Hungarian Empire and were coping with new boundaries and the lack of mechanisms for the collection of taxes.[37] The scenario is similar to that of Paul Erdman's thriller *The Crash of '79*, in which the financial crisis is handled by printing massive amounts of currency and letting the US dollar float, which led to a surge in the demand for goods.[38]

Most governments in Western Europe in the early 1920s sought to stabilize their currencies at their 1914 parities for gold because of tradition – restoring the status quo *ante bellum*. These countries often used stabilization loans, similar in some respects but not identical to loans from a lender of last resort, to dampen the depreciation of their currencies. The French franc was subject to a speculative attack in 1924. Many foreigners bought francs as the currency depreciated in 1919–20 in anticipation of large revaluation gains from the subsequent appreciation toward its pre-war parity – but they eventually gave up and sold.[39] Speculators in Amsterdam, Vienna, and Berlin (perhaps stimulated by the German government) sold francs in anticipation that they would be able to purchase them later at much lower prices.[40] The story was that speculators who had profited from their short positions in the German mark as it depreciated sharply in the hyperinflation of 1923–24 then turned their attention to the French franc. Hundreds of thousands of Frenchmen with liquid securities denominated in the franc watched signals like the advances from the Bank of France to the French government approach the legal ceiling.

On 4 March 1924 panic broke out. The franc, which had been 98 to the pound on 17 February and 104 on 28 February, went to 107 on 4 March. The French government and the Bank of France met in emergency sessions. J.P. Morgan & Co. was willing to help if certain conditions were satisfied. The loan, a six-month revolving credit, would be for $100 million, not $50 million, which the bank, represented by Thomas W. Lamont, judged too small. The Bank of France resisted pledging its gold as collateral, and yielded only after a face-saving formula was found. In turn, the Bank regents, including Rothschild and de Wendel, exacted a conservative financial program from the government. The program was worked out on Sunday 9 March and within three days the rout of the speculators began. The franc appreciated from 123 francs to the pound to 78 by 24 March. The Bank of France then intervened to limit further appreciation. The squeeze against the speculators was successful.[41] Lamont wrote that 'there has never been an operation that has given us more satisfaction'.[42]

Success was short-lived. In 1926 there was another attack on the franc, and the currency depreciated to 240 francs to the pound in July 1926. Subsequently the franc appreciated to 125 in response to conservative Poincaré reforms and

a surge in interest rates that was designed to reassure the French holders of wealth and induce them to bring their money back to Paris.

Stabilization loans were arranged in the 1920s for Austria and Hungary under League of Nations auspices, and for a number of new central banks in Eastern Europe under arrangements with lenders in London, Paris, and New York. The most widely known were the Dawes and Young loans undertaken to recycle German reparations. The Dawes loan stimulated US purchases of foreign bonds. During the 1930s the suggestion was frequently put forward, with only partial irony, that what France needed was a stabilization loan in gold, since the French franc and the currencies of the other gold bloc countries had become overvalued as a consequence of successive devaluations of the British pound, the Japanese yen, and the US dollar, and the imposition of exchange controls on foreign payments by Germany and Austria. The gold, it was said, should be mounted in a vehicle like a glass hearse and paraded through the streets of every town and village in France to convince the people that the authorities had an abundance of gold and thereby encourage them to dishoard from the *bas de laine* (wool sock) in which they kept their *louis d'or*.

The rolling deflation of 1931 underlines the need for an international lender of last resort in a way that differs in scale and therefore in kind from previous episodes. The issues involved included the need for loans of the appropriate magnitude, the political character of the transactions, and the need for some country or group of countries to accept responsibility for the stability of the system.[43] Those who wrote along similar lines included Jørgen Pedersen and R.G. Hawtrey and other British economists whose advice to their government is set out in Susan Howson and Donald Winch's *The Economic Advisory Council, 1930–1939*.

Hawtrey presented a cogent analysis:

> The crisis of 1931 differed from earlier crises in its international character. Earlier crises were international in that the fall of prices and forced sales affected world markets. But only an unimportant part of debts were due to foreign creditors. In 1931 the outstanding characteristic of the panic was that foreign creditors of Germany and Eastern European debtors feared that the foreign exchange market would break down *even if the debtors remained solvent*. Against a panic-stricken withdrawal of foreign balances from London, it [raising Bank rate] was too tardy a remedy. Such a panic-stricken withdrawal *had never occurred before*... the underlying cause of the trouble has been *monetary instability*. The industrial depression, the insolvencies, the bank failures, budget deficits and defaults, are all the natural outcome of a falling price level... The need arises for an *international lender of last resort*. Perhaps some day the Bank for International Settlements... But as things are,

the function can only be undertaken by a foreign central bank or by a group of foreign central banks in co-operation.[44]

Hawtrey's theoretical insight into the wisdom of limits on these stabilization loans was penetrating:

> As a general rule, if credits are to be granted to a central bank in difficulties at all, they should be granted up to the full amount needed. There should be no limit. If the amount is inadequate and the exchange gives way after all, the sums lent are completely wasted... It can be argued in favor of unlimited credits, that if they *had* been granted, there would have been no withdrawal of funds at all... either no credits to the Bank of England or unlimited. But there is some risk. Unlimited credits would have enabled the country to remain on the gold standard, prolonging conditions that were rapidly becoming intolerable... the lesson: if the country can maintain the monetary standard without undue strain, then grant unlimited credits; but if the effort of maintaining parity is excessive, no credits and allow the currency to depreciate.[45]

Howson and Winch set out a series of reports written by British economists to their government in the 1930s, including one from a Committee of Economists chaired by J.M. Keynes that included Hubert Henderson, Alfred Pigou, Lionel Robbins, and Sir Josiah Stamp as members, with Hemming and Richard Kahn as secretaries. The reports advocated cooperation among central banks, preferably through the Bank for International Settlements, both to fund short-term claims and to form a pool of currencies for loans that could be used to prevent currency debacles like those experienced in the early 1920s and to restore 'confidence in the financial stability of those many countries which are now the subject of distrust'.[46]

In July 1932, after Britain had gone off gold but before the World Economic Conference of 1933, the Cabinet Committee on Economic Information chaired by Stamp and including Citrine, Cole, Keynes, Sir Alfred Lewis, and Sir Frederick Leith-Ross as members, with Henderson and Hemming as secretaries, issued a report that discussed the 'international financial crisis'. The document quoted Bagehot, cited the crises of 1825 and 1847, and noted that Britain could no longer act as a lender of last resort; the report recommended that this function be performed by the Bank for International Settlements by the issue of 'paper gold', to be called International Certificates.[47]

The first opportunity to halt the international disintermediation came in May 1931 with the collapse of the Credit Anstalt, the leading Austrian bank. The Austrian central bank had maintained high interest rates to keep money in Vienna which in turn contributed to the softness in the economy and large

losses as asset prices declined. The publication of the Credit Anstalt's statement on 11 May revealed that it had lost 140 million schillings, about three-fourths of its capital. The Austrian government asked the League of Nations for financial assistance because the League had organized the stabilization loans in the 1920s and the League turned to the new Bank for International Settlements that had been established under the Young Plan of 1930 to assist in the transfer of German reparations. The Austrian government wanted to borrow 150 million schillings ($21 million). During the last two weeks of May the BIS arranged for a loan of 100 million schillings from eleven central banks. By 5 June the money from the loan were exhausted, and the Austrian National Bank requested another loan which was arranged by 14 June, subject to the condition that the Austrian government get a two- to three-year loan of 150 million schillings. The French demanded that the Austrian government renounce its customs union agreement with Germany that had been announced two months earlier, but the Austrian government refused and fell. The Bank of England then offered a loan of 50 million schillings ($7 million) for a week. The Austrian government then stopped pegging the schilling to gold, and the currency depreciated.

The run shifted to Germany. The German banking position was weakened by excessive speculation, large loan write-offs, fraud, quarrels among the bankers, banks that had been buying their own shares and depleted liquid reserves – the full range of classic troubles.[48] The outsider was Jacob Goldschmidt of the Danatbank,[49] the product of the merger of the Darmstädter and the National banks. Other bankers, including Oskar Wasserman of the Deutsche Bank, detested Goldschmidt and his aggressive tactics. In 1927 the Berlin Handelsgesellschaft had stopped making loans to the Norddeutsche-Wolkämmerei (Nordwolle), an aggressive firm in the woolen industry. The Danatbank took on Nordwolle as a client. Nordwolle's failure, on 17 June 1931, brought down Danatbank; other bankers were unwilling to support the bank because of their dislike of Goldschmidt. There were political complications, and the internal financial turmoil led to massive withdrawals that were only briefly interrupted by the Hoover moratorium. On 25 June a loan for $100 million was arranged, including $25 million each from the Bank of England, the Bank of France, the Federal Reserve Bank of New York, and the Bank for International Settlements, for the period to 16 July. Hans Luther, president of the Reichsbank, had wanted a larger loan and had asked that the exact amount of the loan be concealed, so the communiqué said only that discount facilities had been arranged in sufficient amount. When the amount of the loan became known and the Reichsbank's statement of 23 June showed that its reserve cover was at 40.4 percent, just above the minimum requirement of 40 percent, the motto became 'Devil take the hindmost'.[50]

New loans were discussed but were not forthcoming. The Germans wanted to borrow $1 billion. The French were willing to consider a loan of $500

million, but they wanted to impose political conditions. The US government was worried about its prospective budget deficit of $1.6 billion and thought it highly unlikely that the US Congress would approve lending more money to Germany. The US government was willing to consider stabilization of existing loans to Germany, which the Reichsbank wanted as well as a further loan. In Britain the foreign secretary Arthur Henderson was attracted to the idea of a loan but Montagu Norman, then the governor of the Bank of England, held that the Bank had 'already lent quite as much as is entirely convenient'.[51] One argument against foreign loans was that the crisis was believed to have been caused by a flight of domestic capital rather than by withdrawals of funds by foreign investors. By 20 July the idea of a loan had been tacitly abandoned, 'brushed aside as impractical'.[52] Instead, the Germans relied on internal measures to halt the disintermediation at home and on a Standstill Agreement, imposed upon reluctant foreign bankers, to halt the external drain.

The speculative pressure then turned to Britain. The run began in mid-July 1931, stimulated partly by losses on the Continent, but also fed by the May and Macmillan reports of the large prospective domestic budget deficits and the unexpectedly high estimate of foreign money in London that might be withdrawn. Since 1927 the Bank of France had converted British pounds to gold by a roundabout device; British pounds were sold in the spot foreign exchange market, and bought in the forward foreign exchange market, and then the Bank of France asked for gold on the dates when the forward exchange contracts matured.[53] The idea was to make it seem that the Bank of France was converting only its newly acquired British pound balances into gold, not its previously acquired balances. In the summer of 1931, the Bank of France cooperated fully and did not sell any of its British pound deposits and securities. At the end of July, the Federal Reserve Bank of New York and the Bank of France each loaned $125 million to the Bank of England. When these funds were depleted, the British government contemplated a one-year loan from the New York and Paris markets. The Bank of England reported that foreign bankers would not lend funds to Britain while it had such a large budget deficit. The British trade unions opposed a reduction in relief payments to the unemployed and withdrew support from the Labour government, which fell on 24 August. Four days later, after the formation of a new 'national government' with Ramsay MacDonald again prime minister and Philip Snowden as Chancellor of the Exchequer, $200 million was borrowed from a Morgan syndicate in New York and another $200 million from a French syndicate in Paris. On one showing, the bankers held up the British government; their own explanation, echoing the Morgan statements to the French in the 1920s, was that they were not imposing political conditions but instead indicating the economic circumstances in which they felt they were justified in risking their own and their depositors' money.

On 5 August 1931, Keynes wrote to Prime Minister MacDonald, at the latter's request, to present a series of proposals for devaluation of the British pound and for the formation of a gold-based, fixed-exchange currency unit at least 25 percent below the current parity, which all member countries of the British empire, together with South America, Asia, Central Europe, Italy, and Spain – in fact, all countries – would be invited to join. The letter pointed out that if the British pound could not be successfully defended, it would be foolish to continue to borrow foreign currencies to support it.[54]

Loans of $400 million in addition to the $250 million were not enough, and the British stopped supporting the pound on 21 September. The Bank of France did not show the restraint towards the United States that they had shown to the British, and together with other members of the gold bloc converted $750 million of US dollar deposits into gold. The deflationary pressure exerted by this reduction in US gold holdings, and by the depreciation of the British pound and of the currencies pegged to the pound, weakened the US banking system. The New York Fed did not ask for help or even for forbearance in the conversion of US dollar deposits into gold. The code of the central banker calls for a stiff upper lip, reminiscent of Walter Mitty refusing the blindfold before the firing squad. In 1929, when Harrison asked Norman whether the pounds that the Federal Reserve Bank of New York had bought would be convertible into gold, he received the curt reply: 'Of course, the sterling is repayable in gold. This is the gold standard.'[55] In 1931 Harrison in turn offered to assist Moreau in converting any or all of the Bank of France's dollars into gold.[56]

Five aspects of the 1931 story are especially striking: (1) the inability of Britain to act as a lender of last resort; (2) the unwillingness of the United States to act as a lender of last resort apart from the (inadequate) loan to Britain (the country of the 'special relationship'); (3) the desire of France to advance its political objectives with respect to Austria and Germany; (4) the paranoia of Germany after 1923, preferring anything to a hint of inflation; and (5) the irresponsibility of the smaller countries.

This analysis has been questioned in a number of quarters. One analyst thought that something more far-reaching was required to restore stability to the world economy after World War I, perhaps something on the order of the Marshall Plan after World War II.[57] Another considered that the German economy could not have been righted by a lender of last resort in 1931 since the authorities were determined to undo the Versailles Treaty and especially its reparations clauses.[58]

Bretton Woods and the international monetary arrangements

There was an extended debate in the early 1940s about the design of multilateral economic institutions that would provide for greater economic stability

than in the twenty years after World War I. An international credit institution would be established to help countries finance short-term balance of payments deficits, which became the International Monetary Fund. Another lending agency would be established to help countries finance their economic rebuilding, which became the International Bank for Reconstruction and Development (IBRD or World Bank). An international trade organization would be established to resolve trade disputes and provide a forum for the reduction of tariffs and other trade barriers. The advent of the planned international trade organization was long delayed, but its preamble, the General Agreement on Tariffs and Trade (GATT), provided the basis for an organization to deal with trade policy issues in a less ambitious way. Eventually, more than fifty years later, the GATT morphed into the World Trade Organization (WTO).

The debate about the international credit institution that became the International Monetary Fund (IMF) was primarily between the British and the Americans, who held different views about both its financial structure and its financial resources. The 'Keynes plan' provided for an institution that would have its own money or unit of account. Member countries would be endowed with deposits in this institution which they could transfer to other countries to finance their payments deficits. The American view, the 'White plan', was that each member country would transfer gold and its own currency to the institution to endow its capital. Each member country would have a quota based on the volume of its trade and its gold holdings; each member's quota would determine its capital subscription and the amounts of gold and of its own currency that it would transfer to the institution as part of this subscription. A member country that wanted to finance a balance of payments deficit would 'buy' one of the currencies owned by the institution with its own currency; the amount of currencies that the member could buy would depend on its quota.

The Americans had all the money, and the US view prevailed. The British got the Americans to agree to double the initial capital of the institution.

The Fund's Articles of Agreement contained a set of rules for the management of the values of member country currencies. Each member country would be required to state a parity for its currency either in terms of gold or in terms of the US dollar; each would be required to limit the variation in the value of its currency to a narrow band around its parity; and each would be required to obtain the approval of the IMF before changing the value of its parity by more than 10 percent relative to its initial parity.

Despite the motivation to establish a credit institution that would help avoid a repetition of the experiences of the 1930s, the structure of the Fund meant it could not act as a lender of last resort because it did not have its own money. Instead the IMF could lend money to countries to help them finance their current account deficits. Countries could retain exchange controls to limit capital flows that might complicate their ability to retain their parities. Loans from

the IMF would have to be repaid. There were narrow limits on the amount of loans to each country set by country quotas; the quota was divided into four tranches and no more than one tranche could be drawn on in a given twelve-month period. Drawing on the first tranche was more or less automatic. Thereafter access to credit was a matter of grace on the part of the IMF rather than a matter of right as Keynes's plan had proposed.

During the first ten to fifteen years after the end of World War II, the IMF and the World Bank were on the sidelines because most of the finance for economic reconstruction was provided under the Marshall Plan, and primarily in the form of grant aid. The Bretton Woods system did not come into full operation until 1958, after the British pound was convertible into other currencies on capital account, and restrictions on money flows were abandoned.

The inattention to financial flows soon had to be modified. It proved impossible to maintain convertibility on current accounts and controls over capital movements, since large money transfers could take place through changes in the 'leads and lags', changes in the timing of payments associated with exports and imports. A country called upon to pay cash for imports instead of getting three months' credit, and forced to extend six months' credit for exports instead of three months, could quickly lose international reserve assets that would be equal to the value of six months of the average of its exports and imports.

In 1960 the General Arrangements to Borrow (GAB) was adopted to increase the funds available to the IMF; ten large financial economies, the Group of Ten, pledged a total of $6 billion that the Fund could use when its own resources proved too small to finance the payments deficits of its members. Many countries believed the Fund resources would still be too small. Moreover, the IMF's decision-making was ponderous and time-consuming. Decisions were taken by weighted voting by directors, many of whom represented more than one country. To frame a proposal, obtain instructions, and arrive at a decision to help a country in crisis could take three weeks.

Cross-border money flows increased in the 1950s and the 1960s. Economic distance was declining because of technological developments and the sharp declines in the costs of communication and storing information. Richard Cooper emphasized this development:

A crude quantitative indicator of these developments is provided by contrasting the maximum daily speculation of under $100 million against the pound sterling, in the 'massive' run of August 1947, with the maximum daily speculation of over $1.5 billion in favor of the German mark in May 1969, and the movement of over $1 billion into Germany in less than an hour in May 1971. Moreover, as the barriers of ignorance fall further, there

is no reason why $1.5 billion should not rise to $15 billion, or even $50 billion a day.[59]

The daily turnover in the currency market had reached $1 trillion by the late 1990s. The increase in the transactions in the currencies of the emerging market economies had been even more rapid.[60]

The increase in international money flows challenged the ability of central banks to retain their parities. Some central banks experimented by selling their currencies in the forward foreign exchange market, which enabled them to maintain their parities without depleting their reserves – at least not until the forward exchange contracts matured. And on those dates the central banks might 'roll over' or renew these contracts. A few observers believed that these forward exchange transactions would relieve central banks of the need to hold international reserve assets or at least reduce the amount of reserves they would hold. The experience of the Bank of England from 1964 to 1967 suggests the limits to this view; by late 1967 its commitments to deliver US dollars in the spot foreign exchange market on the dates that the forward exchange contracts matured were several times larger than its holdings of international reserve assets, and market participants became increasingly reluctant to renew maturing forward exchange contracts. In 1964 the British had been able to postpone the devaluation of the British pound, but the devaluation became inescapable in 1967.

One of the major international financial innovations in the 1960s was the Basel Agreement, which established a network of currency swaps among central banks. The United States had taken a leadership role in the development of this network, which was a series of bilateral credit lines between pairs of central banks – each central bank wrote up an amount of foreign currency as an asset and an equivalent amount of its own currency as a liability. Once the swap lines had been established the foreign currency would be immediately available, although the money drawn under the swap lines would have to be repaid. The first swap involved $50 million between the Bank of France and the Federal Reserve Bank of New York in March 1962. In June the New York Fed entered into $50 million swaps with both the Dutch and Belgian central banks and a $250 million swap with the Bank of Canada. In July the New York Fed established a $250 million line with the Swiss National Bank. By October 1963 the swap network had been increased to $2 billion, by March 1968 to $4.5 billion, and by July 1973 to $18 billion.[61]

In 1961, when the British pound was under attack, representatives of the major central banks at a meeting of the Bank for International Settlements committed a total of $1 billion for credits. Charles Coombs, a senior vice-president of the Federal Reserve Bank of New York, called this development a major breakthrough in international postwar finance.[62] The number was large

enough to convince speculators that they could not break the parity of a currency. Most of the loan would be repaid with the funds obtained from the repatriation of flight capital.

The money available under the swap lines would be a country's first line of defense; the second line would be access to money from the IMF. A number of countries used their swap lines: Canada used more than $1 billion in June 1962; Italy $1 billion in March 1963; and Britain $2 billion in the autumn of 1964. A $1.3 billion package was available for France in July 1968, which would have been extended to $2 billion in November for the defense of the devalued franc.[63] The French decision not to help the British in September 1965 has been characterized as a 'shocking repudiation of the central-banking free masonry'. The pressure to conform to the club, and the will to be different as a matter of foreign policy, were doubtless both intense. 'It cut no ice. The British got the support they wanted.'[64]

The approximate cause of the breakdown of the Bretton Woods arrangements were the efforts by the Federal Reserve and the Bundesbank to follow divergent monetary policies, even though their money markets were closely linked through the offshore deposit market. Any effort to follow divergent monetary policies would lead to massive money flows. The Federal Reserve embarked on a policy of monetary expansion six to eight months before the presidential election of 1972, while the Bundesbank maintained high interest rates. There was a massive shift of money from the dollar area to the mark area.

The decline in interest rates on US dollar securities led to the relaxation of lending standards, as in 1825, 1853, 1871, and 1885. Banks headquartered in New York, London, Tokyo, and other financial centers began to lend freely to Mexico, Brazil, Argentina, and other developing countries as well as to the Soviet Union and the countries in the Eastern bloc.

Further, the large flow of funds from New York to Germany and elsewhere made it impossible to sustain the pegged currency arrangement. The United States made no effort to defend the new parity for the US dollar that had been set in the Smithsonian Agreement. By February 1973 speculation that the mark would be revalued led to increasingly large money flow to Frankfurt; the Bundesbank stopped buying the US dollar, and the mark immediately appreciated.

Most economists had thought that the adoption of floating exchange rates would severely dampen the movement of interest-sensitive money and that once currencies were floating central banks would be able to follow independent monetary policies without any untoward external effects. Economists differed about whether speculative money movements would be stabilizing or, occasionally, destabilizing in a serious way; the general view was that fear of currency losses would deter cross-border money flows. That view proved mistaken; many banks regarded currencies as a new asset class that could be profitably traded.

Under these new circumstances, the initial need for a lender of last resort proved to be domestic, though related to international finance – at least for a time. The changes in currency values encouraged speculators, including some who worked for banks. The Herstatt Bank of Cologne and the Franklin National Bank of New York lost money trading currencies and each failed in June 1974. The closing of the Herstatt in the middle of the trading day created a new problem because Herstatt had collected the money due to it on currency transactions but was closed before it had paid out the German mark counterpart that was due the other banks. For a time, the consortium that guaranteed its liabilities was interested only in domestic German obligations and was prepared to let the sums due to foreign banks go unrequited. Second thoughts prevailed, however: the total liabilities of the bank were covered, foreign as well as domestic, and there was no shockwave felt abroad. The Federal Deposit Insurance Corporation took over Franklin National deposits up to the limit of $40,000, and the Federal Reserve System, acting as lender of last resort, guaranteed the remaining liabilities.

The arrangements worked out at the Bank for International Settlements in the so-called Basel Protocol of March 1975 were supposed to settle the issue of national responsibility in the case of bank failure. A new problem developed in 1982 when the Luxembourg subsidiary of the Banco Ambrosiano of Milan defaulted on $400 million of liabilities to other European banks. The Bank of Italy refused to make good on these liabilities on the legal ground that the Luxembourg unit of Ambrosiano was a subsidiary of the Milan bank which was incorporated under Luxembourg law and so its assets and liabilities were distinct from those of the bank's head office. This placed the Basel Protocol under a cloud.

The sharp increases in the price of oil in 1973 and again in 1979 led to a surge in the export earnings of the oil-producing countries and in their holdings of international reserve assets. The export earnings of the developing countries as a group increased sharply, and their growth rates trended up. Bank loans to the developing countries surged.

The combination of the increase in oil production in the North Sea and Mexico and elsewhere and the decline in demand due to the global recession in the early 1980s led to reductions in oil prices. By mid-1982 Mexico and many of the smaller oil-producing countries were in financial difficulties.

Credits under the swap agreements were limited to the major industrial countries and were not available to the developing countries which depended on credits from the IMF when they incurred financial crises. But since many of these countries had a strong aversion to the conditions demanded by the IMF, they attempted to reschedule debt with the lending banks. It was a bit like Catch-22; often these banks insisted that the developing countries obtain a seal of approval from the IMF before they would re-finance maturing loans.

When financial conditions became acute, a 'bridge' loan might help the borrower during the period of negotiation.

The Federal Reserve Bank of New York extended a $1 billion bridge loan to Mexico in 1982, and the US Treasury bought a billion dollars' worth of oil with current payments for future delivery to the US Strategic Petroleum Reserve. These ad hoc loans enabled Mexico to muddle through and avoid default. The debtors liked this arrangement because otherwise they would have been cut off from international capital markets for a generation (if indications from history were any guide) and because they wanted more foreign capital to enhance their rates of economic growth. The creditors liked these arrangements because otherwise defaults by the borrowers would force them to write off the loans and recognize large loan losses. Whereas default in earlier periods was less of a concern to the credit system of developed countries, since it affected bonds held by individuals and not loans extended by banks, default on Third World debt in the 1980s might led to runs on the major international banks because of concern about the adequacy of their capital.

Conditionality

The populist view is that the IMF has been excessively restrictive about the choice of monetary policies and fiscal policies by the emerging-market countries. Most lenders, both domestic and international, stipulate conditions on their loans. Some French analysts were irked by the requirements laid down by J.P. Morgan & Co. in its 1924 stabilization loan,[65] although the standard response is that the lenders have an obligation to their depositors to ensure that their loans can be repaid. French conditions for loans to Austria and to Germany in 1931 were political. American and French loans to Britain later that summer were regarded by the Labour Party as a 'bankers' ramp' (in American, 'racket') as the lenders thought that the recommendations of the British May Committee to balance the budget and reduce unemployment benefits should be carried out.

Conditions have not been attached to the credits extended under swaps although there are 'understandings'. When the Bank for International Settlements makes a loan to Hungary under a swap, it knows that Hungary will obtain the funds to repay the loan from access to credit at the IMF. Conditions will be attached to that loan. When a G-7 member borrows from the IMF, as Britain did in 1976, conditions also apply.[66]

The Mexican crisis of 1994–95

The Mexican financial crisis of 1994 began with a peasant revolt in January and the assassination of the leading presidential candidate of the dominant

political party. The inflow of foreign money slowed dramatically, and the ability of the Bank of Mexico to support the peso in the currency market to finance the large current account deficit rapidly became exhausted.[67] In April 1994, the United States and Canada came to the rescue, largely because of the 'special relationship' involved in the North American Free Trade Agreement Act of November 1993. A credit line of $6.7 billion was put together – the United States provided $6 billion, Canada $700 million. Mexican troubles continued. There was another attack on the peso in December 1994, partly the flight of Mexican money and partly the withdrawal of money by disenchanted American and other foreign investors. In January 1995, the US Treasury organized a $50 billion rescue fund including $20 billion from the US Exchange Stabilization Fund, $18 billion from the IMF, $10 billion from European central banks organized by the BIS, and $2 billion from Canada.[68] The amount proved persuasive. The capital outflows stopped and money began to flow back to Mexico. Only $12.5 billion of the US credit line was drawn, and in the fall of 1995 repayment started with the help of money loans placed privately.

Several questions remain about the Mexican rescue operation in 1994–95. One is whether the loans made in the early 1980s led the Mexicans to believe that they would be helped if they again encountered difficulties – the moral hazard question. Another is whether the large amounts of the credits created a precedent that would lead to future problems. Moreover, did the financial authorities forestall contagion and prevent the crisis from spreading to Brazil and Argentina, as contagion had spread from Austria in May 1931?

Bagehot enunciated the prescription that a lender of last resort should lend freely because low limits excite. Loans on a scale that seems beyond any possible need can be seen as lending freely.

The East Asian crisis of 1997

The financial crisis in East Asia that started in July 1997 involved euphoric lenders in developed countries keen on diversifying their portfolios, and developing-country borrowers that wanted to increase their investments and their growth rates and that had been pushed by industrial countries to deregulate their financial markets. Other factors included crony capitalism in Indonesia, weak government in Thailand, enormous conglomerates (*chaebol*) in South Korea, and many bad bank loans. The catatonic state of economic policy in Japan robbed the area of what had been a strong source of demand, and the expansion of Japanese direct investment into lower-wage areas and of Japanese bank loans contributed to the increases in the current account deficits in these countries. European banks also increased their loans. The currencies of most

Table 12.1 Official finance commitments (last-resort lending) ($US billions)

Country	IMF	World Bank	Asian Development Bank	Bilateral Commitments	Total
Thailand	3.9	1.9	2.2	12.1	20.1
Indonesia	10.1	4.5	3.5	22.0[a]	30.0
South Korea	21.0	10.0	4.0	22.0	57.0
Total	35.0	16.4	9.7	56.1	117.1
Memo item					
Mexico	17.8	1.5	1.3[b]	21.0	51.6

[a] Including the use of a $5 billion Indonesian contingency reserve.
[b] InterAmerican Development Bank.

of the countries had become overvalued in response to these money inflows – and countries with large current account deficits and overvalued currencies are vulnerable to any shock that leads to a decline in money inflows.

The devaluation of the Thai baht in early July 1997 triggered the contagion effect, and the flow of money to other Asian countries slowed and then was reversed. The President of Malaysia, Dr Mahathir, blamed foreign speculators and specifically George Soros in the United States, who in turn asserted that he had not sold the ringgit short while Malaysia was constructing two new buildings taller than those anywhere else in the world. Malaysia did not seek loans from the IMF but instead imposed controls on outflows of money and interest payments to foreign creditors. Thailand, Indonesia, and South Korea borrowed from the IMF. The amounts, on top of that for Mexico in 1994–95, are set forth in Table 12.1 from the Bank for International Settlements.[69]

The report of the Bank for International Settlements noted that the $50 billion credit lines for Mexico had a 'psychological impact on the markets' and halted the erosion of domestic confidence.[70] These credit commitments also depleted the funds available for rescue operations in the future. At the end of 1998, the US Congress agreed to the proposal of the Clinton Administration to increase IMF quotas and capital. The IMF was thus able to help Brazil when it needed assistance.

The IMF is not a central bank and hence it cannot create money; instead it can lend the money it has obtained from the capital subscriptions of its members and its own borrowing.

A world central bank would be a more efficient lender of last resort than the IMF, but an improbable one. Outside the European Monetary Union, most large countries regard the issue and control of money as a mark of sovereignty; in the United States, these functions are enshrined in the Constitution.

The United States and the dollar

The United States took the lead in the 1940s, the 1950s, and the 1960s in establishing new international financial arrangements – the IMF and the World Bank, the General Arrangements to Borrow, the Special Drawing Rights, the swap network, and the gold pool. Moreover the United States took the initiative at the end of the 1980s in assisting developing countries to write down the value of their loans to major international banks so they might again be creditworthy. And the United States took a leadership role when individual countries like Mexico and Russia and South Korea had major international payments problems.

One of the major surprises at the beginning of the 1980s was the development of a persistent US trade deficit, after one hundred years of trade surpluses. The persistent US trade deficit that began about 1980 led to a dramatic change in the US international financial position. In 1980, the United States was the world's largest creditor country; its net creditor position was larger than the combined positions of all other net creditor countries. By 2000 the United States had become the world's largest debtor, and its net debtor position was larger than the positions of all other net debtors as a group. The US net debtor position has continued to increase.

There is an analogy between the persistent US payments deficit in the 1950s and the 1960s and the persistent US trade deficit in the 1980s and the 1990s and subsequently. The US payments deficit developed because the demand for international reserve assets on the part of other countries was larger than the increases in the supply of new reserve assets from non-US sources. The United States developed a persistent trade deficit because of the demand in other countries for US dollar securities and US real assets.

At the end of the 1960s and throughout the early 1970s, the willingness of foreign official institutions to acquire reserve assets declined, which might reflect, if present indications are correct, the crises of the last two-thirds of the 1990s. The US had also been the most successful economy during the 1990s, with some economic analysts suggesting that with low inflation, low unemployment, government budget surpluses in the making, and technical progress, it had entered a 'New Era' in which the business cycle has been dampened and financial crises, with rapid Federal Reserve responses to trouble, tamed. Not all was completely solid, however. Prosperity relied on increases in US consumer spending and a decline in the household saving rate, and after 2002, a bubble in the housing market. Credit card debt reached new heights. So did household bankruptcies. The increase in US net international indebtedness may pose problems if foreign confidence in the value of the US dollar declines. A strong, perhaps overstated warning by a British economist is entitled 'Seven Unsustainable Processes', although the author is unwilling to discuss timing

beyond saying that downturns are likely within the next five to fifteen years.[71] There is a possibility that the United States and its dollar may lose their positions in 'world economic and financial primacy'.[72]

Some political scientists place faith in what they call 'regimes', habits of cooperation built up during periods of leadership (that they in turn call 'hegemony').[73] Such cooperation worked remarkably well in the 1980s, especially under the initiatives of Secretary of the Treasury James Baker, who abandoned the policy of benign neglect of the value of the dollar and worked out the Plaza Agreement of September 1985 and the Louvre Agreement of January 1987. Few observers placed much reliance on the system of summit meetings among the G-7. These had strong overtones of ceremony and posturing. More suited to the working out of effective agreements was the G-5 among France, Germany, Japan, Britain, and the United States.

Can the IMF and the World Bank offer an alternative to declining US leadership? These institutions, established at Bretton Woods not to help the United States but to solve the problems of others, work slowly, a disability in a time of crisis that may require decisions in hours, not weeks. Moreover their funds may still prove to be too small to cope with market forces when those markets get the bit between their teeth and need the supplement of those of the G-7 central banks. This is particularly the case if the currency in difficulty is the dollar, rather than the euro, the pound, or the yen, with smaller financial markets.

While the dollar has troubles, it continues to be used as a world unit of account, if decreasingly as a medium of exchange, for lack of an adequate substitute. Japan and Germany were good followers of the American lead, but held back from challenging it. Under President de Gaulle, France continually challenged US policies and the dominance of the dollar without, however, great success, and is now fully involved with domestic, European and other international issues. The European Union may grow in economic and financial strength and take over world economic primacy. At the moment, however, the world depends on US leadership for lack of a better alternative; but the United States is holding back, preoccupied with its own political and economic problems, and reluctant to pay the cost of providing international public goods. Regimes work smoothly in quiet times, but something more decisive in the way of leadership is called for in crisis, and the likelihood of escaping economic and financial crises in the years ahead seems small.

13
The Lehman Panic – An Avoidable Crash

The bankruptcy of Lehman Brothers Holdings, the fourth largest US investment bank, in mid-September 2008 triggered the most severe financial panic in a century. Lehman had been aggressive buyer of mortgage-related securities using money obtained from short-term loans; its assets were more than thirty times its capital. Lehman was very profitable during the bubble years because of its high leverage and its aggressive posture toward risks. The rates of return to investors – to individual homeowners, property developers, and financial firms – that are extensively leveraged are very high as long as real estate prices increase. The paradox is that the leverage of some of these investors may decline even as their indebtedness increases because their net worth increases even more rapidly.

The earliest manias reviewed in this book occurred nearly four hundred years ago, and many of the manias and panic took place more than one hundred years ago. The 2008 crisis occurred in 'real time' – as the sixth edition of this book was under preparation.

When a bubble implodes, the market participants that are most vulnerable are those that have expanded most rapidly and have the greatest leverage. As real estate prices declined in the summer of 2008, Lehman was subject to two runs – investors and hedge funds sold their shares, both those they owned and those they could borrow, and owners of Lehman's IOUs were reluctant to renew their loans to the firm as they matured. Both groups were concerned that Lehman seemed slow in raising capital to counter the decline in its net worth from the fall in the value of its mortgage-related securities.

When Lehman failed, the credit markets 'froze' – liquidity disappeared. Trading volumes declined sharply. There was the traditional rush to cash that occurs when the financial environment suddenly becomes highly uncertain; the firms that were in a position to extend credit were extremely reluctant to do so because they wanted to increase their liquidity and their holdings of cash and the firms like Lehman that relied extensively on borrowed short-term

funds to finance their assets found it exceedingly difficult – and extraordinarily expensive – to re-finance their maturing IOUs. The risk premium embedded in the structure of interest rates soared; the interest rates in the interbank market increased by a factor of 30 to 40 relative to the interest rates on US Treasury bills, and the interest rates on high-risk bonds surged relative to those on US Treasury bonds.

The policy problem for the US government was the familiar one – should new measures be adopted to restore liquidity or instead should the government continue with a hands-off approach, even though asset prices would decline further, more financial institutions would fail, and the recession would be steep and perhaps prolonged. Eventually markets would stabilize; the presumption is that the next crisis would be less severe because the lenders would be much more cautious. In contrast, if the government intervened to stabilize the credit markets, the economic slowdown would be moderated but the next crisis would be more severe because lenders and investors would conclude that they would be 'bailed out' – whatever that means.

The bubble in US house prices

Bubbles result from the growth of credit that proves too high to be sustainable; when the credit growth slows, some asset prices begin to decline. US house prices peaked at the end of 2006. Between 2002 and 2006, the market value of the US residential real estate had increased from $16,000 billion to $23,000 billion, or from 110 percent of US GDP to 150 percent of US GDP – which was 15 percent higher in 2006 than in 2002. The sharp increase in property prices resulted from the ready availability of mortgage credit and led to a boom in the construction of new homes and condominium apartments; housing starts were one-third larger than the number associated with the growth of the population and the replacement of the homes lost to fires, storms, and highway construction.

The increase in the credit available for mortgages was facilitated by securitization, which had begun in the 1970s and initially involved depositing mortgages with similar attributes in terms of maturity in a trust, essentially its own corporation, and then issuing new securities (collateralized mortgage obligations or CMOs) that were claims on the interest income of the trust. Securitization accelerated after 2000; the second round involved 'slicing and dicing' groups of CMOs into four or five different tranches that differed in terms of whether they had the first, last or an intermediate priority to the interest income of the CMOs. The tranches with the first priority to the interest income of the CMOs were placed in one trust, the tranches with the second priority to the interest income were placed in a second trust, and so on. Conservative investors could buy the CDOs (collateralized debt obligations) that had the first claim to the

interest income because they were the least risky, while investors that wanted higher yields could buy those that had lower priority to the interest income of the CMOs – and hence were riskier.

A priori it would seem that the CDOs that had the first claim to the income of the CMOs would be less risky than the CMOs, while the CDOs that had the last claim to the their income would be riskier. Somehow the credit-rating agencies concluded from their models that even those CDOs that had the last claim to the income of the CDOs were nevertheless less risky than the CMOs. This outcome seemed counter-intuitive, and after the collapse of the mortgage market, it appeared that the credit-rating agencies had relied on information about the riskiness of various CDOs that was supplied by the investment banks that were eager to satisfy the investors' demand for securities that had high anticipated rates of return and yet were deemed not especially risky. The demand for free lunches is unending.

One consequence of securitization was that the amount of money available for mortgages increased because these mortgage-related securities were more liquid than individual mortgages, and the interest rates on mortgages accordingly were lower. The increase in the supply of money available for mortgages meant that some borrowers with less attractive credit histories qualified for mortgages because 'the money was there'. Moreover there was substantial innovation about both the minimal down payments required of home buyers and the interest payments on mortgages – there were adjustable-rate mortgages or ARMs, and interest-only mortgages, and negative amortization mortgages (part of the interest payment that the borrower was scheduled to make to the lender was added to the principal of the mortgage). Moreover many of these mortgages had 'teaser interest rates' – the interest rates that the borrowers were scheduled to make in the first two or three years were exceptionally low and then would increase by two or three percentage points.

These innovations in the terms of mortgages were motivated because the demand for mortgage-related securities was significantly larger than the supply of loans from creditworthy borrowers – typical of the last several years of an expansive credit cycle.

Another consequence of securitization was 'financial deepening', the supply of mortgage-related securities increased more than twice as rapidly as the supply of mortgages. During this deepening process the investment banks carried large inventories of mortgages and mortgage-related securities in their 'production and distribution pipelines' awaiting sale to the investors; some of these pipelines were six months long. The investment banks carried the credit risk attached to these mortgages as long as they were in the pipelines, although some may have hedged this risk. Some of the investment banks 'sold' some of these mortgage-related securities to their captive affiliates if they could not be readily sold to non-related investors.

Securitization of mortgages also led to an anonymous relationship between the home-owner borrowers and the mortgage lenders that replaced the traditional one where the lenders knew the borrowers because they shared the same neighborhood. The new relationship invited fraud in the mortgage-origination process; some of the fraud was on the part of borrowers, who provided incomplete or misleading information about their incomes, assets and credit histories (which were known as 'liar's loans'). Some of the fraud was by the mortgage brokers, who encouraged home owners to take on indebtedness that was extremely high relative to their incomes. Some mortgage brokers were not concerned about the ability of the borrowers to repay because the credit risk would be transferred to others in the credit distribution pipeline. The eagerness of the credit rating agencies to be asked to provide their ranking of individual mortgage-related securities 'clouded' their objectivity. Moreover this anonymity complicated the 'workout arrangements' when borrowers had problems making their monthly payments.

At some stage it was inevitable that the booms in real estate in the United States, Britain, Ireland, Spain, and other countries would end, property prices would decline, new home construction would fall, and probably sharply, and recessions would follow. There was significant disagreement about the impacts of the recession and the slowdown in construction on house prices and whether the upward momentum would stall or whether instead real estate prices would decline. The view of many lenders was that property prices would not decline because, well, 'US real estate prices had never before declined on a national basis'. (A factually incorrect statement since these prices had declined in the Great Depression.) Most of those involved in the real estate industry believed that there would not be a significant reduction in home prices. A second much smaller group – members of the 'reversion-to-the-mean school' – believed that prices would decline until the traditional ratio of the market value of US real estate to US GDP, plus or minus 10 percent, had been restored. A third view was that house prices would decline 10 to 15 percent below this traditional ratio because of the large excess supply of houses that had developed during the bubble years; these declines below the trend line would be larger in those regions with the largest excess supply.

Housing starts began to decline toward the end of 2006, which contributed to the slowdown in the US GDP growth rate. Still, the decline in US GDP in 2007 and the first eight months of 2008 was modest. The number of people employed peaked in December 2006, and thereafter the number declined continuously but modestly. During the first half of 2008 it appeared as if the United States would have a gradual slowdown that would be prolonged until the excess supply of homes and apartments was nearly absorbed – and that the slowdown would not be especially severe. Then the bankruptcy of Lehman on 15 September triggered the most severe crisis since the Great Depression.

The Japanese experience in the 1990s suggests that declines in asset prices need not led to a crisis or panic, even though the metric of the increase in the ratio of household net worth to GDP in Japan was more than ten times larger than that in the United States – both stock prices and real estate prices in Japan had increased by a factor of five to six while US real estate prices had increased by 40 percent. Although the banks and most of the other financial institutions in Japan had negative net worth in any meaningful mark-to-market test, there never was a run on a major Japanese bank, despite very modest levels of deposit insurance – at least in a formal sense.

Japan was different because it was 'understood' that bank depositors would not incur losses even if a major financial institution were closed. The tradition in Japan is that many losses are socialized that would fall on individuals in the United States and in other western industrial countries – the losses are made good by the government and ultimately by the taxpayers. In effect Japan had 100 percent deposit insurance and the Japanese institutions were 'too big to fail' even though the shareholders in many of these banks eventually would lose most or all of their money.

The first group impacted by the decline in US real estate prices was those who had purchased homes in 2005 and 2006, often first-time homeowners with spotty credit histories and modest if any down payments. If these borrowers had difficulties in making their monthly payments, they would sell their properties and repay their mortgage indebtedness – as long as property prices continued to increase.

The share of subprime mortgages in total mortgages increased from 6 or 7 percent of total mortgages in 2003 and 2004 to 20 percent in 2005 and 2006. Many of these subprime mortgages required very low interest payments in the first several years. Once property prices began to decline, many individuals that had purchased properties with minimal down payments would have 'upside-down mortgages' – their loan indebtedness would be greater than the market value of their homes. These homeowners then would have three choices – they could continue to make the monthly mortgage payments, they could engage in a short sale, that is sell the property and hope that the lender would 'forgive' the excess of the indebtedness over their sales proceeds or they could get involved in 'jingle mail' – return the keys to the lenders and walk, leaving the lenders with the costs and burdens of foreclosure. The larger the decline in house prices, the greater the decline in the net worth of home owners, and the higher the number of mortgage loans that would become non-performing.

The second group impacted by the decline in property prices were the brokers that bought the mortgage loans from the homeowners with the intent to sell them to the investment banks, who had twelve to eighteen months to put these mortgages back to the brokers if the borrowers stopped making their monthly payments. A few mortgage brokers that had extended loans to

subprime borrowers failed in 2005, many more failed in the last few months of 2006. The failure of these brokers meant that the investment banks were stuck with non-performing loans that they had recently acquired and would have returned to the brokers if they had not gone bankrupt.

The investment banks were the third group affected by the defaults of the homeowners, and in several additional ways beyond being stuck with the non-performing mortgages that they had recently acquired. Once real estate prices began to decline, it became much more difficult to sell mortgage-related securities because they were much harder to value. As mortgage defaults increased, the investment banks incurred losses on their holdings of tens of billions of mortgages and mortgage-related securities, including some CMOs and CDOs that they had been unable to sell.

Once an investment bank was a distress seller of mortgage-related securities, their prices declined sharply. The 'mark to market' convention was that firms that owned identical securities had to value them at the recent market price. There was an immense anomaly in that the market prices of these securities were much lower than the prices based on their current and prospective cash flows.

Subsequently the commercial banks that had acquired significant mortgages in a few states failed because of large losses. The Federal Deposit Insurance Corporation closed those banks that had exceptionally large losses relative to their capital; their deposits were sold to banks that had a secure capital position. Nearly two hundred commercial banks had failed by the end of 2009. The money that had been collected by the FDIC from the insurance premium paid by the banks was used to ensure that no depositor incurred a loss. The reserves that they had accumulated from these premium payments were being depleted at a rapid rate.

The FDIC had a $500 billion credit line at the US Treasury; it had rarely drawn on this line. If this line were exhausted, the Congress almost certainly would vote more funds to the FDIC. In anticipation that the losses would exceed its premium payments the FDIC wanted banks to make advance payments of the premiums.

Several of the hedge funds managed by Bear Stearns failed in March 2007; these funds owned mortgage-related securities and were extensively leveraged. Then in mid-August Countrywide Financial, the largest mortgage lender in the United States, had a liquidity problem; the firm had sought to rapidly increase its share of the national market for mortgages, and had relied on money borrowed in the commercial paper market to finance about one-third of the mortgages that it had acquired. Countrywide was unable to renew its commercial paper loans as they matured, and initially it drew on the credit lines that it had with commercial banks. Then Countrywide sold a new issue of preferred shares to Bank of America, and a few weeks later Bank of

America purchased Countrywide. About the same time Northern Rock, the largest mortgage lender in Britain, began to borrow from the Bank of England because it also was not able to re-finance its IOUs in the commercial paper as they matured.

The Bagehot doctrine and insolvent banks

The Bagehot doctrine discussed in Chapter 11 was that a central bank should lend freely – but at a penalty interest rate – to forestall the likelihood that a liquidity crisis would cascade into a solvency crisis because a commercial bank would sell assets to get the cash to meet the demands of their depositors that wanted their money. The prices of these assets would decline, and some of the banks that previously had been well-capitalized might then have too little capital as the value of the assets in their portfolios fell.

The extension of credit from the central bank to a bank that has a shortage of liquidity is a 'bail-out' of its shareowners; otherwise this bank might be forced to close and its shareholders would lose all of their money.

Bagehot probably would have advised against the extension of financial assistance to an insolvent bank. Assume, however, that Alpha Bank – or Lehman or the Royal Bank of Scotland – is insolvent. Should the government provide financial assistance to ensure that Alpha Bank can continue to operate even if it has a negative net worth – as the US Treasury did for Fannie Mae and Freddie Mac? If Alpha Bank goes bankrupt, the legal procedure is that its assets would be sold to get the money to satisfy its creditors, the prices of these assets would decline, and some previously solvent institutions then might have too little capital.

The logic that Bagehot used to justify why the central bank should provide credit to an illiquid bank that is solvent also justifies why the government should provide financial assistance to an insolvent bank. Both arguments centered on the impacts of the failure of one financial institution on the solvency of other financial firms.

Early in 2008 Bear Stearns, the sixth largest US investment bank, was subject to two runs, one on its share price by investors who thought the firm might go bankrupt and another by the holders of its short-term IOUs who were reluctant to renew their loans to Bear. Bear was acquired by JPMorganChase, but only after the Federal Reserve had agreed to take up to $29 billion of losses on some of the 'toxic waste' that had been in Bear's portfolio. Then in the first week of September Fannie Mae and Freddie Mac were taken over by the US Treasury in what was announced as a conservatorship; it was estimated that it would cost the US Treasury at least $100 billion and perhaps $200 billion to re-capitalize each of these firms.

In September 2008, Washington Mutual (WaMu) which was one of the largest and fastest growing mortgage lenders, was subject to a run; the FDIC intervened and detached the bank from its holding company; the bank was sold to JPMorganChase without cost to the FDIC.

The financial position of Lehman Brothers was well known, which made it difficult for the firm to obtain more capital; potential investors had no way to value many of its assets and hence how much capital the firm needed. The US government tried to facilitate the sale of Lehman to other financial institutions – but the statement was that no US government money would be involved. The Secretary of the Treasury subsequently said that the arrangement to sell Lehman to Barclay's Bank had been prevented by the financial authorities in London, who were concerned that the British government might have to bail out Barclay's if the losses of Lehman and its need for capital had been underestimated.

One day after the failure of Lehman, the US government intervened to prevent the collapse of AIG, then the world's largest insurance company. AIG had sold several hundred billion dollars worth of credit default swaps (CDSs), essentially an insurance policy to the buyers of bonds that they would not incur losses if the firms that had issued bonds defaulted. Many of these CDSs were purchased by the firms that owned the bonds. Some were purchased by third parities, who had no direct insurable interest; from their point of view, the premium required to buy a credit default swap was low relative to the payoff if the borrower defaulted.

Financial history would have been different if Lehman Bros had not failed. The US government officials said that they did not have the authority to prevent the failure of Lehman – that they could not support an institution that was bankrupt. Perhaps, but they might have been able to come up with a temporary arrangement until they had the authority. And many in the government were concerned that preventing the failure of Lehman would have meant that there would be many more Bear Stearns and Lehmans and Countrywides that would be on the ropes in the next financial crisis in 2015 or 2020.

'Too big to fail' – the costs and curse of the cliché

The implosion of every real estate bubble leads to the failure of a Lehman or a Northern Rock, usually one of the firms that had expanded most rapidly during the bubble years.

The traditional debate is between those who believe that the government should not intervene to save firms that are financially challenged; those who advance this view believe that if asset prices decline sufficiently sharply, the markets eventually will stabilize and investors who had been waiting on the sidelines will then buy the institutions that had lost most of their capital. Those who support this view subscribe to the moral hazard position that the next financial crisis will be more severe because the lenders would believe that they will be bailed out.

The challenge to this view is that each of these crises is different, and that the economic costs of not stabilizing the system are severe. Moreover, the moral hazard view fails to distinguish between government intervention to stabilize a single failing firm, and intervention to stabilize a group of firms as a real estate bubble implodes. Is the failure or near failure a result of faulty management decisions at a time when the economic environment is stable; in this case, the failure is idiosyncratic and reflects the problems of management. Alternatively, if a large number of financial institutions are on the ropes at more or less the same time, the failures are systemic, and result from a dramatic change in the financial environment, and what might charitably be called the mismanagement of the economy by the monetary authorities.

There are two responses to the view that public money should not be used to stabilize the financial system. One is that the costs in terms of unemployment and business failures and the increase in the government's indebtedness are high. A second is eventually the government will change its policy in the effort to stabilize the system, and it would have been less costly if the government had intervened earlier in the crisis.

If the government decides to intervene to stabilize the financial system before property prices have declined to their long-term equilibrium levels, then there may be need for a series of injections of government money as the decline in the property prices leads to continuing shrinkage in bank capital. If in contrast the intervention is long delayed, then the recession in the economy will lead the banks to incur significant loan losses. The timing of intervention is a bit of an art form.

Banks traditionally have borrowed at short term and lent at long term – one of the risks that they acquired was that short-term interest rates might increase significantly relative to long-term interest rates and another was the liquidity risk that depositors might withdraw their money and that their creditors might be reluctant to purchase more of their IOUs when they matured. The prospect was that the depositors and the creditors would discipline the lending practices of the banks.

The Federal Deposit Insurance Corporation was established in 1934 to reduce runs on banks by owners of small deposits who were concerned that they might lose their money if their bank failed. In effect these depositors were provided by a US government guarantee against loss if the banks failed. Initially the ceiling on insured deposits was $2,500, which was soon raised to $5,000, and then to $10,000 in 1950, $20,000 in 1969, and $100,000 in 1980. Individuals could readily circumvent this ceiling by setting up accounts in the same bank under the names of different family members and different combinations of family members, and by opening accounts in different banks. In effect deposit insurance was unlimited to anyone who wanted to break up a large amount of money into $100,000 units.

The institutional innovation of deposit insurance eliminates the 'discipline' that small depositors might impose on banks, based on the threat that they might rush to get their money, which would mean that the managers of the banks would be super-cautious; the belief was that these small depositors are unsophisticated and would be trigger-happy. There was no need for insurance of larger depositors because their owners were believed to be sophisticated, and much less likely to trigger a run.

The practical issue involved in the decision whether to 'save Lehman' was which of the stakeholders in the firm would avoid losses if government assistance had enabled the firm to avoid bankruptcy. Lehman was significantly larger than Bear. Lehman was not a commercial bank, and hence it was regulated by the Securities and Exchange Commission rather than by the Federal Reserve, although after the rescue of Bear the Fed had opened its discount window to investment banks.

One interpretation of 'too big to fail' is that a faltering bank is so large that there is no other institution has sufficient excess capital to acquire the firm, much as Bear had been purchased by Morgan; hence the government must provide financial assistance to any of the five to ten very large firms because its demise would have costly consequences both for its competitors and the economy.

One of the concerns implicit in the 'too big to fail' cliché is that the various stakeholders in the bank benefit from the proverbial 'free lunch' because they are protected against losses that their counterparts in smaller firms would have incurred when they failed. The largest bank failure prior to the 2008 crisis was that of Continental Illinois Bank in 1984, then the seventh largest US bank. Continental Illinois had incurred large loan losses when the oil price declined sharply. Continental differed from most other banks in that it had a relatively small domestic deposit base. There had been a run on Continental Illinois in 1982, but that had been stanched; the run in 1984 started after a report of very weak earnings. The run started on its offshore branches. Initially the bank borrowed from the Federal Reserve, and then it was rescued by the FDIC.

The owners of the common stock of Continental Illinois lost all their money, and their shares became virtually worthless. The top executives and senior managers lost their positions and a large part of their personal wealth that was invested in the bank's stocks – and they lost a significant part of their professional reputations. The careers of many managers as bankers were shortened. The directors of Continental Illinois were publicly embarrassed – and most lost money since they owned shares in the firm.

Who, then, was 'bailed out'? The owners of the insured deposits did not incur any losses because of the FDIC guarantee – but these depositors had been charged premiums for deposit insurance. The owners of the uninsured deposits also avoided losses because they withdrew virtually all of their money before

Continental Illinois was taken over by the FDIC. The owners of the subordinated debentures which had been issued by the bank holding company for Continental Illinois were bailed out. (There is general agreement that the FDIC made a mistake by providing money so that the owners of the bonds issued by the bank's holding company did not incur losses.)

For all practical purposes, Continental Illinois failed even though it remained in business. Presumably the only way to ensure that the uninsured depositors took a 'haircut' was to have had the bank closed while it was still liquid.

Compare the costs and the benefits of 'saving Lehman' with the costs and benefits of 'letting Lehman go'. In August and September 2008 the Secretary of the Treasury encouraged other financial firms to acquire Lehman, to invest in Lehman, or to develop a cooperative arrangement that would acquire some of the dodgy assets that Lehman owned – but he was adamantly against providing any US government financial assistance to Lehman because he did not want to be known as 'Mr Bailout'.

There were several different models of how Lehman might have been saved. One was Morgan's purchase of Bear; a buyer would have been found for Lehman after the Federal Reserve agreed to acquire $40 or $50 or $60 billion of its mortgage-related securities that were believed most sensitive to default. The owners of Lehman shares would receive a token $1 or $2 a share from the acquiring firm. Some of the senior executives of Lehman would be given positions in the acquiring firm, most would not. The owners of Lehman's short-term IOUs would be made whole, and perhaps the owners of its bonds would be made whole like those of Fannie and Freddie; alternatively these bondholders might have been given haircuts. The Lehman name would disappear from the market place. The financial cost to the Federal Reserve might have been $50 billion or $60 billion or $70 billion that it would have used to buy the toxic securities – but these securities would have had some value after the economy stabilized.

A second model is provided by the US Treasury's investment of Troubled Assets Relief Program (TARP) money in Citibank and Bank of America; the US government would buy 60 or 80 million new shares in Lehman at $1 a share – the amount needed to re-capitalize the firm. The immediate financial cost would be the amount that the Treasury would have to lay out to acquire these shares.

A third model is provided by the US Treasury's takeover of Fannie Mae and Freddie Mac; Lehman also would be placed in a conservatorship and the US Treasury would become its owner. The US Treasury would lend money to Lehman. At some stage this loan would be converted to newly issued shares, and then after the crisis, these shares would be sold to the public. The amount that the Treasury would need to invest would be determined by the size of the 'hole' in Lehman's capital structure.

Estimating the cost to the US government of saving Lehman is conjectural and involves the valuation of its mortgages and mortgage-related securities in the post-crisis period. (During the crisis, the market values of these securities would be highly uncertain, and almost certainly far below their intrinsic value based on the cash flows in a stable environment.). Before its collapse, Lehman had a net worth of $30 billion, and a leverage of thirty, so its assets were about $900 billion and its liabilities $870 billion. If its loan losses were three times its capital – then the cost to the US Treasury would have been $60 billion.

As Lehman approached bankruptcy, it was 'two firms'; one firm continued to generate operating profits of $10 billion to $15 billion a year while the other firm had negative capital as a result of the massive decline in the value of the mortgages and mortgage-related securities that it owned. One advantage of the use of government funds to re-capitalize the firm is that the good bank would continue to generate $10 billion or $15 billion of annual profits from its trading and investment activities, while the bad bank would contain most or all of the low-grade mortgage-related paper. The advantage of the second and the third approaches is that the US government would benefit from the operating income of Lehman, which would offset some of the losses.

The 'additional cost' – always noted by those concerned with the moral hazard argument – is that the next crisis will be more severe because Lehman would have been 'bailed out'. Regardless of the approach taken to prevent Lehman's bankruptcy, the Lehman shareholders would have been wiped out, its top executives would have been replaced and lost much of their personal wealth, and its directors would have been humiliated by their failure to protect the shareholders. Two groups of stakeholders would have benefited if Lehman had not failed – one was employees, perhaps some of those with bonuses, and the other was the counterparties and the owners of Lehman bonds.

AIG – one of the world's largest insurance companies – also was experiencing a run as efforts were made to avoid the bankruptcy of Lehman. AIG had written tens of billions of credit default swaps. In effect AIG acquired the credit risk attached to these bonds in exchange for the premium income; the buyers of the credit default swaps believed that premiums that they paid were modest relative to the credit risk. The run on AIG was like the run on Lehman once removed, since AIG would have to reimburse those who had bought these swaps to protect themselves against losses on their holdings of Lehman bonds when the firms went bankrupt.

The counterfactual is to determine how the crisis would have evolved if Lehman had been saved; there is some likelihood that the run might have been deflected to other firms, perhaps Merrill Lynch or Morgan Stanley. One possibility is that that run on AIG would have continued, perhaps less rapidly. Another possibility is that US government would have made it known at the time that Lehman was saved that it stood ready to assist large firms that had

liquidity problems, or had depleted capital – which was essentially the policy that was adopted when AIG was bailed out. The terms of the 'bail-outs' would have included that the common and preferred shareholders would have lost all of their money and the managements and the directors replaced.

If AIG had not been saved then it is highly likely that other financial firms would have been subject to runs. Asset prices would have declined, and other financial firms would have failed. Lenders would be extremely cautious in making new loans, and in renewing maturing credits. The US recession would have become much more intense, and the US economy might have fallen into the debt-deflation trap.

Thus the amounts needed to save Lehman – the counterfactual – would have been the $50 billion or $100 billion that would have been the financial outlay by the US Treasury minus any recovery and the incremental cost of financial crisis in the future because the lenders would have realized that there was an increased likelihood that they would be bailed out if they followed a more risky investment strategy, The additional cost would be the cost of saving other large financial institutions that had invested too aggressively.

This cost must be adjusted for the additional outlays that the US Treasury and the Federal Reserve incurred when they intervened to save AIG, Citibank, Bank of America, and twenty or thirty other firms. Estimating this amount is extremely difficult – the runs on individual institutions would continue until it was clear beyond the shadow of a doubt that no institution would fail because of a lack of liquidity. The US Treasury invested more than $400 billion in other financial institutions, and the data suggest that virtually all of this money will be recovered. (The investment of TARP money in GM and Chrysler is another issue.)

The immediate impact of the decision not to save Lehman was a panic and crash. Some of the costs of not saving Lehman can be estimated in terms of the increase in unemployment, the decline in US GDP, and the increase in the US government's fiscal deficit. In the first eight months of 2008, employment had declined by an average of forty thousand a month, in the six months after the failure of Lehman, employment declined by six hundred thousand a month.

The striking phenomenon was the surge in interest rates on risky assets – on LIBOR relative to the US Treasury bills. Those banks with a relatively strong cash position no longer had trust in the credit standing of the banks that had incurred large loan losses, which then had to sell assets and shrink; they hoarded cash. Those banks in a weak cash position were then obliged to raise cash in other ways and to shrink their assets.

The primary costs of not saving Lehman are those associated with the incremental decline in US GDP because of the surge in interest rates. In August 2008, the non-partisan Congressional Budget Office projected that the US fiscal deficit would be $450 billion in fiscal 2009; six months later, the projection was $1500 billion, which was ten or fifteen times larger than the highest estimate of the costs of saving Lehman. Most of the increase in the projected

fiscal deficit was due to the sharp decline in economic activity because of the dramatic reduction in the willingness to lend, and because households and firms hunkered down – as employment declined sharply, spending declined – individuals were less confident about their jobs.

'Too big to fail' and moral hazard

A significant number of large financial institutions had failed prior to the run on Lehman; the US government had intervened with both Bear Stearns and Fannie Mae and Freddie Mac. Why was Lehman different?

Banks, mortgage brokers, insurance companies and other financial institutions can be placed in one of four groups in terms of how they fared in the 2008 crisis. One group consists of those firms that went out of business because they lost all or virtually all of their capital; the shareowners in these firms lost all of their money and these firms have disappeared from the marketplace. This group includes Indy Mac, Washington Mutual, and Lehman Brothers. Nearly three hundred mortgage banking firms failed. Nearly two hundred commercial banks were closed by the Federal Deposit Insurance Corporation. CIT is in this group. The top executives lost their positions, and a great deal of their personal wealth; the careers of many of these individuals were shortened. The boards of directors of these firms were disgraced.

A second group comprises those firms that that incurred such large losses that they were acquired by other, better capitalized firms, either with or without financial assistance of the US government. Countrywide Financial was in this group, as was Wachovia Bank and Bear Stearns. Merrill Lynch was in this group. These firms disappeared from the marketplace and their shareholders lost most but not all of their money. Similarly most of the top executives in these firms lost their positions, and their careers were truncated. In some cases, however, the top officials still walked away with tens of millions of dollars.

The third group includes the firms that probably would have been forced to close their doors if they had not had assistance from the US Treasury or the Federal Reserve. AIG, Citibank, and Bank of America are in this group. So are some of the regional banks. The prices of the shares of these firms declined by eighty to ninety percent, although these prices subsequently increased once the US government became a significant shareholder and the likelihood that the firms would fail diminished sharply.

The fourth group consists of the firms that did not take any funds from the US Treasury, or else took these funds because they were instructed to do so by Secretary of the Treasury Paulson even though they did not want the US government's money. JP Morgan Chase is in this group, as is Goldman Sachs, both in what had been the top tier of US investment banks. Several of the firms in this group were able to raise capital in the private markets; Morgan Stanley sold new shares to Mitsubishi Bank.

Why did the failure of Lehman lead to the unprecedented panic in the credit markets? One view is that the initiatives of the US government to prevent the bankruptcy of Bear Stearns and subsequently of Freddie and Fannie followed by the unwillingness to commit government money to save Lehman meant that there appeared to be no coherent government policy toward providing financial assistance to weakened large institutions. If two points determine a line, then measures adopted by the US government to deal with the Bear problem and the Freddie and Fannie problem set the policy. But then the 'policy' was reversed by not saving Lehman. And while the initiatives toward AIG represented a reversal of the reversal, market participants were shaken by the inconstancy of the approach of the US government.

Haircuts for all?

The principal stakeholders that have benefited from the use of government money to stabilize the banks and other financial institutions have been the bondholders and other counterparties, the uninsured creditors, and the senior managers. If these firms had gone bankrupt like Lehman, the bondholders would have taken a haircut, and depending on the severity of the losses, they might have lost 30 or 50 percent of their money. The employment contracts of the senior managers would no longer be binding if the firms failed, and those who had large amounts due because of their success as traders would find that they were just another group of unsecured creditors.

If the bondholders are subject to loss if the firm closes, then many would sell their bonds in anticipation of these losses. Interest rates on these bonds would increase sharply, and the banks would not be able to sell new bonds when the existing bonds mature. The banks would be forced to shrink their assets. Similarly the counterparties would be reluctant to re-finance existing positions. The likelihood of losses by these groups could trigger runs.

One position is that bond instruments should be designed so that the bondholders would incur losses if the firms failed; then they would exercise more discipline on the senior managers so that they would be more cautious in undertaking risky investments. Perhaps. The credit-rating agencies already evaluate the riskiness of the bonds, and it is not obvious that the prospect of haircuts might induce the investors to be more cautious in buying the bonds of the banks.

The hidden cost of too much discipline is that bankers would be extremely cautious lenders, and many worthwhile projects would find it difficult to obtain financing.

The design challenge is to ensure that large firms are allowed to fail when the causes of failure are idiosyncratic to one or two institutions in a stable financial environment. When the failures are systemic and involve a large number of institutions, the policy problem differs; failure has many more costly externalities. The capital of financial institutions as a group declines, and there may be too little capital relative to the liabilities of these institutions. Either asset values will decline until the ratio of bank assets to bank capital is at a satisfactory level, or the government must provide financial investments to the banks. The nub of the problem is that the value of bank capital is highly sensitive to changes in asset prices.

Assume that there had been no rescue program – no TARP money, no Fed investments, and so on. Then other banks that had been dependent on borrowed short-term money to carry long-term assets would have failed. Other banks that had been conservative in their lending policies would have encountered large loan losses because the recession would have been much more severe.

Fast forward to 2020 or 2025, the approximate time of the next major period of rapid credit expansion. How many of the top executives of the major US financial firms at that time would look back at the government intervention in the 2008 crisis and conclude they can afford to take greater risks in the allocation of credit because so many large financial institutions had been 'bailed out'.

These executives would have observed that the stock prices and market value of financial firms as a group had declined by ninety percent or more. They would have observed that the senior executives of most of these institutions were replaced, and that the directors also lost their positions. They would have noticed that many managers lost their careers.

The US investment banking industry had been decimated, with one dominant firm as a survivor; another of the major firms had to sell about twenty percent of its shares to a Japanese firm.

'Too big to fail' is a catchy cliché. But the slogan obscures more than it reveals. And because of the reluctance of government officials to provide financial assistance to Lehman so that the firm could have avoided bankruptcy even as its shareholders lost all their money, a market adjustment cascaded into the most severe crisis in one hundred years.

14
The Lessons of History

The last four hundred years have been replete with financial crises, which often followed increases in the supplies of credit, greater investor optimism, and more rapid economic growth. More and more individuals purchased securities and assets for short-term profits from the increases in their prices. Asset prices increased. Household wealth rose and consumption spending grew. Business firms became more upbeat and invested more. Economic booms followed and then euphoria developed. The increases in the supplies of credit led to bubbles in prices of real estate and of stocks; these prices were too high to be sustainable in the long run.

The uniqueness of the last forty years is that there have been four waves of financial crises; each wave was preceded by a wave of credit bubbles that involved three, four, or in some episodes eight or ten countries. The similarity of increases in the prices of assets in so many countries at about the same time suggests a common cause. The pattern has been that the increases in the flows of money to countries usually led to increases in the values of their currencies and to increases in the prices of their bonds, stocks, and real estate.

The likelihood that these four waves of bubbles were independent and unrelated events is low. Instead it seems plausible that there was a systematic relationship among several of these successive waves, and that the money flows associated with the implosion of one wave of bubbles contributed to the next wave.

The increases in the supplies of credit generally were provided by banks. However, the buyers of bulbs during the Dutch 'tulipmania' of the 1630s could pay the increasingly high prices only because the sellers provided credit. As bulb prices increased, rational exuberance morphed into irrational exuberance, the Dutch economy boomed – and then when bulb prices tumbled, the growth rate slowed. Both the South Sea Bubble in London and the Mississippi Bubble in Paris in 1720 resulted from rapid increases in the supplies of credit from newly established banks. The coincidence of these two bubbles reflects that financial innovations in London were mimicked in Paris.

The 'mother of all bubbles' in Tokyo in the second half of the 1980s resulted from the rapid growth in the supply of credit and the relaxation of regulations that had limited bank loans for purchases of real estate. Once property prices began to increase by 20 to 30 percent a year, there was a surge in the demand for real estate, since the anticipated returns from further increases in the prices of apartment buildings and office buildings and land were so much higher than the returns in other sectors of the economy. Real estate was the collateral for bank loans; as prices increased, the value of the collateral increased, and the supply of credit for purchases of real estate increased in response to the increase in demand.

The bubbles in the real estate markets in the United States, Britain, Spain, Ireland, and Iceland that began about 2002 resulted from the increases in the flows of money to each of these countries, which led to increases in the supplies of credit for the purchases of homes and commercial properties. The increases in the flows of money to these countries generally led to higher values for their currencies.

The increases in the prices of assets in the United States, Britain, Mexico, Iceland, and other countries during each of these waves were an integral part of the adjustment process induced by the money flows, which required a corresponding increase in net imports of goods. Higher asset prices led to increases in household wealth and to more spending on both domestic and foreign goods.

Financial crises developed once some borrowers were no longer able to pay the interest on their outstanding indebtedness. The crises in the eighteenth and nineteenth centuries often followed a sharp decline in commodity prices, which handicapped the ability of farmers to pay the interest on the loans that had enabled them to purchase seeds and real estate. At other times the crises developed in response to sharp declines in stock prices. Crises generally were infrequent, often about once a generation – the United States had one crisis in 1907, and then a severe one in the early 1930s; however in the first half of the nineteenth century a crisis occurred almost every ten years. Four waves of crises in the thirty years since the early 1980s, each involving three, four or more countries, are exceptional.

The bubble in US stock prices in the late 1920s followed an economic boom that resulted from the surge in the production of automobiles and the expansion of highways, the electrification of millions of households, and the rapid expansion of telephones; there was a remarkable increase in investor optimism. Stock prices increased threefold between 1927 and 1929, in part because the Federal Reserve made credit more readily available to reduce the likelihood that the surge in productivity would lead to declines in the price level. Real estate prices trended upward, although not as rapidly as stock prices.

The increase in interest rates on US dollar securities in 1928 led to a reduction of American purchases of foreign bonds and complicated the ability of central banks in Europe and in Latin America to maintain the parities for their currencies because they lacked the money to finance their trade deficits. A series of currency crises followed in the 1930s; initially pressure was directed at the Austrian schilling. The depreciation of the schilling triggered the contagion effect, which led to speculative pressures directed at the German mark and soon thereafter at the British pound, and then at the US dollar. After the establishment of a new US dollar price of gold at $35, the speculative pressure was directed at the currencies of the 'gold bloc' countries – the French franc, the Swiss franc, the Italian lira, and the Dutch guilder, whose convertibility into gold was suspended in 1936.

There were episodic currency crises between the late 1940s and the early 1970s as investors became increasingly skeptical that governments would incur the costs required to maintain the parities for their currencies, which had been set at the end of World War II. Patterns of production and trade had been disrupted by massive destruction of factories and of infrastructure during the war. Many European countries devalued their currencies in September 1949 to adjust for more rapid increases in their price levels than in the US price level during and after World War II. The French franc was devalued by 10 percent in 1959. The British pound was subject to speculative pressure in 1964, and was eventually devalued in the autumn of 1967. There was speculation that the German mark would be revalued in 1969 and that the French franc would be devalued at about the same time. Then the speculative pressure was directed at the US dollar; investors anticipated that either that the currencies of other industrial countries would be revalued or appreciate, or that the US dollar price of gold would be increased.

Most of the currency crises in the 1950s and 1960s were preludes to the devaluations of currencies by 10 to 20 percent; none was associated with a banking crisis. Speculative pressure did not lead to excessively large changes in the values of currencies relative to the values projected from differences in inflation rates.

The uniqueness of the period since the early 1980s is that nearly all of the banking crises have been associated with currency crises. The principal exception is that the financial crisis in Japan in the first half of the 1990s was not associated with a crisis about the value of the yen. Another exception is that the banking crisis in Ireland in 2008 and 2009 was not associated with a currency crisis because Ireland was a member of the European Monetary Union. The financial crisis that involved the debt of the Greek government in the spring of 2010 was centered on the ability of the Greek government to reduce its fiscal deficit and make its debt service payments in a timely way and thus was a credit crisis; however the euro depreciated because some investors were concerned about its viability.

The waves of cross-border money flows that began in the early 1970s were too large to be sustained, because the external indebtedness of individual countries was increasing much more rapidly than their GDPs. When the flows of money to these countries slowed, their currencies depreciated. Often the depreciations were extremely sharp as investors sought to move money from the countries that had formerly received money from abroad.

The coincidence of the waves of banking and of currency crises since the early 1980s reflects that the money flows to individual countries led to the appreciation of their currencies and contributed to increases in their asset prices. Domestic credit often expanded at a rapid rate, since the appreciation of their currencies dampened inflationary pressures in the goods market and enabled the central banks to continue their expansive policies.

When the flows of money to countries slowed and their currencies depreciated, some of the borrowers in these countries became distress sellers of assets because they could no longer obtain the money to pay the interest by borrowing more money. Their sales depressed asset prices. Moreover the reduction in money inflows led to increases in interest rates, which contributed to the declines in the prices of real estate and stocks. Households and firms that had loans denominated in the US dollar, the Swiss franc, or some other foreign currency often went bankrupt because the depreciation of their currencies led to large revaluation losses on their foreign loans. The failures of banks – the financial crises – followed from the large losses incurred by extensively indebted firms and individuals; the value of the collateral pledged to the banks declined sharply relative to the value of the loans.

Domestic financial crises have become less frequent in the last one hundred years after central banks were established to act as lenders of last resort – they were able and willing to provide cash to banks in response to sharp declines in investor demand for speculative assets.

The feature of crises in the twentieth century was that central banks often had to choose between whether they should give priority to maintaining the value of their currencies or instead to give priority to domestic financial stability. Initially many central banks increased interest rates to convince investors that they had an unshakable commitment to their parities; higher interest rates often led to sharp declines in asset prices and economic activity. Then the central banks reversed their priorities. The contagion effect in the 1930s was triggered by the initial mis-alignment of currencies and price levels; the sequence of currency crises reflected that investors became concerned that central banks would not maintain their parities after one of their major trading partners had devalued. Each central bank then would conclude that the costs of retaining its parity was too high in terms of business failures and unemployment.

The contagion effect of the late 1990s was triggered by the sharp depreciation of the Thai baht; investors suddenly realized that money flows to Thailand's neighbors and to other emerging-market countries would decline sharply and that the central banks in these countries would no longer be able to support their currencies because money would flow from these countries in anticipation of the subsequent depreciations. Immediately speculative pressure shifted to one or several other currencies that still retained their pre-crisis values.

This sequence of large declines in the values of the currencies both in the 1930s and then in the second half of the 1990s leads to the question whether an international lender of last resort would have enabled countries to avoid the deflate-and-devalue cycle. Domestic lenders of last resort can provide abundant credit to reduce the likelihood that liquidity crises will cascade into a solvency crisis. The problem in both the 1930s and the 1990s was that currencies were not appropriately aligned with national price levels, and that once one country devalued or allowed its currency to depreciate, speculative pressure was deflected to the currencies of its trading partners.

The counter-factual question is whether an increase in the supply of international reserve assets in the early 1920s, when the potential shortage of gold was first recognized, would have obviated some of the changes in the currency values in the interwar period. National price levels had increased much more rapidly in some countries than in others during and after World War I, with the result that currency values were not aligned with price levels. A larger volume of international reserve assets would have made it possible for the adjustment of currency values to occur over a longer period. But neither a larger volume of international reserve assets nor an international lender of last resort would have obviated the need for the adjustments of currency values. But a larger volume – or perhaps a much larger volume – might have reduced the frequency and severity of speculative attacks.

During the interwar period central banks were committed to pegging their currencies to gold, initially at their pre-war parities, while the general view since the early 1970s is that currencies should float – although central banks in some countries in Asia have intervened extensively to limit the appreciation of their currencies.

The large variability in asset prices and in currency values and in the number and severity of banking crises since the early 1980s suggest that the lessons of the interwar period have been overlooked – if they have been remembered. One of the lessons was that cross-border movements of money often distort national competitiveness in the short and medium run – for three, four, or more years. Another was that currencies 'overshoot' and 'undershoot' their long-run average values unless central banks are committed either to parities for their currencies or to limiting the variability in currency values. A third is that some countries

will manage the values of their currencies to secure competitive advantages in the goods market unless they are constrained by international agreement or rules from doing so. 'Permanent undershooting' is a form of currency protectionism and invites critical responses from trading partners.

The three decades since the early 1980s have been the most tumultuous in monetary history in terms of the number, scope, and severity of financial crises. More national banking systems collapsed than in any previous comparable period; the loan losses of banks in Japan, in Sweden and Norway and Finland, in Thailand and Malaysia and Indonesia, and in Mexico (twice), in Brazil and Argentina, and in the United States and Britain and Iceland and Ireland ranged from 10 to 50 percent of their assets. In some countries the costs to the taxpayers of providing the money to fulfill the implicit and explicit government guarantees on bank deposits and other bank liabilities amounted to 15 to 20 percent of their GDPs. The loan losses in many countries were much greater than those in the United States in the Great Depression of the 1930s.

Occasionally the failure of a bank was firm-specific; Franklin National Bank of New York City and then Herstatt AG of Cologne lost the large bets that they had made on changes in currency values. However most of the bank failures since the early 1980s have been systemic events that have occurred in waves and involved large numbers of lenders in different countries because of dramatic changes in the international financial environment. In some episodes virtually all of the banks in a country failed. Usually these countries experienced sharp declines in real estate values after periods when these prices had increased rapidly.

The dominant pattern, evident in Mexico, Argentina, Thailand, Malaysia, and Russia and many other countries was that banking and currency crises occurred together. These crises followed periods when money flows to these countries led to increases in the value of their currencies and in the prices of domestic assets. Often the subsequent sharp depreciations of the currencies led simultaneously to deflationary shocks as interest rates surged, and to inflationary shocks as the prices of tradable goods soared.

In the early 1980s many banks in Texas, Oklahoma, and Louisiana failed when oil prices declined sharply while small banks in Iowa, Kansas and other states in the agricultural Midwest also went under because the sharp declines in the prices of cereals led to even larger percentage declines in the value of farm real estate. The failure of these lenders resulted from the reversal of US monetary policy in the autumn of 1979 and the resulting sharp decline in anticipated US and global inflation rates. Several thousand US thrift institutions failed in the early 1980s when short-term interest rates surged. Many banks in Thailand, South Korea, and Indonesia failed in the Asian Financial Crisis; in the late 1990s, soon after the financial debacle in Russia, Long-Term Capital Management, then the largest US hedge fund, collapsed and would

have failed if the Federal Reserve had not mandated that its largest creditors invest equity capital in the firm.

The range of movement in currency values since the early 1970s has been much larger than in any previous period. Moreover this range was much larger than the proponents of floating currencies had anticipated; they believed that allowing the values of currencies to adjust to differences in national inflation rates and other shocks was preferable to forcing adjustment of national economies so that the established parities could be retained. These large swings in currency values initially were attributed to the lack of familiarity of market participants with freely floating exchange rates. Then the large changes in currency values were explained because of the inherent lags before the supplies of tradable goods could increase in response to changes in the relationship between domestic and foreign prices and costs.

Moreover the scope of overshooting and undershooting of currency values was much larger than in any previous period. Thus the appreciation of the German mark and the Japanese yen in the late 1970s was greater than the increases that would have been forecast based on the excess of the US inflation rate over the inflation rates in Germany and in Japan. Then in the early 1980s both currencies depreciated sharply even though the US inflation rate remained higher than their inflation rates. In the late 1990s the US and European inflation rates were similar but the newly established euro – the successor to the German mark, the French franc, the Italian lira and the currencies of the other countries that joined the European Monetary Union – depreciated by 30 percent following its establishment at the beginning of 1999.

Overshooting and undershooting are an integral part of the adjustment process in response to cross-border movement of money when currencies are floating. Changes in the cross-border movement of money must be accompanied by corresponding changes in the cross-border movement of goods and in trade balances. Countries that experience inflows of money must develop larger trade deficits, which result from increases in consumption spending attributable to higher levels of wealth and the appreciation of their currencies.

The increases in the flows of money to countries are like increases in the savings available in these countries, which can adjust by counterpart increases in household consumption, business investment, and larger government deficits. Household consumption accounts for 60 to 70 percent of GDP in most countries; as a result most of the adjustment to an increase in money inflows involves increases in household consumption – or what is the same thing, reductions in household savings. The increases in the availability of foreign savings lead to higher asset prices and more consumption spending and hence to declines in domestic saving.

The large swings in the values of the currencies result from the reversal in the direction of cross-border money flows. Increases in the flows of money to countries often lead to real appreciations of 20 or 30 percent and hence to overshooting. Then when the shock occurs and money leaves the countries, these currencies may depreciate by 40 or 50 percent. Hence the initial money flows to various countries and then the inevitable reversals lead to bubbles in currency values.

The increases in the flows of money to countries often followed from an institutional innovation, including the relaxation of financial regulations. These movements may have been reinforced because the appreciations of the currencies enhanced the anticipated rates of return available to investors that were willing to incur the cross-border risks. However, as a group these investors failed to appreciate the end-game, and that the currencies eventually would depreciate because the borrowers' indebtedness had been increasing too rapidly.

Thus the financial crises followed from reversals in the pattern of cross-border money flows following one, two, or more years when these flows increased rapidly. When the cross-border movements of money slowed, some of those who had borrowed to finance the purchases of real estate or stocks or some other assets became distress sellers because they needed to pay the interest on their indebtedness. The values of the Mexican peso, the Brazilian cruzeiro, the Argentinean peso, and the currencies of many other developing countries plummeted in the early 1980s when the money inflows diminished sharply. The depreciation of these currencies led to large revaluation losses by firms that had liabilities denominated in the US dollar. When these firms went bankrupt, the banks in these countries incurred large loan losses.

The scope of overshooting and undershooting of the currencies of the emerging-market countries has been even larger than for the industrial countries. The Mexican peso lost nearly half of its value during that country's presidential transition at the end of 1994 and the beginning of 1995. The values of the Thai baht, the Malaysian ringgit, the Indonesian rupiah, and the South Korean won declined by 50 to 70 percent in the last six months of 1997. The Russian ruble depreciated sharply in August 1998. The Argentinean peso lost more than two-thirds of its value in January 2001.

There were more waves of credit and asset-price bubbles between 1970 and 2010 than in any other comparable period. The first wave involved increases in bank loans to the governments and government-owned firms in Mexico, Brazil, Argentina, and other developing countries in the 1970s; their external indebtedness increased from $125 billion to $800 billion. Then real estate prices and stock prices in Japan increased more than fivefold in the 1980s; household wealth surged, and the economy boomed. Finland, Norway, and Sweden also experienced bubbles in their real estate markets and their stock

markets at about the same time. The third wave involved bubbles in the real estate and stock markets in Thailand, Malaysia, Indonesia, and several nearby countries in Southeast Asia in the first half of the 1990s; in most of these countries the bubbles in real estate begat the bubbles in stocks because many of the companies listed on the stock exchanges owned large amounts of property. US stock market wealth doubled in the late 1990s and the market values of the stocks of the firms in the dot.com and information technology industries increased fourfold. The fourth wave involved sharp increases in the market value of real estate in the United States, Britain, Ireland, Spain, Iceland, and South Africa after 2002.

One feature of each of these waves of bubbles was that the indebtedness of a large group – occasionally governments, most often investors in real estate, including home owners – increased by 20 to 30 percent a year for three, four, or more years. Rapid increases in indebtedness enabled borrowers to use the money from new loans to pay the interest on their outstanding loans. The indebtedness of these borrowers could increase more rapidly then their GDPs for a few years, but eventually the lenders became more cautious in extending new loans, and some of the borrowers then had to scramble to get the money to pay the interest. Since most of these waves of bubbles involved cross-border money flows, declines in the flows of money to countries led to the depreciation of their currencies.

The bubbles in real estate and stock markets and the failure of banks was systematically related to the overshooting and the undershooting of currencies, and resulted from changes in the amounts and direction of cross-border money flows. The four waves of crisis followed from the sharp depreciations of the currencies which automatically led to increases in the indebtedness of the domestic firms that had debts denominated in the US dollar, the Swiss franc or some other foreign currency. The almost universal surges in interest rates as the currencies depreciated triggered sharp declines in the values of real estate and of stocks during the crash phase of the financial cycle. Most of these crises were 'predictable', since increasing reliance on money from new foreign loans to pay the interest on the outstanding foreign indebtedness was not sustainable; the uncertainty involved the sharpness of the declines in the values of the currencies when the money inflows diminished. Thus the mania phase of the expansions in Mexico and the other developing countries in the 1970s and again in the 1990s and in Thailand and Malaysia and Indonesia in the 1990s could not continue because the borrowers' indebtedness was increasing too rapidly relative to their incomes. It was inevitable that the lenders would become more cautious in increasing their loans to these borrowers, although the details and the timing of these moves could not have been predicted. As the flow of cash from new loans diminished, the borrowers would need a new source of money for their interest payments. The likelihood that these countries could adjust to

the decline in the inflow of money without the depreciation of their currencies was low. Similarly, at some stage it was inevitable that Japanese real estate prices would stop increasing; many of those that had recently borrowed money to buy properties would encounter a cash bind because the interest payments on their loans would be larger than their rental income.

The causes of financial tumultuousness

The financial tumult since the early 1970s resulted from the impacts of monetary shocks and credit market shocks on the direction and scope of cross-border money flows. The monetary shocks involved unanticipated changes in the rates of money supply growth and the accompanying impacts on anticipated inflation rates and on interest rates. The credit market shocks involved the relaxation of financial regulations that allowed the banks to increase their loans to specific groups of borrowers, who then became more attractive to lenders. In several cases a monetary shock and a credit shock occurred at the same time and had complementary impacts on money flows across national borders.

Increases in the flows of money to countries induced both increases in the values of their currencies and increases in the prices of their assets, which moved above their long-run equilibrium values. The money flows could not be sustained indefinitely; declines in the flows almost always led to a depreciations of these currencies and often triggered crashes in asset prices of 50 or 60 percent or more.

The first major shock in this extended period was the increase in the annual US inflation rate to the range of 5 and 6 percent in the second half of the 1960s; in the previous twenty years this rate was almost always below 3 percent and usually less than the inflation rates in Germany and its neighbors in Western Europe. Annual US payments deficits surged at the end of the 1960s as it seemed increasingly likely that the prices of the European currencies and the Japanese yen would increase. Investors and firms moved money from the United States to avoid losses and profit from these anticipated changes in parities. Because the United States was reluctant to increase the US dollar price of gold and Germany and France and Japan were unwilling to revalue their currencies, the payments imbalances became larger and the international reserve assets owned by Germany and Japan and other countries with payments surpluses increased more rapidly. Then in 1971, when the US economy slowed and the inflation rate declined the Federal Reserve adopted a more expansive monetary policy, and the decline in interest rates on US-dollar securities led to larger flows of money from New York to foreign centers.

The global inflation in the early 1970s was an unprecedented peacetime event that followed from the combination of the more rapid growth of the

US money supply after monetary policy was eased and the more rapid growth in the money supplies in Germany and in Japan in response to their increasingly large payments surpluses. The more rapid increase in the US inflation rate meant that a realignment of the values of national currencies was inevitable. Because the US inflation rate exceeded the inflation rate in Germany by more than 2 percent, the Bretton Woods system of adjustable parities was no longer viable and the abandonment of parities for the German mark, the Japanese yen, and the currencies of other industrial countries became inevitable.

Interest rates on US dollar securities increased as the US inflation rate climbed; however interest rates on US dollar demand deposits were subject to ceilings that the Federal Reserve had adopted to limit competition among banks. These ceilings did not apply to interest rates on US dollar deposits in London and other foreign financial centers, which increased and induced investors to move funds from domestic to offshore financial centers. The rapid increases in money supply growth in the United States and other industrial countries in the early 1970s contributed to sharp increases in demand for primary products, and in the prices of oil, wheat, and other commodities. The rates of growth of GDP in the countries that produced these primary products increased. The Saudi Arabian embargo on oil shipments to the United States and the Netherlands following the Yom Kippur War of October 1973 triggered a sharp increase in the demand for petroleum and the oil price surged. The decline in oil supplies following the Iraqi invasion of Iran in 1979 had a much larger impact on the price of oil and global inflation. Investors increased their purchases of gold and other precious metals, collectibles, real estate, and other 'hard assets' as inflation hedges.

The first wave of credit bubbles

The increase in the rates of economic growth in the primary producing countries in the 1970s led to the growth of bank loans to governments and government-owned firms in Mexico, Brazil, Argentina, and other developing countries of 30 percent for about ten years; the external indebtedness of these countries increased by 20 percent a year. Banks headquartered in Canada, several European countries and Japan used US dollars that they borrowed in the offshore deposit markets in London, Zurich, and Luxembourg to fund these loans and to 'poach' on what had been the turf of US banks. The US banks responded aggressively to minimize the declines in their market shares. Moreover US and foreign banks wanted to circumvent the regulations that limited the growth of their domestic loans and assets. The increase in the flows of money to Mexico, Brazil, Argentina and other developing countries enabled them to finance larger trade deficits.

The change in the operating procedures of the Federal Reserve in October 1979 (the so-called 'Volcker shock') was the next major shock. Previously the Federal Reserve had stabilized interest rates and market forces had determined the growth of credit; under the new policy the Fed sought to limit the growth of credit, and market forces determined interest rates – which surged. The US dollar price of gold peaked ten weeks after this policy change; the anticipations that the inflation rate would continue to increase were shattered. Investment spending fell, a recession followed, and the prices of petroleum and other commodities dropped sharply.

The combination of the much higher interest rates on US dollar securities and the sharp reduction in the anticipated US inflation rate led to an increase in investor demand for US dollar securities, and the German mark, the Japanese yen, and other currencies depreciated. Mexico and other developing countries were squeezed by the scissors-like increase in the interest rates on their foreign loans and the decline in both the volumes and the prices of their exports. Banks in Texas were similarly impacted by higher interest rates and lower oil prices. Interest rates paid by US thrift institutions on their short-term deposits increased rapidly and in many cases began to exceed the interest rates on their long-term mortgage loans.

The second wave of credit bubbles

The appreciation of the Japanese yen that began in the spring of 1985 induced the Bank of Japan to buy US dollars to dampen the strengthening of the currency; the central bank's holdings of US dollar securities surged, which led to a rapid increase in the money supply. The Japanese financial system had been extensively regulated with the intent to provide loans to corporate borrowers at very low interest rates. Moreover administrative guidance required that these banks extend loans to firms in the industries that the government bureaucrats believed were strategically important. One motive for financial liberalization was that the industrial demand for bank loans had declined so that it was not longer necessary to allocate credit among borrowers on a preferential basis; another was that the US authorities demanded that US banks and other US financial firms have access to the banking and capital markets in Tokyo on terms comparable to the access available to Japanese banks in New York.

Financial deregulation enabled the banks headquartered in Tokyo and in Osaka to increase their real estate loans at a rapid rate. Because of building restrictions and the time consumed in assembling larger lots for construction, the increase in the demand for real estate had a much larger impact on the price of land than on the supplies of living space and office space. Many of the firms that were listed on the Tokyo Stock Exchange were real estate investment

companies; the increases in real estate prices contributed to sharp increases in the value of their assets and in the prices of their stocks.

Restrictions on the foreign investments of Japanese banks and firms were relaxed in the effort to limit the upward pressure on the yen. These banks rapidly increased the number of their branches and subsidiaries in London, New York, Zurich, and other financial centers. The flow of savings from Japan to the United States and various European countries surged; the cliché in both New York and Tokyo was 'Where will the US Treasury get the money to finance its fiscal deficit if the Japanese stop buying US government securities?' These newly established foreign branches of Japanese banks increased their loans in their host countries, using the funds obtained in the offshore deposit market; these newly established branches charged lower interest rates than their host-country competitors because they wanted to increase their market shares. Moreover Japanese investors began to purchase real estate – office buildings, apartment buildings, golf courses, and ski resorts – in the United States and in other industrial countries; most of these purchases were funded with money borrowed from the branches of Japanese banks in London, Zurich, and other offshore financial centers.

Financial liberalization in Finland, Norway, and Sweden enabled banks headquartered in these countries to source for domestic loans by borrowing in the offshore market. Stock prices and real estate prices in these countries increased sharply in response to the money inflows. The decision of the newly appointed Chair of the Board of the Bank of Japan at the beginning of 1990 to restrict the growth of bank loans for real estate pricked the asset-price bubble; heavily indebted borrowers could no longer obtain enough money from new loans for their scheduled interest payments. Stock prices declined by 30 percent in 1990 and by 25 percent in 1991. The growth of the Japanese economy slowed dramatically. The Japanese yen appreciated as exports surged relative to imports. Japanese firms responded to the adverse impact of the increase in the value of the yen on their profitability by increasing their investments in manufacturing facilities in China and Thailand and other countries in Southeast Asia that would be used primarily as sources of supply for markets in Japan, the United States and other industrial countries. Japanese banks rapidly increased their loans in these countries.

The third wave of credit bubbles

The development of the Brady bonds in 1989 and 1990 enabled Mexico and other developing countries to convert bank loans that had been in default into long-term bonds that were partially guaranteed by the US government, which effectively ended their financial isolation ('the lost decade') from the global capital market. Mexico began to prepare for membership in the North American Free Trade Agreement; the Bank of Mexico adopted a contractive

monetary policy to reduce the inflation rate, hundreds of government-owned firms were privatized, and government regulations on international trade and business practices were liberalized. Foreign direct investment in Mexico surged as US, European, and Japanese firms rapidly increased their manufacturing facilities in Mexico. US money market funds bought peso securities because the interest rates were high and attractive. US pension funds and mutual funds increased their purchases of stocks of a new asset class – 'emerging market equities'. Mexico's current account deficit increased to 6 percent of its GDP.

Similarly there was a surge in money flows to Brazil, Argentina, Thailand, Malaysia, and other emerging-market countries (which had been re-christened after having been known as developing countries for several decades). Their currencies appreciated in real terms, and their trade and current account deficits increased.

Then at the beginning of 1994 several political incidents – an Indian uprising in the southernmost province and then the assassination of the leading presidential candidate of the dominant political party – led to a decline in the flow of money to Mexico, which meant that the Bank of Mexico could no longer finance its large trade and current account deficits with the money inflows and the peso depreciated sharply. The flow of money to Thailand slowed significantly in late 1996 because the non-bank finance companies that had been established by the banks to circumvent regulations on consumer loans were incurring large loan losses – which in effect were the losses of the banks once-removed. The Bank of Thailand was unable to maintain the value of the Thai baht once its international reserve assets had been depleted at the beginning of July 1997. The depreciation of the baht triggered the contagion effect that rippled through the Asian countries as well as to Brazil, Argentina, and Russia; the imports of this group of countries declined relative to their exports by $150 billion. Several of these shocks were true surprises: the political events in Mexico in the first few months of 1994 could not have been foreseen. However, the Mexican current account deficit was too large to be sustained and some trigger eventually would have led to a decline in the inflow of money. Similarly the current account deficits of Thailand and Malaysia in 1996 were too large to be sustained. Some event would have led to declines in money flows, although the catalyst for these declines could not have been foreseen. The implosion of the bubbles in real estate prices and stock prices Thailand and Malaysia was inevitable when the money flows to these countries declined.

The fourth wave of credit bubbles

The striking feature of the period between 2002 and 2007 was the sharp increases in the prices of residential and commercial real estate in the United States, Britain, Spain, Ireland, Iceland, South Africa, New Zealand, and several

other countries. The prices of residential real estate in several of these countries more than doubled. The increases in prices of residential real estate in the United States were smaller than in other countries; however, these US price increases occurred primarily in one-third of the states, that were in the south and along both seaboards, and the prices in some of these states more than doubled.

Each of these countries experienced an increase in money inflows and their currencies appreciated, except for Spain and Ireland, which were members of the European Monetary Union. Once again the rate of increase in indebtedness was too large to be sustained. When the money flows to the United States and Britain declined, the supplies of credit available for real estate purchases fell, and real estate prices tumbled. The prices of mortgage-related securities declined, and many financial institutions in the United States and in Britain failed. Ireland had experienced a massive construction boom; its banks failed when the bubble burst. Iceland had experienced a spectacular increase in stock prices during the years when there were large inflows of money; when these inflows stopped, stock prices declined sharply and the currency lost nearly half of its value.

US real estate prices began to decline at the beginning of 2007. One of the large investment banks was merged into one of the largest commercial banks with a wedding dowry that might reach nearly $30 billion depending on its loan losses. In mid-September 2008, two of the large government-sponsored mortgage lenders that owned 50 percent of US mortgages were effectively taken over by the US Treasury; the owners of their common stock and their preferred stock lost all their money. The owners of the bonds of these mortgage lenders were 'bailed out' by the US Treasury; otherwise they would have incurred losses of several hundred billion dollars. Several days later, Lehman Brothers, the fifth largest US investment bank, failed; the efforts to negotiate the sale of the firm were not successful. A massive credit panic and crash followed; the interest rates of corporate bonds surged relative to interest rate on US Treasury securities. A few months later there were runs on the bonds of the governments of Greece and Portugal. The fiscal deficit of the Greek government was more than 12 percent of its GDP, and its total indebtedness was more than 125 percent of its GDP. The large international banks had been willing to acquire the bonds of these governments, which enabled them to finance fiscal deficits that were much too large to be sustained. The total indebtedness of the Greek government had surged in part because one or several previous governments had not been fully forthcoming with information about its total indebtedness. Nevertheless the indebtedness of these governments had surged because the 'money was there' – the banks were willing to buy the bonds which enabled the governments to finance large deficits. Then the crises occurred when the bank lenders suddenly became much more cautious, and the governments did

not have enough cash to pay the interest on their indebtedness and meet the government payrolls.

The patterns in the data

The identity of the borrowers in each of the wave of credit bubbles differs. The borrowers in the first wave were governments and government-owned firms in Mexico and ten other developing countries. The borrowers in the second wave were home owners and real estate companies in Japan; similarly it was the banks and home owners in Finland, Norway, and Sweden that were the borrowers. The borrowers in the third wave often were banks in emerging-market countries, who used external sources of funding because they were less costly than the domestic sources. The borrowers in the fourth wave were primarily home owners and property developers.

Several large banks were lenders in more than one wave. Banks headquartered in the United States, Canada, Japan, Britain, and France were among the lenders in the first wave. Japanese banks were the primary lenders in their domestic real estate bubble in the 1980s. US banks as well as banks based in Britain, Ireland, Iceland, and Spain were among the lenders in the fourth wave. Several of the US banks that encountered large losses during this wave also had incurred large losses in the first wave.

Despite these differences in the identities of the borrowers and the lenders in these several waves, there was remarkable similarity in the pattern of cash flows. Most of the countries that were identified with credit bubbles also experienced an appreciation of their currencies and an increase in their current account deficits. Japan is an exception in that it had a current account surplus; however, the yen appreciated and its trade and current account surplus declined as its credit bubble expanded. The increases in the flows of money to these countries contributed to the increases in asset prices.

The appreciation of currencies of these countries and the increase in asset prices were responses to the increase in money inflows and components of the international adjustment problem; domestic consumption had to increase to ensure that the increase in imports more or less matched the money inflows. The higher levels of consumption spending induced by the elevated asset prices led to economic booms. The governments in some of these countries – including the US government in the late 1990s – realized fiscal surpluses in response to the spending booms, although several governments increased their spending almost as rapidly as their revenues increased.

The pattern of cash flows between lenders and borrowers in each wave was Ponzian – the rate of increase in the indebtedness of the borrowers was several times higher than the interest rate on their indebtedness. For three, four, or more years, the borrowers incurred no burden in their debt service payments,

since there was more than enough cash from new loans to pay the interest on their indebtedness. However, the pattern of cash flows was not sustainable; at some stage it was inevitable that the lenders would become more cautious in extending new loans, perhaps because their loans to this group of borrowers were becoming a larger share of their loan portfolios. Any slowing in the growth of cross-border loans to this group of borrowers would lead to depreciation of their currencies – which might contribute to the greater cautiousness by the lenders. The borrowers would incur a real debt-servicing burden for the first time when the rate of growth of indebtedness declined below the interest rate; then they would need to find some other source of money for part of their interest payments. Governments would need to increase their taxes relative to their expenditures to get the money to make the interest payments – or they would default.

In a world with perfect foresight, the lenders would have foreseen that the inevitable depreciation of the borrowers' currencies would lead to massive losses by the borrowers, which might spill over to large losses by the lenders. The losses by the lenders were so large that the reasonable conclusion is that the lenders had less than perfect foresight; they failed to ask 'Where will the borrowers get the money to pay the interest on their outstanding indebtedness if there are not enough new loans to provide the money?'.

The likelihood is high that these several waves of bubbles were not independent events; instead there was a systematic relationship between the implosion of one wave and the beginning of the next wave. When the first wave of bubbles imploded in the early 1980s, the currencies of Mexico and the other developing countries depreciated sharply, and their trade balances morphed into surpluses. The countries that had developed large trade surpluses when the credit bubbles were expanding would experience the real appreciation of their currencies and declines in their trade surpluses.

The Japanese were reluctant to accept the decline in their trade surplus in the second half of the 1980s because of the adverse impacts on profits and employment in their export industries; the intervention by the Bank of Japan to dampen the appreciation of the yen led to the surge in the reserves of the Japanese banks and the increase in the money supply. Regulations that constrained domestic and foreign transactions were liberalized to limit the appreciation of the yen; banks were able to increase the supply of credit available for real estate purchases, which led to higher property prices.

When the bubble in real estate and stock prices in Japan imploded, the yen appreciated; Japanese firms then increased their investments in Thailand, Malaysia, and nearby countries in anticipation that production costs for low-value-added activities would be below the domestic costs. About the same time, money flows to the emerging-market countries increased rapidly because investment banks had discovered 'emerging market equities as a new

asset class'. Many countries were involved in extensive privatization activities, which attracted firms based in the industrial countries.

When the bubbles in Thailand and its neighbors imploded, their currencies depreciated sharply and their trade deficits morphed into trade surpluses. The counterpart was that there was a surge in the flow of money to the United States as they reduced their foreign indebtedness, which contributed to the increase in the US trade deficit and the bubble in US stock prices.

After the US stock price bubble imploded in the spring of 2000, the US dollar began to depreciate. The flow of money to the United States, Britain, and other countries increased from about 2002, and the currencies of these countries appreciated, and asset prices in these countries increased. Thus the large changes in the cross-border money flows led to large changes in currency values and in the prices of assets, and these changes contributed significantly to the bubbles in currencies and in asset prices. The real estate bubbles that developed in the United States, Britain, and other countries after 2002 were a response to the increases in the inflows of money. It is as if there is a pool of international money that can be tapped by borrowers in various countries when they conclude that the net interest costs are lower than the costs of obtaining funds in their domestic markets. Many of these borrowers incur the currency exposure but presumably they have concluded that the reduction in their interest payments more than compensates for this risk.

The impacts of monetary shocks and credit market shocks

The striking feature of the last forty years has been the variability of cross-border flows of money. The German mark and the Japanese yen appreciated throughout most of the 1970s and then depreciated sharply in the first half of the 1980s. When the money was flowing to Mexico in the early 1990s and its economy was booming, the peso appreciated significantly after adjustment for differences in inflation rates, and the Mexican trade deficit reached 6 percent of its GDP. When money was withdrawn from Mexico at the end of 1994, the peso depreciated sharply and the Mexican trade surplus reached 4 percent of its GDP. This change in cross-border money flows was massive and sudden, and had powerful impacts on the value of the peso, the inflation rate in Mexico, the prices of peso securities and of real assets and the solvency of Mexican firms, households, and banks.

Financial market events in Thailand, Norway, Indonesia, Iceland, and many other countries were similar to those in Mexico. The Minsky story of the cyclical variability in the supplies of credit in domestic economies is dramatically evident in the global economy in the variability in the cross-border money flows. In domestic economies, increases in the credit supply contribute to

the economic booms and euphoria, which in turn lead investors to become more optimistic and encourage lenders to extend more credit. Increases in the flows of money to countries led to the appreciation of their currencies and to dramatic increases in the prices of securities and other assets. Investor optimism led lenders to provide more credit to domestic borrowers and economic booms often followed.

Some of the credit market and monetary shocks led to increases in the flows of money to countries and to increases in the prices of commodities, currencies, stocks, and real estate. As long as the currencies appreciated and asset prices were increasing, the rates of return on these currencies and securities were high and increasing; optimism about the economic futures increased. Then another shock would trigger a decline or reversal in cross-border money flows; and the prices of currencies, securities, and other assets crashed.

These manic-type shocks resulted from extensive changes in the preferences of investors for securities and other assets denominated in different currencies. Investors became concerned that the US inflation rate would increase in the 1970s; they sold US dollar securities and bought securities denominated in the German mark, the Swiss franc, and the British pound, and these currencies appreciated by much more than the excess of the US inflation rate over the foreign inflation rates. During the same period the US dollar price of gold increased 'because gold was a good inflation hedge', although the annual percentage increases in the price of gold in the second half of the 1970s were many times larger than the percentage increases in the US price level. Early in 1980 investors became convinced that the US inflation rate would decline; they sold securities denominated in the German mark and other foreign currencies and these currencies depreciated rapidly.

One explanation for the greater variability of cross-border capital flows in the last thirty years is that shocks, and especially those that involved changes in the stance of monetary policy, have been larger than in earlier periods when currencies were pegged or when there was a commitment to parities for national currencies. One of the major arguments in the case for floating exchange rates is that central banks then would have greater independence to change their monetary policies and the rates of growth of their money supplies and their interest rates to achieve their domestic economic objectives. In effect the commitments to parities for national currencies constrained changes in the central banks' monetary policies and especially the adoption of more expansive monetary policies; these commitments meant that national inflation rates could not differ significantly from the inflation rates in the countries' major trading partners. In the absence of a commitment to parities, the policies adopted by central banks have led to changes in the current and anticipated inflation rates that in turn induced large changes in cross-border flows of money. Hence the much greater variability in the cross-border flows of money in part reflects that

changes in national monetary policies, and the changes in anticipated infla-
tion rates have been larger than when currencies have been pegged.

The much greater variability in the cross-border flows of money in part
reflected that the monetary shocks were greater than when currencies had
been pegged; these shocks led to changes in investor estimates of the infla-
tion rates in different countries and hence in the anticipated currency values
at various future dates. The expansive US monetary policies of the late 1960s
and the early 1970s led investors to revise upward their estimates of the US
inflation rate and also to revise upward their estimates of the anticipated
values for the German mark and the Japanese yen. Investors sold US dollar
securities and bought securities denominated in the German mark, the Swiss
franc, and the British pound, which led to the sharp appreciation of these
currencies.

If, as a group, investors increase the proportion of non-dollar securities in
their portfolios, then the United States would develop a larger current account
surplus (or a smaller current account deficit), which would require that the
German mark – or the euro as its successor – the Japanese yen, and other foreign
currencies appreciate less rapidly than would be inferred from the difference in
national inflation rates. These foreign currencies overshoot the values inferred
from the differences in national inflation rates as long as investors increased
the rate at which they acquired foreign securities.

The adoption of a more contractive US monetary policy in the autumn of
1979 led investors to reduce their estimates of the US inflation rate and to revise
downward their estimates of the values of the German mark, the Japanese yen,
and other foreign currencies. Their purchases of US dollar securities led to the
depreciation of the mark and the yen.

Overshooting was inevitable whenever investors wished to increase their
holdings of securities denominated in a particular currency; similarly, under-
shooting was inevitable whenever investors wished to decrease their holdings
of securities denominated in this currency. The earlier clichés applied to large
and rapid deviations between the currency values and those that were consist-
ent with the differences in national inflation rates – the 'vicious and virtuous
cycle' and 'destabilizing speculation' – reflected the impacts of sudden changes
in the direction of cross-border money flows. Changes in anticipated inflation
rates – more precisely changes in the differential in national inflation rates –
lead to overshooting and undershooting because the impact of the changes in
these differentials on the anticipated currency values induce changes in cross-
border money flows.

Because the bubble in Japanese stocks and real estate attracted more money
from abroad, the yen appreciated. Credit market shocks impact the values of
individual currencies as they affect investor demand for securities denominated

in different currencies. The credit market shocks in the last thirty years have had major impacts on the values of the Mexican peso, the Thai baht, and many other currencies because they led to changes in the amounts of the securities denominated in these currencies that investors wished to hold.

A second, complementary, explanation for the greater variability in cross-border money flows is that when currencies are not pegged, a shock of a given magnitude in the form of an increase in demand for securities denominated in a currency has a larger immediate impact on the country's GDP as a result of the increase in the prices of its securities and real estate. When currencies were pegged, the immediate impact of increases in the flows of money to countries was that their central banks' holdings of international reserve assets increased and their monetary liabilities increased correspondingly. The price of securities available in the country also increased in response to the purchases by the foreign investors.

When currencies were not pegged, comparable increases in the foreign demand for securities denominated in these currencies initiated the adjustment process to ensure that the countries' trade balances changed by the amounts that corresponded to the increases in the money inflows. The invisible hand operated to ensure that the immediate impacts of the increases in the flows of money was that domestic investment spending increased as the cost of capital declined and consumption spending increased in response to higher levels of household wealth. Household consumption spending is three or four times larger than investment spending in most countries, and accounted for most of the increase in total spending. The inevitable outcome of the adjustment process was that domestic saving declined in the countries that received large inflows of cross-border money.

The invisible hand led to increases in the values of the currencies of the countries that experienced larger money inflows and to increases in the values of assets in these countries. Moreover the invisible hand led to higher rates of growth of GDP as asset prices increased and consumption spending climbed. Thus the increases in the variability in the ratios of the changes in the trade balances to GDP that resulted from the initial increase in the flows of money to countries induced changes in the adjustment process that led to higher rates of return on securities and other assets in these countries. The increase in wealth contributed to the economic boom. In effect there was a feedback mechanism from the initial increases in the flows of money to the higher rates of return that induced further money inflows. The economic booms were prolonged and pervasive; many of the participants may have failed to recognize that the cross-border pattern of flows of money could not be sustained. One of the patterns in the data is that increases in the flows of money to countries were associated with economic booms, which were

evident in Mexico and other developing countries in the 1970s, in Mexico, Thailand, and other Asian countries in the first half of the 1990s, and in the United States in the second half of the 1990s. The appreciation of their currencies reduced the inflationary pressures associated with a robust economic expansion and the increase in export prices relative to import prices led to higher rates of economic growth. The money flows were also associated with a non-sustainable pattern of cash flows because some of the borrowers in these countries obtained the cash to pay the interest to their creditors from new loans. The continuation of the economic booms may explain why the lenders – at least a large number of them – failed to recognize that eventually there would be costly adjustments.

Could an international lender of last resort have made a significant difference?

Might the financial tumultuousness of last thirty years have been mitigated if there had been an international lender of last resort? Central banks were established as lenders of last resort to reduce the likelihood that the increase in the demand for money would lead to distress sales of assets and trigger solvency crises; these banks would increase the supply of money in response to the exceptional increase in demand. During the expansion phase of the cycle, investors took on greater risks in anticipation of higher returns – and they reduced their estimates of the risks attached to various securities. And at the same time, many lenders became less cautious in the effort to increase their share of the markets for mortgages, and for loans to business and governments.

Could an international lender of last resort have made a significant difference and reduced the frequency and scope of financial crises since the early 1980s? Or could a more active international lender of last resort have made a significant difference? The source of the problem – one source of the problem – is that the cross-border money flows have been much larger than when currencies were attached to parities, and more highly variable. Would an international lender of last resort have moderated the volume of cross-border money flows that have led to exceptional increases in the supplies of credit, or is it possible that investors' demand for foreign securities would have been even larger because they would have greater confidence that the central banks would be more ambitious in their efforts to moderate the changes in currency values? The answer depends on the responsibilities and the financial resources available to the international lender of last resort and its willingness to extend credit to central banks whose currencies may depreciate sharply, perhaps because of the decline in money inflows. Would the international lender of last resort

have the authority and be willing to caution central banks about the volume and variability of cross-border money flows?

This variant of the moral hazard argument is that efforts to stabilize the changes in currency values in response to the sharp variations in cross-border movements of money might lead investors to incur larger cross-border risks because they would have greater confidence that the range of movement in the currency values would be smaller. The lender of last resort might seek to moderate the flows of money to various countries – but the likelihood that countries would agree that the international lender might place limits on money inflows seems low. The lender might be empowered to provide cautionary statements that the cross-border flows are non-sustainable. National governments would seem more likely to agree to an arrangement that would enable to them to have access to more money from an international lender to moderate the undershooting when the direction of the money flows is reversed – if they could agree on the source of the money for the international lender.

Most of the currency crises since the early 1980s have resulted from the reversal in cross-border money flows. In virtually all of these country cases the pace of the money flows was too high to be sustained because the borrowers' indebtedness was increasing too rapidly. During the manic phase the currencies appreciated – and overshot – in response to the increase in money inflows. The inevitable first sign of a modest slackening of the pace of money inflows would have led to the depreciation of the currencies, which would feed back to much sharper declines in the inflows and then to massive depreciations.

Overshooting and undershooting were an inevitable part of the adjustment process to the changes in the volume and direction of the cross-border movement of money. In some countries the temporary severe depreciation of the currency associated with its undershooting would imperil the solvency of domestic firms that had debts denominated in foreign currencies because of the sharp increases in the domestic currency equivalent of their interest payments. The bankruptcy of these firms could imperil the solvency of domestic banks and other financial institutions.

A domestic lender of last resort might on occasion make some public statements that stock prices were increasing too rapidly and that the market was characterized by irrational exuberance. Similar statements might be made about prices in the markets for residential and commercial real estate; the central bank might advise banks to limit the increases in their real estate loans. The counterpart is that an international lender of last resort might provide private and even public statements that the increases in the external indebtedness of one or several countries were too rapid to be sustained and that the eventual adjustment to sustainable values could be costly and perhaps messy. Investors

and other market participants would be left to draw their own conclusions about the implications of these statements.

The International Monetary Fund was established in the 1940s to assist countries in financing their balance of payments deficits and thus to reduce the pressure on countries to devalue their currencies because of a shortage of funds to finance their cyclical or otherwise temporary payments deficits. The motive for establishing the IMF was the belief that much of the financial instability in the 1920s and especially in the 1930s could have been avoided or mitigated if there had been an international lender of last resort. Fund staff visit each of the member countries once or twice a year to discuss the country's economic policies. The Fund is a large repository of data and country experiences about the changes in asset prices and changes in international indebtedness of individual countries. It has rarely sounded the alarm that the external indebtedness of member countries was increasing too rapidly – that their current account deficits were too large, and that the transition to sustainable values for their current account balances would be costly in terms of economic stability – and that there was more likely to be a 'hard landing' than a 'soft landing'. Nor has the IMF been able to provide the credits at the time of the crash to avoid extensive and debilitating undershooting.

Many of these countries used money from the Fund to limit the depreciation of their currencies after the advent of the currency crisis. Some countries were reluctant to accept money from the IMF because the conditions attached to acceptance would deflate their economies.

The US Treasury became a lender of last resort to Mexico at the time of that country's financial crisis at the end of 1994, because the financial resources from the IMF were too small relative Mexico's needs; the IMF and several foreign governments also provided assistance. The announcement that this money would be available limited the further depreciation of the peso. If a comparable initiative had been taken one or several weeks earlier, the undershooting of the peso would have been smaller, and the adverse impacts of the depreciation of the peso on the Mexican economy would have been less severe.

Whether the shortfall in the performance of the IMF relative to the ambitions that led to its establishment has resulted from the failures of analysis or policy or member country truculence is a topic for another book.

15
Epilogue 2010–2020

The global economic crisis that began in 2008 led to extensive discussion of changes in the regulations that would forestall a similar train of events and reduce the likelihood that the US government and the governments of other countries would need to 'bail out' their large banks in 2020 or 2025. Most of the reform proposals reflect a view – implicit if not explicit – about the cause or causes of the crisis. The standard themes include that the bubble in real estate and subprime mortgages resulted from greed, reckless lending, skewed compensation arrangements for bankers, or regulatory failures.

The prelude to the panic and crash that began in September 2008 was a rapid expansion in credit that began in 2002 that financed both purchases of real estate in the United States, Britain, South Africa, and Iceland and large fiscal deficits in Greece, Portugal, and Spain. US real estate prices peaked at the end of 2006; the subsequent decline in housing starts led to a recession that began in January 2008. Some financial institutions began to fail in the summer of 2007; both Countrywide Financial and Northern Rock needed financial assistance in August 2007.

Subsequently there were runs on Bear Stearns in February 2008 and on Lehman Brothers in August and September 2008. After Lehman failed in mid-September, the credit markets froze, and other financial institutions were subject to runs, which abated only after the public credit of governments was committed to support these firms. The government money extended to these firms was viewed as a 'bail out' and the cliché 'too big to fail' dominated the discussion, even though the shareholders of these banks as well as the shareholders of AIG lost 90 to 95 percent of their money, the firms became wards of the government, and their managements were replaced. US government assistance was intended to prevent the implosion of the US and global payments arrangements, which would have led to the failure of many otherwise solvent banks and firms and a sharp increase in unemployment. The immediate beneficiaries of the government financial assistance were the bondholders and other unsecured creditors of the banks – as well as the general public.

One group of reform proposals seeks to limit the amounts and types of assets that the firms that benefit from government deposit insurance can acquire. For example, these firms would be restricted in their investments in hedge funds and in private equity firms. Another group of proposals would limit the transactions by banks for their own accounts in various types of derivatives. Still another proposal is that bank transactions in derivatives could occur only on an organized exchange much like the futures markets in currencies and in gold, rather than in the over-the-counter market.

A large set of proposals was directed at the capital structure of banks. One generic proposal was to increase the bank capital requirement, which is the thrust of the next update to the Basel Accords ('Basel III'). A variant is that the required capital of each bank would depend on the composition of the bank's assets, the higher the share of 'risky assets', the higher the required capital. Another variant is that the capital requirements would be keyed to the rate of growth of bank assets; the higher the rate, the higher the requirement. A different approach is that the banks have 'contingent capital' – if loan losses lead to a decline in capital below a regulatory minimum, then some of the bonds issued by the banks would be converted into equity. Hence each bond holder – or the holders of certain types of bonds that had been issued by the banks – would be a contingent shareholder.

A third group of proposals focuses on changes in the institutional structure of the banking and financial services industry. One generic recommendation (the 'Volcker Plan') is that the investment banking activities of banks be separated from their conventional commercial banking activities, the distinction that had been legislated in the Glass-Steagall Act of 1933. Another proposal is that the very largest banks be 'downsized', so that thereafter, no bank would be too big to fail because of the concern that closing the bank might trigger runs on its competitors.

One set of proposals centers on the need to change the institutional location of the bank regulators. The premise is that the regulators failed to enforce the rules, either because they had been 'captured' by the firms that they were supposed to regulate or because of bureaucratic indolence – regulation seems superfluous in good times and so individuals who want a quiet life become regulators.

The regulation of banks began when they were first chartered by governments; they were subject to various constraints in exchange for the privilege of producing money. The pattern is incremental regulation; each crisis leads to the adoption of new regulations to forestall the likelihood that the same type of crisis might recur. Banks are subject to portfolio regulations, capital requirements, liquidity requirements, branching restrictions, interest rate ceilings, and deposit insurance. Regulations limit self-dealing by bankers and loans to bank owners and officers.

In the US, the Dodd-Frank 'Wall Street Reform and Consumer Protection Act' requires that banks hold more capital and that the same requirement apply to bank holding companies. There are constraints on compensation payments for mortgage originators, who are required to hold 5 percent of mortgages that they originate. There are limits on the amount of shares in hedge funds and private equity firms that the banks can own, and there are also limits on their participation in some risky derivative transactions. Moreover the Office of Thrift Supervision is to be merged into the Office of the Controller of the Currency.

Would the financial history of the last ten years have been significantly different had the regulations mandated in the Dodd-Frank legislation been adopted in 2000? Would the extraordinary increase in the prices of US residential and commercial real estate that occurred over the next seven years have been significantly smaller?

The bubble in the US housing market resulted from an extraordinary increase in the supply of credit, or what is the same thing, an extraordinary increase in the demand for mortgages and for mortgage-related securities. Part of this demand was from foreign firms, including central banks in Asia, and part of the demand was from the US government-sponsored lenders, Fannie Mae and Freddie Mac, and the Federal Home Loan Banks. (Agents of Freddie and Fannie encouraged central banks to believe that their IOUs were effectively guaranteed by the US Treasury – which proved to be true.) Part of the increase in the supply of credit was from US pension funds and US insurance companies. Moreover US investment banks had acquired tens of billions of dollars of mortgages that they would eventually securitize, which were held in inventories until they could be transferred to the trusts that would issue the mortgage-backed securities.

The US commercial banks accounted for no more than 20 percent of the increase in the supply of credit for real estate. The massive loan losses that the banks incurred between 2007 and 2010 resulted because their judgments about the trend of real estate prices were wrong; these firms – and the financial regulators – underestimated the credit risks attached to mortgages and mortgage-related securities because they did not believe that the price of residential real estate could decline, perhaps because they believed that prices had never before declined on a national basis. (The lenders did not realize that real estate prices had declined on a national basis in the Great Depression.) The losses that the banks incurred on their transactions in esoteric derivatives – except for those associated with mortgage-related securities – were small; these transactions had a significant impact on the solvency of only one firm, AIG, which was primarily an insurance conglomerate (although AIG owned a small bank). AIG needed increasingly large amounts of cash as margin for the credit default swaps that it sold as the credit-rating agencies downgraded its risk profile. The rationale for

massive government assistance to AIG was that if it failed, many of the counterparties that had bought the swaps that it had sold would also fail.

The supply of credit available for the purchase of mortgages and the scope of the real estate bubble that began in 2002 may have been modestly stronger because of the belief that the bonds of Fannie Mae and Freddie Mac were effectively guaranteed by the US Treasury. Similarly the supply of credit may have been modestly larger because the credit-rating agencies were corrupted by the investment banks in their assessments of the risk attached to various mortgage-related securities. Moreover, the supply of credit may have been modestly larger because some of the firms that originated mortgages that would be securitized were less meticulous in the appraisal of the credit risks than if they planned to hold the mortgages in their own portfolios – hence the requirement that the originators hold 5 percent of the credit risk.

One of the stylized facts is that as credit bubbles expand, the lenders extend credit to borrowers who are increasingly less attractive in terms of their repayment histories and their ability and willingness to adhere to the contracts. The increase in the share of subprime loans from 6 per cent to 20 percent of the total market for residential real estate mortgages that occurred between 2004 and 2005 resulted because there weren't enough prime mortgages to satisfy the demand for mortgage-related securities. Some of the lenders wanted the higher rates of return associated with mortgage-related securities. The borrowers on the borderline of being creditworthy were able to obtain money because there were no binding macro constraints on the increases in the supply of credit. Many firms, including Countrywide Financial, Washington Mutual, and Northern Rock, took on more credit risk because they wanted to increase their market shares while many of their competitors were reluctant to cede market share with the result that credit standards declined.

The Dodd-Frank legislation will increase the costs of banks. One of the standard features of financial market development is that when the costs of regulation imposed on various firms become significant, new institutions are developed that circumvent these regulations. The parallel financial system developed alongside the traditional system in response to the regulations imposed on traditional banks; thus money market funds, offshore banking, and special investment vehicles are components of this system. Fannie Mae and Freddie Mac had a cost advantage relative to traditional mortgage lenders because of the implicit government guarantee of their bonds, but they also had an advantage because they had much lower capital requirements. During the financial crisis, the money market funds were brought under the umbrella of the federal deposit insurance guarantees to staunch incipient runs – a free lunch for the owners of money market funds and for the firms that issue these IOUs.

It is too soon to determine whether the increase in the costs of regulation that will follow from the Dodd-Frank Law will lead to a significant expansion in the

roles of firms in the parallel financial system. Some of these new unregulated firms might be owned by the regulated institutions. There is nothing in the Dodd-Frank Act that would prevent a surge in the supply of credit available for real estate or an increase in real estate prices comparable to the one that occurred after 2002 – or to even more extensive increases. The share of credit from regulated banks would be somewhat smaller because of the regulations. Nevertheless when the real estate bubble of 2020 implodes, some banks will incur such large loan losses that they will be forced to close or to seek a well-capitalized partner for a merger; others will be protected because of the higher level of required capital. How many banks will be forced to close will depend on the scope of the increase in real estate prices during the bubble relative to the increase in bank capital. When real estate prices begin to decline, it is inevitable that there will be runs on the firms in the parallel banking system, much like the runs on the investment banks and the money market funds and the large insured banks in 2007 and 2008 and 2009. The regulators will have said during the bubble years and perhaps earlier: 'We will not bail out or otherwise support the lenders that have circumvented regulation.' Perhaps the authorities at the time of the crisis will adhere to this commitment – but the lesson of history is that they will not, and that institutions that were developed to circumvent regulation will receive government financial assistance during the crisis because their failure would be costly for the regulated institutions, because the prices of most assets – except for government bonds – would decline.

The premise for many of the proposed changes in financial regulation and for the major features of the Dodd-Frank legislation is that the cause of the financial crisis was on the demand side and the reckless lending by individual banks and other firms. The competing view is that the cause of the crisis was that the supply of credit had increased at too rapid a rate in the United States, Britain, Ireland, Spain, Iceland, and South Africa; similarly the supply of credit available for the purchase of bonds of the governments of Greece and Portugal had increased at too rapid a rate. Reckless lending and greed had only a minor impact on the supply of credit; the lenders were responding to the rapid increase in the supply of credit. The banks were one of the channels for credit flows. The supply of credit would have been satisfied in some other way if the banks had been more cautious in buying mortgages and mortgage-related securities.

The shortcoming of the proposals for financial reform and the Dodd-Frank legislation is that they do not affect the supply of credit and the increase in the demand for mortgage-related securities. The likelihood is high that if the US financial regulatory regime had been restructured in 2000 along the lines of the Dodd-Frank legislation, the monetary history of the next ten years would not have been significantly different.

Appendix

A Stylized Outline of Financial Crises, 1618 to 2008

Year	1618–1623	1636–1637	1690–1696
Countries (city)	Holy Roman Empire	Dutch Republic	England
Related to	Thirty Years' War	Boom in war against Spain	Glorious Revolution 1688; war with France 1689–1697
Preceding speculation in	Subsidiary coin, exchanging bad for good	Shares of Dutch East India Company, real estate, exotic tulip bulbs, common tulip bulbs, canals	East India Company, treasure, new companies, lotteries
Monetary expansion from	Debasement of coins by weight, fineness, denomination	None (?), down payments in kind	Coin debasement, Bank of England established 1694
Speculative peak	February 1622	February 1637	1695
Crisis (crash, panic)	February 1622	February 1637	1696
Lender of last resort	None	None	None

Year	1720		1763	1772	
Countries (city)	England	France	Amsterdam	Britain	Amsterdam
Related to	Treaty of Utrecht, 1713	Death of Louis XIV, 1715	End of Seven Years' War	Seven Years' War (10 years after)	
Preceding speculation in	South Sea Company stock; government debt	Mississippi Company, Banque Générale, Banque Royale	Commodities, esp. sugar (?)	Housing, turnpikes, canals	East India Co.
Monetary expansion from	Sword Blade Bank	John Law banks	*Wisselruitij* (chain of accommodation bills)	Ayr Bank; country banks	*Wisselruitij*; Bank of Amsterdam
Speculative peak Crisis (crash, panic)	Apr. 1720 Sept. 1720	Dec. 1719 May 1720	Jan. 1763 Sept. 1763	June 1772 Jan. 1773	
Lender of last resort	Bank of England (??)	None	Bank of England	Bank of England	City of Amsterdam

Year	1792	1793	1797	1799	1810	1815–16
Countries (city)	United States	England	England	Hamburg	England	England
Related to	Constitution adopted, 1789	Reign of Terror (France)	Collapse of *assignats*; French landing, Fishguard	Break in Continental blockade	Wellington's peninsula campaign	End of Napoleonic War
Preceding speculation in	Recovery of US bonds	Canal mania	Securities, canals	Commodities	Exports to Brazil (and Scandinavia)	Export commodities, Continent and United States
Monetary expansion from	Treasury accepted public securities at par for stock of Bank of the US	Capital flows from France	Country banks	*Wechselreiterei*	Country banks	Banks
Speculative peak	Jan. 1792	Nov. 1792	1796	1799	1809	1815
Crisis (crash, panic)	March 1792	Feb. 1793	Feb.–June 1797	Aug.–Nov. 1799	1810, Jan. 1811	1816
Lender of last resort	US Treasury open market purchases, deferred customs receipts	Exchequer bills	Exchequer bills, abandon gold	Admiralty bills	Exchequer bills	?

Year		1819	1825	1828	1836	1837	1838
Countries (city) Related to	England Waterloo (five years after)	United States	England Success of Baring loan; decline in interest rates	France Decline in interest rates	England Textile boom	United States Jackson presidency	France July 1830 Monarchy
Preceding speculation in	Commodities, securities	Manufacturing behind embargo	Latin American bonds, mines, cotton	Canals, cotton, building sites	Cotton, railroads	Cotton, land	Cotton, building sites
Monetary expansion from	Banks generally	Bank of the US	Bonds sold in installments, country banks	Paris banks	Joint-stock banks	Wildcat banks; retention of silver	Regional banks
Speculative peak	Dec. 1818	Aug. 1818	Early 1825		Apr. 1836	Nov. 1836	
Crisis (crash, panic)	None	Nov. 1818 to June 1819	Dec. 1825	Dec. 1827	Dec. 1836	Sept. 1837	June 1837
Lender of last resort	None needed	Treasury specie deposits	Bank of England	Paris, Basel banks, Bank of France banks,		Bank of France and Bank of Hamburg helped Bank of England	

Year	1847	1848	1857		
Countries (city)	England	Continent	United States	England	Continent
Related to	1846 potato blight, wheat failure			End of Crimean War	
Preceding speculation in	Railways, wheat	Railways, wheat, building (Cologne)	Railroads, public lands	Railroads, wheat	Railroads, heavy industry
Monetary expansion from	Installment sale of railway securities	Regional banks	Gold discoveries, clearing-house	Bank mergers, clearing-house	Crédit Mobilier, new German banks
Speculative peak	Jan. 1847	Mar.–Apr. 1848	End of 1856	March 1857	
Crisis (crash, panic)	Oct. 1847	Mar. 1848	Aug. 1857	Oct. 1857	Nov. 1857
Lender of last resort	Suspension of Bank Act of 1844	Bank of England loan to Bank of France; Russian purchase of French *rentes*	Capital inflow from England	Suspension of Bank Act of 1844	*Silberzug* (Hamburg)

Year	1864	1866		1873	1882
Countries (city) Related to	France End of Civil War	England/Italy General limited liability	Germany/Austria Franco-Prussian indemnity	United States Fraud exposed in 1872 campaign	France Expansion into southeastern Europe
Preceding speculation in	Cotton,	shipping companies generally	Building sites, railroads, securities, commodities	Railroads, homesteading, Chicago bldg.	Stocks of new banks, Lyons
Monetary expansion from	Crédit Mobilier	Joint-stock discount houses	New industrial banks, broker banks, construction banks	Short-term credit, inflow of European capital	Securities bought on margin
Speculative peak	1863	July 1865	Fall 1872	March 1873	Dec. 1881
Crisis (crash, panic)	Jan. 1864	May 1866	May 1873	Sept. 1873	Jan. 1882
Lender of last resort	Maturities of bills extended	Suspension of Bank Act; Italy abandoned fixed parity	None	None	Limited help from Paris banks

Year	1890	1893		1907		1920–21
Countries (city)	England	United States	Australia	United States	France/Italy	Britain United States
Related to	Argentine clearing of southern lands; Brazil, coffee; Chile, nitrates; South Africa, gold	Sherman Silver Act, 1890	Growth of cities	Russo-Japanese war (?), San Francisco earthquake (??)		End of postwar boom
Preceding speculation in	Argentine securities, private companies going public	Silver, gold	Land, gold mines	Coffee, Union Pacific	Industrial borrowing from banks	Securities, ships, commodities, inventories
Monetary expansion from	Goschen conversion	Contraction	Capital inflow	Trust companies	Società Bancaria Italiana	Banks
Speculative peak	Aug. 1890	Dec. 1892	1891	Early 1907	March 1906	Summer 1920
Crisis (crash, panic)	Nov. 1890	May 1893	Spring 1893	Oct. 1907	Aug. 1907	Spring 1921
Lender of last resort	Baring liabilities guaranteed; Bank of France, Russian gold loans to Britain	Repeal of Sherman Silver Act of Aug. 1893	None	$100 million inflow from Britain	Bank of Italy	None

Year	1929	1931–33	1950s, 1960s	1974–75
Countries (city) Related to	United States End of extended postwar boom	Europe Cut-off of US foreign lending	Worldwide Convertibility without macroeconomic coordination	United States Worldwide Collapse of Bretton Woods: OPEC 1973 price rise
Preceding speculation in	Land to 1925, stocks 1928–29	Not applicable	Foreign exchange	Stocks, REITs, office buildings, tankers, Boeing 747s
Monetary expansion from Speculative peak	Stocks bought on margin Sept. 1929	US lending 1929	Not applicable Speculation in currencies of	Eurodollar market flooding in 1970–71 1973
Crisis (crash, panic)	Oct. 1929	Austria, May 1931; Germany, June 1931; Britain, Sept. 1931; Japan, Dec. 1931; United States, Mar. 1933	France, 1958; Canada, 1962; Italy, 1963; Britain, 1964; France, 1968; United States, 1973, etc.	1974–75
Lender of last resort	FRBNY open-market operations (inadequate)	Feeble efforts in United States, France	BIS swap network	BIS swap network

Year	1979–82	1982–87
Countries (city)	United States World	United States
Related to	Third World syndicated bank loans, OPEC 1979 price rise in oil, real estate in southwest United States, US farmland, dollar	Stock market, luxury housing, office buildings, dollar
Monetary expansion from speculative peak	1979	Capital inflow dollar, 1985; stocks, 1987; real estate, 1987
Crisis (crash, panic)	Dollar, 1979; farmland, 1979; oil, 1980; Third World debt, 1982	Stocks, Oct. 19, 1987
Lender of last resort	IMF; FRBNY, US govt. for Mexican debt, Farm Loan Bank Board	FRBNY open-market operations banks, FDIC, FSLIC dollar, swaps

Year	1990	1994–95	1997–98	2007–08
Countries (city)	Japan	Mexico	Thailand, Indonesia, Malaysia, Korea, Russia, Brazil	United States, Britain, Iceland, Ireland, Spain, Greece
Related to	Nikkei shares index; real estate	Deregulation; capital inflow and outflow; domestic boom	Deregulation, capital inflow and outflow; borrowing abroad	Real estate, stocks (Iceland), Government borrowing (Greece)
Monetary expansion from	Interest-rate reduction 1986	Capital inflow, bank lending, domestic new banks 1991, nationalized banks privatized 1991	Bank lending; construction boom; crony capitalism	Money inflows
Speculative peak	First half 1989	1994–95	1997–98	Late 2006
Crisis (crash, panic)	January 1990	1994–95	1997–98	September 2008
Lender of last resort	Ministry of Finance and Bank of Japan slow to react	US Stabilization Fund; IMF; IADB	IMF; World Bank; ADB; bilateral country loans	Central banks, International Monetary Fund

Notes

1 Financial Crisis: a Hardy Perennial

1. Ezra Vogel, *Japan as Number One: Lessons for America* (Boston: Harvard University Press, 1979).
2. C.P. Kindleberger, *The World in Depression, 1929–1939*, 2nd edn (Berkeley: University of California Press, 1986).
3. See Robert D. Flood and Peter W. Garber, *Speculative Bubbles, Speculative Attacks and Policy Switching* (Cambridge, Mass.: MIT Press, 1994), who believe in 'fundamentals' as determining economic behavior, unless governments change the rules. One particular change in the last quarter of the twentieth century was deregulation of financial markets.
4. Edward Shaw, *Financial Deepening in Economic Development* (New York: Oxford University Press, 1973); and Roland I. McKinnon, *Money and Capitalism in Economic Development* (Washington, DC: Brookings Institution, 1973). A detailed study of regulation in developing countries is 'A Survey of Financial Liberalization' by John Williamson and Molly Mohar, *Essays in International Finance*, no. 221 (Princeton, NJ: International Finance Section, November 1998).
5. *Recent Innovations in International Banking* (Basel: Bank for International Settlements, 1986).
6. See Kindleberger, 'Panic of 1873', in *Historical Economics* (New York: Harvester Wheatsheaf, 1990), pp. 310–25; idem, 'International Propagation of Financial Crises'; Henrietta M. Larson, *Jay Cooke, Private Banker* (Cambridge, Mass.: Harvard University Press, 1936); and Matthew Simon, *Cyclical Fluctuations in the International Capital Movements of the United States, 1865–1897* (New York: Arno, 1979).

2 The Anatomy of a Typical Crisis

1. Joseph A. Schumpeter, *Business Cycles: a Theoretical, Historical and Statistical Analysis of the Capitalist Process* (New York: McGraw-Hill, 1939), vol. 1, chap. 4, esp. pp. 161ff.
2. Hyman P. Minsky, *John Maynard Keynes* (New York: Columbia University Press, 1975); and idem, 'The Financial Instability Hypothesis: Capitalistic Processes and the Behavior of the Economy', in C.P. Kindleberger and J.-P. Laffargue, eds, *Financial Crises: Theory, History and Policy* (Cambridge: Cambridge University Press, 1982), pp. 13–29. For a view of the work of Hyman Minsky in historical context, see Perry Mehrling, 'The Vision of Hyman P. Minsky', in the *Journal of Economic Behavior and Organization*, vol. 39 (1999), pp. 125–58.
3. See R.C.O. Matthews, 'Public Policy, and Monetary Expenditure', in Thomas Wilson and Andrew S. Skinner, eds, *The Market and the State: Essays in Honour of Adam Smith* (Oxford: Oxford University Press, Clarendon Press, 1976), p. 336.
4. See James B. Stewart, *Den of Thieves* (New York: Touchstone Books [Simon & Schuster], 1991, 1992), p. 97: 'What really fueled the takeover boom [in the 1980s] was the sight of other people making money, big money, by buying and selling companies.'

5. See C.P. Kindleberger, *The World in Depression, 1929–1939*, 2nd edn (Berkeley: University of California Press, 1986), pp. 1–3.
6. Robert D. Flood and Peter W. Garber, *Speculative Bubbles, Speculative Attacks and Policy Switching* (Cambridge, Mass.: MIT Press, 1964), pp. 73–4, 85, 96, 98, etc.
7. Alvin Hansen, *Business Cycles and National Income* (New York: W.W. Norton, 1957), p. 226.
8. Newspaper accounts state that George Soros's Quantum Fund made a profit of $1 billion going short of the British pound and the Italian lira in 1992–93 and lost $600 million shorting the yen in the spring of 1994.

3 Speculative Manias

1. John F. Muth, 'Rational Expectations and the Theory of Price Movements', *Econometrica*, vol. 29 (July 1961), pp. 313–35.
2. Harry G. Johnson, 'Destabilizing Speculation: a General Equilibrium Approach', *Journal of Political Economy*, vol. 84 (February 1976), p. 101.
3. Milton Friedman, 'The Case for Flexible Exchange Rates', in *Essays in Positive Economics* (Chicago: University of Chicago Press, 1953). On one occasion, Friedman moved to a different position: 'Destabilization speculation is a theoretical possibility, but I know of no empirical evidence that it has occurred even as a special case, let alone as a general rule.' Milton Friedman, 'Discussion' of C.P. Kindleberger, 'The Case for Fixed Exchange Rates, 1969', in Federal Reserve Bank of Boston, *The International Adjustment Mechanism* (Boston: Federal Bank of Boston, 1979), pp. 114–15.
4. See Fernand Braudel, *The Structures of Everyday Life*, vol. 1 of *Civilization and Capitalism: the Limits of the Possible*, trans. Siân Reynolds (New York: Harper and Row, 1981), pp. 220, 221, 281, 315, 318, 335, etc.
5. H.M. Hyndman, *Commercial Crises of the Nineteenth Century* (1892; 2nd edn [1932], reprinted, New York: Augustus M. Kelley, 1967), p. 96.
6. Walter Bagehot, *Lombard Street: a Description of the Money Market* (1873; reprint edn, London: John Murray, 1917), p. 18.
7. Sir John Clapham, *The Bank of England: a History* (Cambridge: Cambridge University Press, 1945), vol. 2, p. 326.
8. Adam Smith, *An Inquiry into the Nature and Causes of the Wealth of Nations* (1776; reprint edn, New York: Modern Library, 1937), pp. 703–4.
9. Alfred Marshall, *Money, Credit and Commerce* (1923; reprint edn, New York: Augustus M. Kelley, 1965), p. 305.
10. More and more economic theorists are moving away from unswerving reliance on the assumption that market participants are uniformly intelligent, informed, and independent in thought, introducing such concepts as asymmetric information (different knowledge available to different participants), cognitive dissonance (unconscious suppression of information that fails to fit a priori views), herd behavior, procrastination that results in failure to act in timely fashion, and so on. Those interested should consult the work especially of George Akerlof and Richard Thaler. For relevant studies, see Frederic S. Miskin, 'Asymmetric Information and Financial Crises: a Historical Perspective', in R. Glenn Hubbard, ed., *Financial Markets and Financial Crises* (Chicago: University of Chicago Press, 1991), pp. 69–108; and Thomas Lux, 'Herd Behavior, Bubbles and Crashes', *Economic Journal*, vol. 105 (July 1995), pp. 881–96.

11. Gustav LeBon, *The Crowd: a Study of the Popular Mind* (London: T. Fischer, Unwin, 1922).

12. Charles Mackay, *Memoirs of Extraordinary Delusions and the Madness of Crowds* (1852; reprint edn, Boston: L. C. Page Co., 1932).

13. John Carswell, *The South Sea Bubble* (London: Cresset Press, 1960), p. 161.

14. David Cass and Karl Shell, 'Do Sunspots Matter?' *Journal of Political Economy*, vol. 91, no. 2 (April 1983), pp. 193–227. This concept of a completely extraneous event was included in the first edition more or less randomly. Since 1983, however, 'sunspots' has become a word of art to cover general uncertainty as opposed to the 'fundamentals' that feature in rational expectations.

15. Irving Fisher, *The Purchasing Power of Money: Its Determination and Relation to Credit, Interest and Crises*, 2nd edn (New York: Macmillan, 1911), esp. chap. 1, dealing with crises; Knut Wicksell, *Interest and Prices* (London: Macmillan, 1936) (first published 1898).

16. Henrietta M. Larson, *Jay Cooke, Private Banker* (Cambridge, Mass.: Harvard University Press, 1934).

17. John Berry McFerrin, *Caldwell and Company: a Southern Financial Empire* (Chapel Hill: University of North Carolina Press, 1939; reprint, Nashville: Vanderbilt Press, 1969).

18. Peter Temin, *Did Monetary Forces Cause the Great Depression?* (New York: W.W. Norton, 1976), pp. 90–3.

19. Joan Edelman Spero, *The Failure of the Franklin National Bank: Challenge to the International Banking System* (New York: Columbia University Press, 1980).

20. Bagehot, *Lombard Street*, pp. 131–2.

21. George W. Van Vleck, *The Panic of 1857: an Analytical Study* (New York: Columbia University Press, 1953), p. 31.

22. R.C.O. Matthews, *A Study in Trade-Cycle History: Economic Fluctuations in Great Britain, 1832–1842* (Cambridge: Cambridge University Press, 1954), pp. 49, 110–11; and M.C. Reed, *Investment in Railways in Britain: a Study in the Development of the Capital Market* (London: Oxford University Press, 1976).

 The ladies and clergymen – in American parlance, 'widows and orphans' – may more properly belong to a third stage when the securities have become seasoned in the market. The French call such investments suitable for 'the father of a family'. Charles Wilson, in *Anglo-Dutch Commerce and Finance in the Eighteenth Century* (Cambridge: Cambridge University Press, 1941), produces a number of variations on investor groups: in the Netherlands 'spinsters, widows, retired naval and army officers, magistrates, retired merchants, parsons and orphanages' (p. 118); 'hundreds of other merchants…as well as thousands of civil servants, magistrates, widows and orphans and charitable institutions' (p. 135); 'widows, parsons, orphanages, magistrates and civil servants' (p. 162); 'country gentry, wealthy burghers and officials of Amsterdam, widows and wealthy spinsters' (p. 181); 'spinsters, theologians, admirals, civil servants, merchants, professional speculators, and the inevitable widows and orphans' (p. 202).

 In the quotation from Bagehot that constitutes one of the epigraphs of this book, owners of the blind capital who lacked the wisdom to invest it properly were characterized in the excised portion as 'quiet ladies, rural clergymen and country misers' and again as 'rectors, authors, grandmothers'. See Bagehot, 'Essays on Edward Gibbon', quoted in Theodore E. Burton, *Financial Crises and Periods of Industrial and Commercial Depression* (New York: Appleton, 1902), pp. 321–2.

In his essay on Lord Brougham (1857), Bagehot quotes his subject on the crisis of 1814:

> The frenzy, I can call it nothing less...descended to persons in the humblest circumstances, and the farthest removed by their pursuits, from commercial cares...Not only clerks and labourers, but menial servants, engaged the little sums which they had been laying up for a provision against old age and sickness...

> The great speculators broke; the middling ones lingered out a precarious existence, deprived of all means of continuing their dealings either at home or abroad; the poor dupes of the delusion had lost their little hoards and went on the parish. (Norman St John-Stevas, ed., *Bagehot's Historical Essays* [New York: New York University Press, 1966], pp. 118–19)

Another British expression for the naive and innocent who were drawn into the last phases of a bubble is 'greengrocers and servant girls'. The American 1929 categories were 'bootblacks and waiters', whereas a more modern characterization is 'house painters and office girls' (John Brooks, *The Go-Go Years* [New York: Weybright and Talley, 1973, p. 305]). Classes of current well-to-do amateurish and sometimes badly advised investors in the United States include successful doctors and dentists and professional athletes.

23. Max Wirth, *Geschichte der Handelskrisen*, 4th edn (1890; reprint edn, New York: Burt Franklin, 1968), p. 480.

24. Ilse Mintz, *Deterioration in the Quality of Bonds Issued in the United States, 1920–1930* (New York: National Bureau of Economic Research, 1951).

25. Benjamin Stein, 'The Day Los Angeles's Bubble Burst', *New York Times*, 8 December 1984.

26. 'For Investors, Condo Craze Ends: Once Hot Market Makes Do Without Speculators', *Boston Globe*, 14 February 1988.

27. Homer Hoyt, *One Hundred Years of Land Values in Chicago* (Chicago: University of Chicago Press, 1933), p. 136.

28. Johnson, 'Destabilizing Speculation', p. 101.

29. Larry T. Wimmer, 'The Gold Crisis of 1869: Stabilizing or Destabilizing Speculation under Floating Exchange Rates', *Explorations in Economic History*, 12 (1975), pp. 105–22.

30. Christina Stead, *House of All Nations* (New York: Knopf, 1938).

31. Carswell, *South Sea Bubble*, pp. 131, 199.

32. Ibid., p. 120.

33. Clapham, *Bank of England*, vol. 2, p. 20. Hyndman, a socialist, sarcastically ascribes this example to the 1820s: 'The most ridiculous blunders were made by the class which was supposed to be carrying on business for the general benefit. Warming-pans were shipped to cities within the tropics, and Sheffield carefully provided skaters with the means of enjoying their favorite pastime where ice had never been seen. The best glass and porcelain were thoughtfully provided for naked savages, who had hitherto found horns and cocoa-nut shells quite hollow enough to hold all the drink they wanted.' (See H.M. Hyndman, *Commercial Crises of the Nineteenth Century* [1892; 2nd edn, 1932, reprinted, New York: Augustus M. Kelley, 1967], p. 39.) Clapham is right and Hyndman wrong. The source for both is J.R. McCullough, *Principles of Political Economy*, 2nd edn (Edinburgh, 1830), which refers to 1810 not 1825.

The announcement of the formation of the South Sea Company in May 1711 produced expectations of a strong demand for British goods in Latin America that

would provide 'a triumphant solution to the [British] financial problem and need for expansion for the support of our way of life'. Booming markets were anticipated in 'Colchester bays [a type of cloth], silk handkerchiefs, worsted hose, sealing wax, spices, clocks and watches, Cheshire cheese, pickles, scales and weights for gold and silver' (see Carswell. *South Sea Bubble*, p. 55).

34. William Smart, *Economic Annals of the Nineteenth Century* (1911; reprint edn, New York: Augustus M. Kelly, 1964), vol. 2, p. 292.

35. Matthews, *Trade-Cycle History*, p. 25.

36. D. Morier Evans, *The History of the Commercial Crisis, 1857–1858, and the Stock Exchange Panic of 1859* (1859; reprint edn, New York: Augustus M. Kelley, 1969), p. 102.

37. Max Wirth, 'The Crisis of 1890', *Journal of Political Economy*, vol. 1 (March 1893), p. 230.

38. P.L. Cottrell, *Industrial Finance, 1830–1914: the Finance and Organization of English Manufacturing Industry* (London: Methuen, 1980), p. 169. Cottrell notes that the Guinness flotation was for £6 million, was handled by Baring, and was oversubscribed many times.

39. A.C. Pigou, *Aspects of British Economic History, 1918–25* (London: Macmillan, 1948).

40. J.S. Mill, *Principles of Political Economy, with some of their Applications to Social Philosophy* (1848; 7th edn, reprint edn, London: Longmans, Green, 1929), p. 709.

41. Maurice Lévy-Leboyer, *Les banques européennes et l'industrialisation internationale dans la première moitié du XIX^e siècle* (Paris: Presses universitaires de France, 1964), p. 715.

42. Charles Wilson, *Anglo-Dutch Commerce and Finance in the Eighteenth Century* (Cambridge: Cambridge University Press, 1941), p. 25. For a series from the early seventeenth century, see J.G. Van Dillen, 'The Bank of Amsterdam', in *History of the Principal Public Banks* (The Hague: Martinus Nijhoff, 1934), p. 95.

43. For 1822 and 1824, see Smart, *Economic Annals*, vol. 2, pp. 82, 215. For 1888, see W. Jett Lauck, *The Causes of the Panic of 1893* (Boston: Houghton Mifflin, 1907), p. 39.

44. A. Andréadès, *History of the Bank of England* (London: P.S. King, 1909), pp. 404–5, see also p. 249.

45. O.M.W. Sprague, *History of Crises under the National Banking System* (1910; reprint edn, New York: Augustus M. Kelly, 1968), pp. 35–6.

46. Great Britain, *Parliamentary Papers* (*Monetary Policy, Commercial Distress*), 'Report of the Select Committee on the Operation of the Bank Acts and the Causes of the Recent Commercial Distress, 1857–59' (Shannon: Irish University Press, 1969), vol. 4; Consular report from Hamburg, no. 7, 27 January 1858, p. 438.

47. This statement appears in italics in Donald H. Dunn's fictionalized book, *Ponzi, the Boston Swindler* (New York: McGraw-Hill, 1975), p. 98.

48. *The Collected Works of Walter Bagehot*, ed. Norman St John Stevas (London: *The Economist*, 1978), vol. 11, p. 339.

49. Wirth, *Handelskrisen*, p. 109.

50. Wirth, 'Crisis of 1890', pp. 222–4; Alfred Pose, *La monnaie er ses institutions* (Paris; Presses universitaires de France, 1942), vol. 1, p. 215. Lauck puts the cost of the rescue operation at 25 million francs from the leading banks and 100 million francs from the Bank of France; see *Causes of the Panic of 1893*, p. 57.

51. Ronald I. McKinnon, *Money and Capital in Economic Development* (Washington, DC: Brookings Institution, 1973).

52. Carlos F. Diaz-Alejandro, 'Goodbye Financial Repression, Hello Financial Crash', *Journal of Development Studies*, vol. 18, no. 1 (Sept–Oct 1985), pp. 1–24.
53. Ronald I. McKinnon and Donald J. Mathieson, 'How to Manage a Repressed Economy', *Essays in International Finance*, no. 145 (Princeton, NJ: International Finance Section, Princeton University, 1981).
54. Wirth, *Handelskrisen*, p. 519. An apparent parallel can be found in a major exhibition in Melbourne, Australia, to celebrate the hundredth anniversary of European settlement, which boosted the city's economy briefly. See Geoffrey Searle, *The Rush To Be Rich: a History of the Colony of Victoria* (Melbourne: Melbourne University Press, 1971), pp. 285–7.
55. J.W. Beyen, *Money in a Maelstrom* (New York: Macmillan, 1959), p. 45.
56. The expression is that of Gerald Malynes in 1686, quoted in Violet Barbour, *Capitalism in Amsterdam in the 17th Century* (Ann Arbor: University of Michigan Press, 1963), p. 74.
57. William R. Scott, *The Constitution and Finance of English, Scottish and Irish Joint-Stock Companies to 1720*, 3 vols (London, 1922), as summarized by J.A. Schumpeter, *Business Cycles* (New York: McGraw-Hill, 1939), vol. 1, p. 250.
58. Carswell, *South Sea Bubble*, p. 139.
59. Hans Rosenberg, *Die Weltwirtschaftskrise von 1857–59* (Stuttgart-Berlin: W. Kohlhammer, 1934), p. 114.
60. David Divine, *Indictment of Incompetence: Mutiny at Invergordon* (London: Macdonald, 1970).
61. C.P. Kindleberger, 'The Economic Crisis of 1619 to 1623', *Journal of Economic History*, vol. 51, no. 1 (March 1991), pp. 149–75.
62. Wirth, *Handelskrisen*, p. 92.
63. Ibid., p. 458.
64. Bertrand Gille, *La Banque et le crédit en France de 1815 à 1848* (Paris: Presses universitaires de France, 1959), p. 175.
65. Lévy-Leboyer, *Banques européennes*, p. 673.
66. Gille, *Banque et crédit*, p. 304.
67. Honoré de Balzac, *César Birotteau* (Paris: Livre de Poche, 1972), esp. pp. 13–14.
68. Leland H. Jenks, *The Migration of British Capital to 1875* (New York: Knopf, 1927), p. 34.
69. Rosenberg, *Weltwirtschaftskrise*, pp. 50, 100–1. See also Stewart L. Weisman, *Need and Greed: the Story of the Largest Ponzi Scheme in American History* (Syracuse: University of Syracuse Press, 1999).
70. Carswell, *South Sea Bubble*, p. 171.
71. Ibid., pp. 140, 155.
72. Ibid., p. 159.
73. Guy Chaussinand-Nogaret, *Les financiers de Languedoc au XVIIIᵉ siècle* (Paris: S.E.V.P.E.N., 1970), p. 146. This author observed that the financiers were much more realistic than John Law, stimulating the speculation (*agiotage*) but keeping themselves aloof from the fever and ruining the system by converting their notes when they judged the moment to be the most favorable. Ibid., p. 129.
74. Clapham, *Bank of England*, vol. 2, p. 239.
75. T.S. Ashton, *Economic Fluctuations in England, 1700–1800* (Oxford: Oxford University Press, Clarendon Press, 1959), p. 151.
76. Ibid., p. 127.
77. See the review of Johannes van der Voort, *De Westindische Plantage van 1720 tot 1795* (Eindhoven: De Witte, 1973), in *Journal of Economic History*, vol. 36 (June 1976), p. 519.

78. Wilson, *Anglo-Dutch Commerce*, pp. 169–87; Ashton, *Economic Fluctuations*, pp. 127–9; Clapham, *Bank of England*, vol. 1, 242–9; Martin G. Buist, *At Spes non Fracta, Hope & Co., 1770–1815: Merchant Bankers and Diplomats at Work* (The Hague: Martinus Nijhoff, 1974), pp. 21ff.

79. Arthur D. Gayer, W.W. Rostow and Anna J. Schwartz, *The Growth and Fluctuation of the British Economy, 1790–1850* (Oxford: Oxford University Press, Clarendon Press, 1953), vol. 1, p. 92.

80. Ruth Benedict, *Patterns of Culture* (Boston: Houghton Mifflin, 1934).

81. See Herman van der Wee, *The Growth of the Amsterdam Market and the European Economy* (The Hague: Martinus Nijhoff, 1963), vol. 2, p. 202; J.A. van Houtte, 'Anvers', in Amitore Fanfani, ed., *Città Mercanti Dottrine nell' Economia Europea* (Milan: A. Guiffre, 1964), p. 311; Simon Schama, *The Embarrassment of Riches: an Interpretation of Dutch Culture of the Golden Age* (Berkeley: University of California Press, 1988), pp. 347–50; Ernest Baasch, *Holländische Wirtschaftsgeschichte* (Jena: Gustav Fischer, 1927), p. 240, quoting Büsch.

82. Schama, *The Embarrassment of Riches*, pp. 503, 505.

83. Clement Juglar, *Des crises commerciales et leur retour périodiques en France, en Angleterre et aux Etats-Unis*, 2nd edn (1889; reprint edn, New York: Augustus M. Kelley, 1967).

84. Theodore E. Burton, *Financial Crises and Periods of Industrial and Commercial Depression* (New York: D. Appleton, 1902), pp. 39–41.

85. The real prices of Australian land rose from 100 in 1870 to 450 in 1895, fell to about 360 before 1900, and then took off again to 600 by 1905. See Kevin H. O'Rourke and Jeffrey G. Williamson, *Globalization and History* (Cambridge, Mass.: MIT Press, 1999), Figure 3.1.

86. Francis W. Hirst, *The Six Panics and Other Essays* (London: Methuen, 1913), p. 2.

87. Michel Chevalier, *Lettres sur l'Amérique du Nord*, 3rd edn (Brussels: Société belge du librairie, 1838), vol. 1, pp. 261–2.

88. Ibid., vol. 2, pp. 151 ff.

89. Andréadès, *Bank of England*, p. 404.

90. Louis Wolowski, testimony before Ministère des Finances et al., *Enquête sur les principes et les faits généraux qui régissent la circulation monétaire et fiduciaire* (Paris: Imprimerie imperiale, 1867), vol. 2, p. 398.

91. Wilson, *Anglo-Dutch Commerce*, p. 77, quoting Isaac de Pinto, *Jeu d'Actions* (eighteenth century).

92. Fritz Stern, *Gold and Iron: Bismarck, Bleichröder, and the Building of the German Empire* (London: Allen & Unwin, 1977), p. 500, quoting Constantin Franz in 1872.

93. Oskar Morgenstern, *International Financial Transactions and Business Cycles* (Princeton, NJ: Princeton University Press, 1959), p. 550.

94. Robert Bigo, *Les banques françaises au cours du XIXᵉ siècle* (Paris: Sirey, 1947), p. 262.

95. Stead, *House of All Nations*, p. 233.

96. Ibid., p. 244.

4 Fueling the Flames: the Expansion of Credit

1. At least as far as I can tell from limited sources. Typically at the height of the bubble, sellers had no bulbs, and some (many?) buyers made down-payments, if at all, in kind, that is, in personal possessions or commodities, presumably because they

lacked cash. The difference between the value of the down-payment and the negotiated price was personal credit. See N.W. Posthumus, 'The Tulip Mania in Holland in the Years 1636 and 1637', *Journal of Economic and Business History*, vol. 1 (1928–29), reprinted in W.C. Scoville and J.C. LaForce, eds, *The Economic Development of Western Europe*, vol. 2, *The Sixteenth and Seventeenth Centuries* (Lexington, Mass.: D.C. Heath, 1969), p. 142; Simon Schama, *The Embarrassment of Riches: an Interpretation of Dutch Culture in the Golden Age* (Berkeley: University of California Press, 1987), p. 358; Robert P. Flood and Peter M. Garber, *Speculative Bubbles, Speculative Attacks and Policy Switching* (Cambridge, Mass.: MIT Press, 1994), p. 60.

2. Peter Temin, *The Jacksonian Economy* (New York: W.W. Norton, 1969), pp. 79–82.

3. Jean Bouvier, *Le Krach de l'Union Générale, 1878–1885* (Paris: Presses universitaires de France, 1960), pp. 129–34.

4. 'Kuwait's Great $70 Bn Paper Chase', *Financial Times*, 25 September 1982; 'Kuwait Aide Says Speculators Own "Price of Follies"', *International Herald Tribune*, 29 October 1982; 'Kuwaitis Try a New Exchange', *New York Times*, 16 December 1984, sec. D.

5. Milton Friedman, *The Optimum Quantity of Money and Other Essays* (Chicago: Aldine, 1969), pp. 1–50.

6. Jacob Viner, *Studies in the Theory of International Trade* (New York: Harper, 1937), pp. 232–3.

7. J.G. Van Dillen, 'The Bank of Amsterdam', in *History of the Principal Public Banks* (The Hague: Martinus Nijhoff, 1934), pp. 79–123.

8. Eli F. Heckscher and J.G. Van Dillen, eds, 'The Bank of Sweden in Its Connection with the Bank of Amsterdam', in ibid., p. 169.

9. Walter Bagehot, 'The General Aspect of the Banking Question', no. 1, a letter to the editor of *The Economist*, 7 February 1857, in *The Collected Works of Walter Bagehot*, ed. Norman St John Stevas (London: *The Economist*, 1978), vol. 9, p. 319.

10. Great Britain, Committee on the Working of the Monetary System, *Report* (Radcliffe Report), Cmnd 827 (London: H.M. Stationery Office, August 1959), pp. 133–4, 391–2.

11. Ibid., pp. 134, 394.

12. James S. Gibbons, *The Banks of New York, Their Dealers, the Clearing House and the Panic of 1857* (New York: D. Appleton, 1859), pp. 376–7.

13. J.S. Mill, in *Westminster Review*, vol. 41 (1844), pp. 590–1, quoted in Jacob Viner, *Studies in International Trade*, p. 246.

14. See Benjamin M. Friedman, 'Portfolio Choice and the Debt-to-Income Relationship', *American Economic Review*, vol. 75, no. 2 (May 1985), pp. 338–43; and idem, 'The Roles of Money and Credit in Macro-economic Analysis', in James Tobin, ed., *Macroeconomics, Prices, and Quantities: Essays in Memory of Arthur Okun* (Washington, DC: Brookings Institution, 1983), pp. 161–89.

15. Samuel L. Clements (Mark Twain) and Charles Dudley Warner, *The Gilded Age: a Tale of Today* (New York: Harper & Brothers, 1873; reprint edn, author's national edn, 10 vols, 1915), vol. 1, p. 263.

16. For an extended discussion see C.P. Kindleberger, 'The Quality of Debt', in D.B. Papadimitriou, ed., *Profits, Deficits and Instability* (Basingstoke: Macmillan, 1992). reprinted in idem, *The World Economy and National Finance in Historical Perspective* (Ann Arbor: University of Michigan Press, 1995), pp. 117–30.

17. See Hyman P. Minsky, 'The Financial-instability Hypothesis: Capitalist Processes and the Behavior of the Economy', in C.P. Kindleberger and J.P. Laffargue, eds, *Financial Crisis: Theory, History and Policy* (Cambridge: Cambridge University Press, 1982), pp. 13–39.

18. See the comments on the Minsky paper by J.S. Flemming, Raymond W. Goldsmith, and Jacques Melitz, in ibid., pp. 39–47.

19. 'Revco Drugstore Chain in Bankruptcy Filing', *New York Times*, 29 July 1988, sec. D.

20. Henry Kaufman, *Interest Rates, the Markets and the New Financial World* (New York: Times Books, 1986).

21. Alfred Marshall noted that paper money was used in China 2000 years before his writing, and that the apt term *flying money* was given there to bills of exchange 1000 years ago. See Appendix E, 'Notes on the Development of Banking, with Special Reference to English Experience', in *Money, Credit and Commerce* (1923; reprint, New York: Augustus M. Kelley, 1965), p. 305n.

 In Europe the bill of exchange was developed by Italian merchants to balance accounts at fairs. Net debtors at the end of trading paid in bills drawn on a fair in a different location or on the next fair at the same place. This 'private money' was needed because there was not enough coin (money of the prince) to square accounts. See Marie-Thérèse Boyer-Xambeu, Ghislain Deleplace, and Lucien Gillard, *Private Money and Public Currencies: the 16th-Century Challenge*, translated from the French (Armonk, NY: M.W. Sharpe, 1984).

22. T.S. Ashton, 'The Bill of Exchange and Private Banks in Lancashire, 1790–1830', in T.S. Ashton and R.S. Sayers, eds, *Papers in English Monetary History* (Oxford: Oxford University Press, Clarendon Press, 1953), pp. 37–8.

23. Francis C. Knowles, *The Monetary Crisis Considered* (1827), referring to the House of Lords Committee on Scottish and Irish Currency of 1826; quoted in J.R.T. Hughes, *Fluctuations in Trade, Industry and Finance: a Study of British Economic Development, 1850–1860* (Oxford: Oxford University Press, Clarendon Press, 1960), p. 267.

24. Ibid., p. 258.

25. Kurt Samuelsson, 'International Payments and Credit Movements by Swedish Merchant Houses, 1730–1815', *Scandinavian Economic History Review*, vol. 3 (1955), p. 188.

26. For an early example of such attitudes, see the hypothetical discussion of the board of directors at a New York bank in the 1850s by James S. Gibbons, *The Banks of New York, Their Dealers, the Clearing House, and the Panic of 1857*, p. 50. A director is pleading the loan application of a Mr Black, 'rich beyond a contingency', who wants to build a new house on Fifth Avenue for $60,000 and to spend $40,000 to furnish it, and proposes expanding his firm's discount line at the bank by the whole amount. Another director objects:

 > Mr. President, my notion is, that we have no right to discount any thing at the Board but a *bona fide* commercial note that will be paid when due. And on top of that the indorser must be able to take it up himself, if the drawer should fail or die. Don't you see that we are discounting this paper to pay for Mr. Black's house and furniture, just for his single enjoyment? This isn't commercial paper, sir! It's accommodation paper in the true sense.

 Gibbon's book, with chapters on the various tasks in a bank, is a mid-nineteenth-century precursor to Martin Mayer's *The Bankers* (New York: Ballantine Books, 1974).

27. R.G. Hawtrey, *The Art of Central Banking* (London: Longmans, Green, 1932), pp. 128–9.

28. See Herman E. Krooss, ed., *Documentary History of Banking and Currency in the United States* (New York: Chelsea House, 1969), vol. 1, p. 31.

29. Viner, *Studies in International Trade*, pp. 245ff., esp. pp. 249–50.

30. Adam Smith, *An Inquiry into the Nature and Causes of the Wealth of Nations* (1776; reprint edn, New York: Modern Library, 1937), pp. 293–7.
31. R.G. Hawtrey, *Currency and Credit*, 3rd edn (New York: Longmans, Green 1930), p. 224.
32. Arthur D. Gayer, W.W. Rostow, and Anna Jacobson Schwartz, *The Growth and Fluctuation of the British Economy, 1790–1850* (Oxford: Oxford University Press, Clarendon Press, 1953), vol. 1, p. 105.
33. Great Britain, *Parliamentary Papers* (*Monetary Policy, Commercial Distress*), 'Report of the Select Committee on the Operation of the Bank Acts and the Causes of the Recent Commercial Distress, 1857–59' (Shannon: Irish University Press, 1969), vol. 4, p. 113, question 1661, and p. 115, question 1679.
34. Albert E. Fr. Schäffle, 'Die Handleskrise von 1857 in Hamburg, mit besonderer Rucksicht auf das Bankwesen', in Schäffle, *Gesammelte Aufsätze* (Tübingen: H. Haupp'schen, 1885), vol. 2, p. 31.
35. Wirth, *Handelskrisen*, p. 91.
36. See C.P. Kindleberger, *The World in Depression, 1929–1939*, 2nd edn (Berkeley: University of California Press, 1986), p. 133 and note.
37. Bouvier, *Le Krach*, pp. 130–1.
38. After the bank's dissolution, the French prosecutor of Bontoux wrote that there had been grave irregularities in the issuance of the shares and in the increases in capital. Subscriptions to the capital had been made by the bank both in its own name and in the names of fictitious client subscribers. A rival bank, the Banque de Lyon et de la Loire, formed in April 1881 with a capital of 25 million francs (raised to 50 million in November) and with one-quarter of its capital theoretically paid in, was similarly a bubble. Of 50,000 shares originally issued, more than half had not paid amounts due, and the bank had less than half of the 6.5 million francs of capital claimed at the outset.

 The Union Générale also bought its own securities in the open market and, as we shall see, loaned money to others to buy them. (Bouvier, *Le krach*, pp. 123, 164–5, 167.)
39. Ibid., p. 131.
40. Ibid. For prices, see tables 7 and 8, pp. 136 and 144; for the shortfall, see p. 144.
41. Ibid., tables 7 and 8, pp. 136, 144, 145.
42. John Carswell, *The South Sea Bubble* (London: Cresset Press, 1960), p.171.
43. Bouvier, *Le krach*, pp. 112, 113.
44. Federal Reserve System, *Banking and Monetary Statistics* (Washington, DC: Board of Governors of the Federal Reserve System, 1943), p. 494.
45. Alexander Dana Noyes, *The Market Place: Reminiscences of a Financial Editor* (Boston: Little, Brown, 1937), p. 353.
46. Peter H. Lindert, *Key Currencies and Gold, 1900–1913*, Princeton Studies in International Finance, no. 24 (August 1969).
47. Jeffrey G. Williamson, *American Growth and the Balance of Payments, 1830–1913: a Study of the Long Swing* (Chapel Hill: University of North Carolina Press, 1964).
48. Alvin H. Hansen, *Business Cycles and National Income* (New York: W.W. Norton, 1957), chaps 13, 15.
49. A.C. Pigou, *Industrial Fluctuations* (London: Macmillan, 1927), pt. 1, chap. 7, and p. 274.
50. Milton Friedman and Anna J. Schwartz, *A Monetary History of the United States, 1867–1960* (Princeton, NJ: Princeton University Press, 1963). Chapter 10 of the book is

published separately as *The Great Contraction, 1929–1933* (Princeton, NJ: Princeton University Press, 1965).

51. This assertion was made by Thomas A. Mayer in a seminar on money and the Great Depression, University of California, Berkeley, 11 May 1977.

52. Peter Temin, *Did Monetary Forces Cause the Great Depression?* (New York: W.W. Norton, 1976), passim.

53. Ben S. Bernanke, 'Nonmonetary Effects of the Financial Crisis in the Propagation of the Great Depression', *American Economic Review*, vol. 73, no. 3 (June 1983), pp. 237–76.

54. Frederic S. Miskin, 'Illiquidity, Consumer Durable Expenditure, and Monetary Policy', *American Economic Review*, vol. 66 (September 1976), pp. 642–54.

55. See Minsky's review of Temin, in *Challenge*, vol. 19, no. 1 (September/October 1976), pp. 44–6; and see Milton Friedman, 'The Monetary Theory and Policy of Henry Simons', in Friedman, *Optimum Quantity of Money*, pp. 81–93.

56. Henry Simons, *Economic Policy for a Free Society* (Chicago: University of Chicago Press, 1948).

57. Friedman, 'Henry Simons', p. 83.

58. See Friedrich A. Hayek, 'Choice in Currency: a Way to Stop Inflation', Occasional Paper no. 48 (London: Institute of Economic Affairs, 1982); Roland Vaubel, 'Free Currency Competition', *Weltwirtschaftliches Archiv*, vol. 113 (1977), pp. 435–59; Richard H. Timberlake, 'Legislative Construction of the Monetary Control Act of 1980', *American Economic Review*, vol. 75, no. 2 (May 1985), pp. 97–102; Leland B. Yeager, 'Deregulation and Monetary Reform', *American Economic Review*, vol. 75, no. 2 (May 1985), pp. 103–7: Lawrence H. White, *Free Banking in Britain: Theory, Experience and Debate* (New York: Cambridge University Press, 1984); George Selgin, *The Theory of Free Banking* (Totowa, NJ: Rowan and Littlefield, 1989). For a defense of central banking, see Charles Goodhart, *The Evolution of Central Banks* (Cambridge: Cambridge University Press, 1989).

59. 'The Post-1990 Surge in World Currency Reserves', *Conjuncture*, 26th year, no. 9 (October 1996), pp. 2–12.

60. Pascal Blanqué, 'US Credit Bubble.com', *Conjuncture*, 29th year, no. 4 (April 1999), pp. 12–21.

61. Graciela L. Kaminsky and Carmen W. Reinhart, 'The Twin Crises: the Causes of Banking and Balance-of-Payments Problems', *American Economic Review* (June 1999), pp. 433–500.

62. Gayer, Rostow, and Schwartz, *Growth and Fluctuation*, vol. 1, p. 300.

63. Hughes, *Fluctuations*, p. 12.

64. Ibid., p. 261.

65. Elmer Wood, *English Theories of Central Banking Control, 1819–1858, with Some Account of Contemporary Procedures* (Cambridge, Mass.: Harvard University Press, 1939), p. 147.

66. A. Andréadès, *History of the Bank of England* (London: P.S. King, 1909), pp. 356–7.

67. Wirth, *Handelskrisen*, p. 463.

68. Ibid., pp. 515–16.

69. E. Victor Morgan, *The Theory and Practice of Central Banking, 1797–1913* (Cambridge: Cambridge University Press, 1943), pp. 184–5.

70. O.M.W. Sprague, *History of Crises under the National Banking System* (1910; reprint edn, New York: Augustus M. Kelley, 1968), p. 241.

71. Gibbons, *Banks of New York*, p. 375.

5 The Critical Stage – When the Bubble Is About To Pop

1. *The Collected Works of Walter Bagehot*, ed. Norman St John Stevas (London: *The Economist*, 1978), vol. 9, p. 273.
2. Milton Friedman, 'In Defense of Destabilizing Speculation', in *The Optimum Quantity of Money and Other Essays* (Chicago: Aldine, 1969), p. 288.
3. Harry G. Johnson, 'The Case for Flexible Exchange Rates, 1969', in Federal Reserve Bank of St Louis, *Review*, vol. 51 (June 1969), p. 17.
4. John Carswell, *The South Sea Bubble* (London: Cresset Press, 1960), p. 139.
5. W.R. Brock, *Lord Liverpool and Liberal Toryism, 1820–1827* (Cambridge: Cambridge University Press, 1941), p. 209.
6. R.C.O. Matthews, *A Study of Trade-cycle History: Economic Fluctuations in Great Britain, 1832–1842* (Cambridge: Cambridge University Press, 1954), p. 162.
7. Maurice Lévy-Leboyer, *Les banques européennes et l'industrialisation internationale dans la première moitié du XIX^e siècle* (Paris: Presses universitaires de France, 1964), pp. 618–20.
8. Ibid., p. 713.
9. Wladimir d'Ormesson, *La grande crise mondiale de 1857: L'histoire recommence, les causes, les remèdes* (Paris-Suresnes: Maurice d'Hartoy, 1933), pp. 110ff.
10. Hans Rosenberg, *Die Weltwirtschaftskrise von 1857–59* (Stuttgart: Verlag von W. Kohlhammer, 1934), p. 210.
11. Max Wirth, *Geschichte der Handelskrisen*, 4th edn (1890; reprint edn, New York: Burt Franklin, 1968), p. 463.
12. Fritz Stern, *Gold and Iron: Bismarck, Bleichröder, and the Building of the German Empire* (London: Allen & Unwin, 1977), p. 242.
13. *The Economist*, 21 April 1888, p. 500. This citation and the following one were brought to the author's attention by Richard C. Marston.
14. Ibid., 5 May 1888, pp. 570–1.
15. M.J. Gordon, 'Toward a Theory of Financial Distress', *Journal of Finance*, vol. 26 (May 1971), p. 348.
16. Carswell, *South Sea Bubble*, p. 170.
17. Sir John Clapham, *The Bank of England: a History* (Cambridge: Cambridge University Press, 1945), vol. 2, p. 257.
18. Edouard Rosenbaum and A.J. Sherman, *M.M. Warburg & Co., 1758–1938: Merchant Bankers of Hamburg* (New York: Holmes & Meier, 1979), p. 129.
19. D.P. O'Brien, ed., *The Correspondence of Lord Overstone* (Cambridge: Cambridge University Press, 1971), vol. 1, p. 368.
20. Michel Chevalier, *Lettres sur l'Amérique du Nord*, 3rd edn (Brussels: Société belge du librairie, 1838), vol. 1, p. 37.
21. Jean Bouvier, *Le krach de l'Union Générale* (Paris: Presses universitaires de France, 1960), pp. 129, 133, 137.
22. Wirth, *Handelskrisen*. p. 508.
23. D. Morier Evans, *The History of the Commercial Crisis, 1857–1858, and the Stock Exchange Panic of 1859* (1859; reprint edn, New York: Augustus M. Kelley, 1969), p. 203.
24. Testimony of Louis Adolphe Thiers, Ministère des Finances et al., *Enquête sur les principes et les faits généraux qui régissent la circulation monétaire et fiduciaire* (Paris: Imprimerie imperiale, 1867), vol. 3, p. 436.

25. Stephen A. Schuker, *The End of French Predominance in Europe, the Financial Crisis of 1924 and the Adoption of the Dawes Plan* (Chapel Hill: University of North Carolina Press, 1976), pp. 87, 104.
26. Arthur D. Gayer, W.W. Rostow, and Anna J. Schwartz, *The Growth and Fluctuation of the British Economy, 1790–1850* (Oxford: Oxford University Press, Clarendon Press, 1953), vol. 1, p. 190.
27. Ibid., p. 312.
28. Bouvier, *Le krach*, pp. 29, 130.
29. James G. Gibbons, *The Banks of New York, Their Dealers, the Clearing House, and the Panic of 1857* (New York: D. Appleton, 1859), p. 94.
30. See Clément Juglar, *Des crises commerciales et leur retour périodique en France, en Angleterre et aux Etats-Unis*, 2nd edn (1889; reprint edn, New York: Augustus M. Kelley, 1967), p. 427.
31. W.T.C. King, *History of the London Discount Market* (London: George Routledge & Sons, 1936), p. 232.
32. O.M.W. Sprague, *History of Crises under the National Banking System* (1910; reprint edn, New York: Augustus M. Kelley, 1968), p. 127.
33. Ibid., p. 33.
34. Ibid., p. 36.
35. Bouvier, *Le krach*, p. 133.
36. Christina Stead may be referring to this episode in *The House of All Nations* (New York: Simon & Schuster, 1938), when she has one of her characters, Stewart, say: ' "My first job. By jove we had fun. At one time they had a short position in Union Pacific which exceeded the floating supply. Were they ruined? Not that time. They came to terms with them...they had to, otherwise a world panic would have resulted".'
37. Sprague, *History of Crises*, pp. 237–53.
38. See C.P. Kindleberger, 'Asset Inflation and Monetary Policy', *Banca Nazionale del Lavoro Quarterly Review*, no. 192 (March 1995), pp. 17–35.
39. Carswell, *South Sea Bubble*, pp. 136–7, 158.
40. Evans, *Commercial Crisis*, p. 13.
41. Gayer, Rostow, and Schwartz, *Growth and Fluctuation*, p. 307.
42. W. Jett Lauck, *The Causes of the Panic of 1893* (Boston: Houghton Mifflin, 1907), pp. 59–60.
43. Oskar Morgenstern, *International Financial Transactions and Business Cycles* (Princeton: Princeton University Press, 1959), p. 523.
44. Part of the reason Germany sold Russian bonds was political, as was the basis for German buying of Italian bonds. The French bought Russian bonds and sold Italian bonds. But Germany did float a Mexican loan of £10.5 million in 1888, so one cannot make the case that the domestic boom in Germany required capital that in nonpolitical circumstances would have gone abroad. See Fritz Stern, *Gold and Iron: Bismarck, Bleichröder, and the Building of the German Empire* (London: Allen & Unwin, 1977), pp. 427, 433, 442. Professor Stern has indicated that there is nothing in the Bleichröder correspondence that bears on German selling of Argentine securities.
45. Johan Åkerman, *Structure et cycles économiques* (Paris: Presses universitaires de France, 1955–57), vol. 2, p. 292.
46. E. Ray McCartney, *Crisis of 1873* (Minneapolis: Burgess Publishing Co., 1935), pp. 58, 71.
47. Wirth, *Handelskrisen*, p. 110.

48. George W. Van Vleck, *The Panic of 1857: an Analytical Study* (New York: Columbia University Press, 1943), p. 68.
49. H.S. Foxwell, introduction to Andréadès, *History of the Bank of England*, p. xvii.
50. E. Victor Morgan, *The Theory and Practice of Central Banking, 1797–1913* (Cambridge: Cambridge University Press, 1943), p. 109.
51. Elmer Wood, *English Theories of Central Banking Control, 1819–1858* (Cambridge, Mass.: Harvard University Press, 1939), p. 183.
52. R.G. Hawtrey, *Currency and Credit*, 3rd edn (New York: Longmans, Green, 1927), p. 28.
53. Clapham, *Bank of England*, vol. 2, p. 153.
54. Leone Levi, *History of British Commerce* (London: John Murray, 1872), p. 233.
55. Joan Edelman Spero, *The Failure of the Franklin National Bank: Challenge to the International Banking System* (New York: Columbia University Press, 1980), pp. 66, 71, 85, 91.
56. Milton Friedman and Anna J. Schwartz, *A Monetary History of the United States, 1867–1960* (Princeton: Princeton University Press, 1963), p. 339.
57. Lauck, *Causes of the Panic of 1893*, chap. 7.
58. Sprague, *History of Crises*, p. 253.
59. Thomas Joplin, *Case for Parliamentary Inquiry into the Circumstances of the Panic, in a Letter to Thomas Gisbourne, Esq. M.P.* (London: F. Ridgeway & Sons, n.d. [after 1832]), pp. 14–15.
60. Robert Baxter, *The Panic of 1866, with its Lessons on the Currency Act* (1866; reprint edn., New York: Burt Franklin, 1969), pp. 4, 26.
61. Clapham, *Bank of England*, vol. 2, p. 101.
62. Ibid., p. 100.
63. Rosenberg, *Weltwirtschaftskrise*, p. 118.
64. Van Vleck, *Panic of 1857*, p. 74.
65. Rosenberg, *Weltwirtschaftskrise*, p. 121.
66. Sprague, *History of Crises*, p. 113.
67. Alvin H. Hansen, *Cycles of Prosperity and Depression in the United States, Great Britain and Germany: a Study of Monthly Data, 1902–1908* (Madison: University of Wisconsin, 1921), p. 13.
68. H.S. Foxwell, 'The American Crisis of 1907', in *Papers in Current Finance* (London: Macmillan, 1919), pp. 202–3.

6 Euphoria and Paper Wealth

1. Peter M. Garber, 'Tulipmania', in Robert P. Flood and Peter M. Garber, *Speculative Bubbles, Speculative Attacks, and Policy Switching* (Cambridge, Mass.: MIT Press, 1994), p. 72.
2. N.W. Posthumus, 'The Tulip Mania in Holland in the Years 1636 and 1637', *Journal of Business and Economic History*, vol. 1 (1928–29), reprinted in W.C. Scoville and J.C. LaForce, eds, *The Economic Development of Western Europe*, vol. 2, *The Sixteenth and Seventeenth Centuries* (Lexington, Mass.: D.C. Heath, 1969), p. 169.
3. Simon Schama, *The Embarrassment of Riches: an Interpretation of Dutch Culture in the Golden Age* (New York: Knopf, 1987), p. 358. Schama's source is Krelage, *Bluemenspekulation*.
4. Jonathan I. Israel, *The Dutch Republic: Its Rise, Greatness and Fall, 1477–1806* (Oxford: Clarendon Press, 1995), p. 533. Jan de Vries and Ad van der Woude state that in

the summer of 1636, speculation in commodity derivatives spread far beyond the circle of tulip fanciers, its flame being fanned by severe outbreaks of bubonic plague that released inhibitions (*The First Modern Economy: Success, Failure and Perseverance of the Dutch Economy, 1500–1815* [Cambridge: Cambridge University Press, 1997], pp. 150–1).

5. Jan de Vries, *Barges and Capitalism: Transportation in the Dutch Economy* (Wageningen: A.G. Bidragen, 1978), pp. 52ff.

6. Israel, *The Dutch Republic*, p. 533.

7. Garber, 'Tulipmania', pp. 71–2.

8. Israel, *The Dutch Republic*, chap. 33.

9. Ibid., p. 869.

10. Homer Hoyt, *One Hundred Years of Land Values in Chicago: the Relationship of the Growth of Chicago to the Rise in Land Values, 1830–1933* (Chicago: University of Chicago Press, 1933).

11. Quoted in Hoyt, *Land Values*, p. 165. This suggests that the 'greater fool theory' – a speculator buying an asset he believes is overpriced because he thinks he can sell it to a greater fool, widely quoted today – goes back at least a century.

12. Fritz Stern, *Gold and Iron: Bismarck, Bleichröder, and the Building of the German Empire* (London: Allen & Unwin, 1977), p. 161.

13. Hoyt, *Land Values*, p. 102. One stimulus to the boom was the huge fire of 6 October 1871, which destroyed somewhat more than one-fourth of the 60,000 houses in the city.

14. Hoyt, *Land Values*, p. 401.

15. 'How to Ruin a Safe Bet: Did Rockefeller Center Financiers Reach Too Far?', *New York Times*, 5 October 1995, pp. D1, D11.

16. Keizai Koho Center, *Japan 1994: an International Comparison* (Tokyo, 1993), Chart 11–3, 'Increase in Land Prices in Japan', p. 83.

17. Koichi Hamada, 'Bubbles, Busts and Bailouts', in Mitsuaki Okabe, ed., *The Structure of the Japanese Economy* (London: Macmillan, 1995), pp. 263–86.

18. Masahiko Takeda and Philip Turner, 'The Liberalization of Japanese Financial Markets: Some Major Themes', *BIS Economic Papers*, no. 34 (November 1992), graph 8, p. 53.

19. Takeda and Turner, 'Liberalization of Japanese Financial Markets', pp. 99–121.

20. Ibid., Table A-1, pp. 120–1.

21. Ibid., pp. 58–65. Hamada calls the revelations 'scandals' (p. 9).

22. Herman Kahn, *The Emerging Japanese Superstate: Challenge and Response* (Englewood Cliffs, NJ: Prentice-Hall, 1970).

23. 'Erosion in Japan's Foundation: Real Estate Crash Threatens the Entire Economy', *New York Times*, 4 October 1995.

24. David Asher and Andrew Smithers, 'Japan's Key Challenges for the 21st Century', SAIS (School for Advanced International Studies) Policy Forum Studies, April 1998.

25. Keizai Koho Center, *Japan 1994*, Charts 5–0, p. 52, and 4–19, p. 44.

26. See C.E.B. Borio, N. Kennedy, and S.D. Prowse, 'Exploring Aggregate Price Formation across Countries: Measurement, Determinants and Monetary-Policy Implications', *BIS Economic Papers*, no. 40 (Spring 1994), p. 46: 'It has been widely accepted that the primary goal of monetary policy should be price stability.'

27. See Armen A. Alchian and Benjamin Klein, 'On a Correct Measure of Inflation', *Journal of Money, Credit and Banking*, vol. 5, no. 1 (February 1973), pp. 172–91.

7 Bernie Madoff: Frauds, Swindles, and the Credit Cycle

1. See Norman C. Miller, *The Great Salad Oil Swindle* (New York: Coward, McCann, 1965).
2. Hyman P. Minsky, 'Financial Resources in a Fragile Financial Environment', *Challenge*, vol. 18 (July/August 1975), p. 65.
3. Martin F. Hellwig, 'A Model of Borrowing and Lending with Bankruptcy', Princeton University Econometric Research Program, Research Memorandum no. 177 (April 1975), p. 1.
4. Daniel Defoe, *The Anatomy of Change-Alley* (London: E. Smith, 1719), p. 8. See also the title of Jean Carper's book on fraud, *Not with a Gun* (New York: Grossman, 1973).
5. Jacob van Klavaren, 'Die historische Erscheinungen der Korruption', *Vierteljahrschrift für Sozial- und Wirtschaftsgeschichte*, vol. 44 (December 1957), pp. 289–324; vol. 45 (December 1958), pp. 433–69, 469–504; vol. 46 (June 1959), pp. 204–31. See also idem, 'Fiskalismus – Mercantilismus – Korruption: Drei Aspecte der Finanz- und Wirtschaftspolitik während der Ancien Regime', ibid., vol. 47 (September 1960), pp. 333–53.
6. E. Ray McCartney, *The Crisis of 1873* (Minneapolis: Burgess, 1935), p. 15.
7. John Carswell, *The South Sea Bubble* (London: Cresset, 1960), p. 13.
8. Maximillian E. Novak, *Economics and the Fiction of Daniel Defoe* (Berkeley: University of California Press, 1962), p. 103.
9. Bray Hammond, *Banks and Politics in America from the Revolution to the Civil War* (Princeton: Princeton University Press, 1957), p. 268.
10. Carswell, *South Sea Bubble*, pp. 222–4.
11. William G. Shepheard, Wall Street editor of *Business Week*, in the preface to Donald H. Dunn, *Ponzi, the Boston Swindler* (New York: McGraw-Hill, 1975), p. x.
12. Milton Friedman, 'In Defense of Destabilizing Speculation', in *The Optimum Quantity of Money and Other Essays* (Chicago: Aldine, 1969), p. 290.
13. Quoted in Max Winkler, *Foreign Bonds, an Autopsy: a Study of Defaults and Repudiations of Government Obligations* (Philadelphia: Roland Swain, 1933), p. 103.
14. E. Victor Morgan, *The Theory and Practice of Central Banking, 1797–1913* (Cambridge: Cambridge University Press, 1943), p. 177.
15. Honoré de Balzac, *Melmoth réconcilié* (Geneva: Editions de Verbe, 1946), pp. 45–50.
16. Dunn, *Ponzi*, p. 188.
17. Introduction by Robert Tracey to Anthony Trollope, *The Way We Live Now* (1874–75; reprint edn, New York: Bobbs-Merrill, 1974), p. xxv.
18. Sir John Clapham, *The Bank of England: a History* (Cambridge: Cambridge University Press, 1945), vol. 1, p. 229.
19. Charles Wilson, *Anglo-Dutch Commerce and Finance in the Eighteenth Century* (Cambridge: Cambridge University Press, 1941), p. 170.
20. George W. Van Vleck, *The Panic of 1857: an Analytical Study* (New York: Columbia University Press, 1943), p. 65.
21. Earl J. Hamilton, 'The Political Economy of France at the Time of John Law', *History of Political Economy*, vol. 1 (Spring 1969), p. 146.
22. Jacob van Klavaren, 'Rue de Quincampoix und Exchange Alley: Die Spekulationsjähre 1719 und 1720 in Frankreich und England', *Vierteljahrschrift für Sozial- und Wirtschaftsgeschichte*, vol. 48 (October 1961), pp. 329ff.
23. Dunn, *Ponzi*, op. cit.

24. Max Wirth, *Geschichte der Handelskrisen*, 4th edn (1890; reprint edn, New York: Burt Franklin, 1968), p. 510.
25. William Robert Scott, *The Constitution and Finance of English, Scottish and Irish Joint-Stock Companies to 1720* (Cambridge: Cambridge University Press, 1911), vol. 3, pp. 449ff.; and D. Morier Evans, *The Commercial Crisis, 1847–48*, 2nd edn, rev. (1849; reprint edn, New York: Augustus M. Kelley, 1969), pp. 33–4. A separate list for the South Sea Bubble, prepared by a contemporary and less detailed, is set out in Wirth, *Handelskrisen*, pp. 67–79.
26. A. Andréadès, *History of the Bank of England* (London: P.S. King, 1909), p. 133.
27. Carswell, *South Sea Bubble*, p. 142.
28. Scott, *Joint-Stock Companies*, p. 450.
29. Hans Rosenberg, *Die Weltwirtschaftskrise von 1857–59* (Stuttgart: W. Kohlhammer, 1934), p. 103.
30. Wirth, *Handelskrisen*, p. 480.
31. Anthony Trollope, *The Three Clerks* (New York: Harper & Brothers, 1860), p. 346.
32. Carswell, *South Sea Bubble*, p. 177.
33. Fritz Stern, *Gold and Iron: Bismarck, Bleichröder, and the Building of the German Empire* (London: Allen & Unwin, 1977), p. 358.
34. Ibid., pp. 396–7.
35. US Senate, Committee on Finance, 72nd Cong., 1st sess., *Hearings on Sales of Foreign Bonds of Securities*, held 18 December 1931, to 10 February 1932 (Washington, DC: US Government Printing Office, 1932).
36. O.M.W. Sprague, *History of Crises under the National Banking System* (1910; reprint edn, New York: Augustus M. Kelley, 1968), p. 341.
37. Quoted in Wirth, *Handelskrisen*, p. 80.
38. Honoré de Balzac, *La maison Nucingen*, in *Oeuvres complètes* (Paris: Calmann Lévy, 1892), p. 68.
39. Carswell, *South Sea Bubble*, pp. 176, 181.
40. Wirth, *Handelskrisen*, p. 491.
41. Emile Zola, *L'Argent* (Paris: Livre de Poche, n.d.), p. 125.
42. Wirth, *Handelskrisen*, p. 491.
43. Zola, *L'Argent*, p. 161.
44. Jean Bouvier, *Le krach de l'Union Générale* (Paris: Presses universitaires de France, 1960), p. 36.
45. *The Economist*, 21 October 1848, pp. 1186–8, quoted in Arthur D. Gayer, W.W. Rostow, and Anna J. Schwartz, *The Growth and Fluctuation of the British Economy, 1790–1850* (Oxford: Oxford University Press, Clarendon Press, 1953), p. 316.
46. Rosenberg, *Weltwirtschaftskrise*, p. 101.
47. Stern, *Gold and Iron*, chap. 10 ('Greed and Intrigue') and p. 364.
48. Novak, *Economics of Defoe*, pp. 14–15 and 160 n. 35.
49. Ibid., p. 16, note 50.
50. Maurice Lévy-Leboyer, *Les banques européennes et l'industrialisation internationale dans la première moitié du XIXᵉ siècle* (Paris: Presses universitaires de France, 1964), pp. 632–3.
51. Ibid., p. 503, note 90.
52. Bouvier, *Le krach*, p. 124.
53. Ibid., p. 161, note 50.
54. Stern, *Gold and Iron*, chap. 11.
55. Theodore Dreiser's *Trilogy of Desire* consists of three novels: *The Financier* (1912), *The Titan* (1914), and *The Stoic* (1947). See *The Titan* (New York: World, 1972), pp. 371–2.

56. Ibid., pp. 515–40.
57. Henry Grote Lewin, *The Railway Mania and its Aftermath, 1845–1852* (1936; reprint edn, rev., New York: Augustus M. Kelley, 1968), pp. 262, 357–64.
58. See Paul W. Gates, *Illinois Central Railroad and Its Colonization Work* (1934; reprint edn, Cambridge, Mass.: Harvard University Press, 1968), pp. 66, 75–6; John L. Weller, *The New Haven Railroad: the Rise and Fall* (New York: Hastings House, 1969), p. 37n; Van Vleck, *Panic of 1857*, p. 58.
59. See Willard L. Thorp, *Business Annals* (New York: National Bureau of Economic Research, 1926), p. 126.
60. Watson Washburn and Edmund S. Delong, *High and Low Financiers: Some Notorious Swindlers and their Abuses of our Modern Stock Selling System* (Indianapolis: Bobbs-Merrill, 1932), p. 13.
61. Ibid., pp. 85, 101, 144, 309.
62. Barrie A. Wigmore, *The Crash and its Aftermath: a History of Security Markets in the United States, 1929–1933* (Westport, Conn.: Greenwood, 1985), pp. 344–8.
63. Ibid., pp. 358–60.
64. James S. Gibbons, *The Banks of New York, Their Dealers, the Clearing House, and the Panic of 1857* (New York: D. Appleton, 1859), p. 104.
65. Ibid., p. 277.
66. John Kenneth Galbraith, *The Great Crash, 1929*, 3rd edn (Boston: Houghton Mifflin, 1972), pp. 133–5.
67. David F. Good, *The Economic Rise of the Hapsburg Empire, 1750–1914* (Berkeley: University of California Press, 1984), p. 165.
68. D. Morier Evans, *Facts, Failures and Frauds* (1839; reprint edn, New York: Augustus M. Kelley, 1968), p. 235.
69. See Robert Shaplen, *Kreuger: Genius and Swindler* (New York: Knopf, 1960).
70. Carswell, *South Sea Bubble*, pp. 225, 265–6.
71. Bouvier, *Le krach*, pp. 211, 219.
72. Herbert I. Bloom, *The Economic Activities of the Jews of Amsterdam in the Seventeenth and Nineteenth Centuries* (Williamsport, Pa.: Bayard Press, 1937), p. 199.
73. Carswell, *South Sea Bubble*, p. 210.
74. Dreiser, *The Titan*, p. 237.
75. Christina Stead, *House of All Nations* (New York: Simon & Schuster, 1938), p. 643.

8 International Contagion 1618–1930

1. Herbert Hoover, *The Memoirs of Herbert Hoover* (New York: Macmillan, 1952), vol. 3, pp. 61–2.
2. Milton Friedman and Anna J. Schwartz, *A Monetary History of the United States, 1867–1960* (Princeton: Princeton University Press, 1963), pp. 359–60.
3. Quoted in Leone Levi, *History of British Commerce* (London: John Murray, 1872), p. 234.
4. S. Saunders, quoted in D. Morier Evans, *The History of the Commercial Crisis, 1857–1858, and the Stock Exchange Panic of 1859* (1859; reprint edn, New York: Augustus M. Kelley, 1969), p. 13.
5. R.C.O. Matthews, *A Study in Trade-Cycle History: Economic Fluctuations in Great Britain, 1832–1842* (Cambridge: Cambridge University Press, 1954), p. 69.
6. Friedman and Schwartz, *Monetary History*, p. 360.
7. Jørgen Pedersen, 'Some Notes on the Economic Policy of the United States during the Period 1919–1932', in Hugo Hegelund, ed., *Money Growth and Methodology: Papers*

in Honor of Johan Åkerman (Lund: Lund Social Science Studies, 1961), reprinted in J. Pedersen, *Essays in Monetary Theory and Related Subjects* (Copenhagen: Samfundsvienskabeligt Forlag, 1975), p. 189.

8. R.T. Naylor, *The History of Canadian Business, 1867–1914*, vol. 1, *The Banks and Finance Capital* (Toronto: James Lorimer & Co., 1975), p. 130.

9. Clément Juglar, *Des crises commerciales et leur retour périodique en France, en Angleterre et aux Etats-Unis*, 2nd edn (1889; reprint edn, New York: Augustus M. Kelley, 1967), pp. xiv, 17, 47, 149, and passim.

10. Wesley C. Mitchell, introduction to Willard L. Thorp, *Business Annals* (New York: National Bureau of Economic Research, 1926), pp. 88–97.

11. Oskar Morgenstern, *International Financial Transactions and Business Cycles* (Princeton: Princeton University Press, 1959), chap. 1. esp. sec. 6; on international stock exchange panics from 1893 to 1931, see table 139, pp. 546–7, and chart 72, p. 548.

12. C.P. Kindleberger, 'The International (and Interregional) Aspects of Financial Crises', in Martin Feldstein, ed., *The Risk of Economic Crisis* (Chicago: University of Chicago Press, 1991), pp. 128–32.

13. Friedman and Schwartz, *Monetary History*, p. 308.

14. See C.E.V. Borio, N. Kennedy, and S.D. Prowse, 'Exploring Aggregate Price Fluctuations across Countries: Measurement, Determinants, and Monetary Policy Implications', *BIS Economic Papers*, no. 40 (April 1994), Graph A.1, p. 74.

15. C.P. Kindleberger, 'The Economic Crisis of 1619 to 1623', *Journal of Economic History*, vol. 51, no. 1 (March 1991), esp. pp. 159–61.

16. Johan Åkerman, *Structure et cycles économiques* (Paris: Presses universitaires de France, 1957), vol. 2. pp. 247, 255.

17. John Carswell, *The South Sea Bubble* (London: Cresset Press, 1960), pp. 84, 94, 100, 101.

18. Ibid., pp. 151, 160–1, 166. There may be some doubt about this. P.G.M. Dickson maintains that the Canton of Berne still held £287,000 of South Sea annuities in 1750 (*The Financial Revolution in England: a Study in the Development of Public Credit, 1688–1756* [New York: St Martin's Press, 1967], p. 90).

19. Charles Wilson, *Anglo-Dutch Commerce and Finance in the Eighteenth Century* (Cambridge: Cambridge University Press, 1941), pp. 103, 124.

20. Carswell, *South Sea Bubble*, p. 167.

21. Ibid., pp. 178, 199. In *The Great Mirror of Folly* [*Het Groote Tafereel der Dwaasheid*], *an Economic Bibliographical Study* (Boston: Baker Library, Harvard Graduate School of Business Administration, 1949), Arthur H. Cole observes that Holland had a full-scale bubble between April and October 1720, stimulated by the excitements of Paris and London. Forty new companies were floated in thirty mostly smaller towns, in the amount of 350 million guilders. Shares of the Dutch East India Company tripled in this period, and those of the West India Company went from 40 to 600 before the bubble burst (pp. 5, 6).

22. T.S. Ashton, *Economic Fluctuations in England, 1700–1800* (Oxford: Oxford University Press, Clarendon Press, 1959), p. 120.

23. George Chalmers, *The Comparative Strength of Great Britain* (London, 1782), p. 141, quoted in Ashton, *Economic Fluctuations*, p. 151.

24. E.E. de Jong-Keesing, *De Economische Crisis van 1763 te Amsterdam* (Amsterdam, 1939), pp. 216–17.

25. Wilson, *Anglo-Dutch Commerce*, p. 168.

26. De Jong-Keesing, *Economische Crisis van 1763*, p. 217.

27. Ernst Baasch, *Hollandische Wirtschaftsgeschichte* (Jena: Gustav Fischer, 1927).

28. Max Wirth, *Geschichte der Handelskrisen*, 4th edn (1890; reprint edn, New York: Burt Franklin, 1968), p. 87.

29. Stephan Skalweit, *Die Berliner Wirtschaftskrise von 1763 und ihre Hintergründe* (Stuttgart/Berlin: Verlag W. Kohlhammer, 1937), p. 50.

30. Wilson, *Anglo-Dutch Commerce*, p. 168; Alice Clare Carter, *Getting, Spending and Investing in Early Modern Times: Essays on Dutch, English and Huguenot Economic History* (Assen: Van Gorcum, 1975), p. 63.

31. William Smart, *Economic Annals of the Nineteenth Century*, vol. 1 (1911; reprint edn, New York: Augustus M. Kelley, 1964), pp. 529–30.

32. Arthur D. Gayer, W.W. Rostow, and Anna J. Schwartz, *The Growth and Fluctuation of the British Economy, 1790–1850* (Oxford: Oxford University Press, Clarendon Press, 1953), vol. 1, p. 159.

33. Smart, *Economic Annals*, vol. 1, chap. 31.

34. Murray N. Rothbard, *The Panic of 1819: Reactions and Policies* (New York: Columbia University Press, 1962), p. 11.

35. Bray Hammond, *Banks and Politics in America from the Revolution to the Civil War* (Princeton: Princeton University Press, 1957), chap. 10, esp. pp. 253–62.

36. Maurice Lévy-Leboyer, *Les banques européennes et l'industrialisation internationale dans la première moitié du XIXe siècle* (Paris: Presses universitaires de France, 1964), pp. 464–79.

37. Åkerman, *Structure et cycles économiques*, p. 294.

38. Maurice Lévy-Leboyer, 'Central Banking and Foreign Trade: the Anglo-American Cycle in the 1830s', in C.P. Kindleberger and J.-P. Laffargue, eds, *Financial Crises: Theory, History and Policy* (Cambridge: Cambridge University Press, 1982), pp. 66–110.

39. R.G. Hawtrey, *Currency and Credit*, 3rd edn (New York: Longmans, Green, 1927), p. 177.

40. Lévy-Leboyer, *Banques européennes*, pp. 570–83.

41. Juglar, *Crises Commerciales*, p. 414.

42. D. Morier Evans, *The Commercial Crisis, 1847–48* (1849; reprint edn, New York: Augustus M. Kelley, 1969).

43. Richard Tilly, *Financial Institutions and Industrialization in the Rhineland, 1815–1870* (Madison: University of Wisconsin Press, 1970), p. 112.

44. Alfred Krüger, *Das Kölner Bankiergewerbe vom Ende des 18. Jahrhunderts bis 1875* (Essen: G.D. Baedeker Verlag, 1925), pp. 12–13, 35, 49, 55–6, 202–3. Reference courtesy of Professor Richard Tilly.

45. This is discussed in a Swedish study of the crisis of 1857: P.E. Bergfalk, *Bidrag till de under de sista hundrade aren inträffade handelskrisershistoria* (Uppsala: Edquist, 1859), referred to in Theodore E. Burton, *Financial Crises and Periods of Industrial and Commercial Depression* (New York: Appleton, 1902), pp. 128–9.

46. Sir John Clapham, *The Bank of England: a History* (Cambridge: Cambridge University Press, 1945), vol. 2, p. 226.

47. Hans Rosenberg, *Die Weltwirtschaftskrise von 1857–1859* (Stuttgart: Verlag von W. Kohlhammer, 1934), p. 136.

48. Åkerman, *Structure et cycles économiques*, p. 323.

49. Clapham, *Bank of England* vol. 2, p. 268. R.G. Hawtrey makes the point that the crisis was not isolated but a sequel to the Continental crisis of 1864. See *Currency and Credit*, p. 177.

50. Shepard B. Clough, *The Economic History of Modern Italy* (New York: Columbia University Press, 1964), p. 53.

51. David S. Landes, *Bankers and Pashas: International Finance and Economic Imperialism in Egypt* (Cambridge, Mass.: Harvard University Press, 1958), p. 287.

52. Wirth, *Handelskrisen*, pp. 462–3.

53. Larry T. Wimmer, 'The Gold Crisis of 1869: a Problem in Domestic Economic Policy and International Trade Theory' (PhD diss., University of Chicago, 1968); US Congress, House, *Gold Panic Investigation*, 41st Cong., 2nd sess., H. Rept. 31, 1 March 1870.

54. Wirth, *Handelskrisen*, p. 464.

55. US Congress, House, *Gold Panic Investigation*, p. 132.

56. C.P. Kindleberger, 'The Panic of 1873', paper presented to the NYU-Salomon Brothers Symposium on Financial Panics, reprinted in *Historical Economics: Art or Science?* (New York: Harvester/Wheatsheaf, 1990), pp. 310–25.

57. Henrietta M. Larson, *Jay Cooke, Private Banker* (Cambridge, Mass.: Harvard University Press, 1936).

58. Homer Hoyt, *One Hundred Years of Land Values in Chicago: the Relationship of the Growth of Chicago to the Rise in Land Values, 1830–1933* (Chicago: University of Chicago Press, 1933), pp. 101–2, 117.

59. R. Ray McCartney, *Crisis of 1873* (Minneapolis: Burgess, 1935), p. 85.

60. Morgenstern, *International Financial Transactions*, p. 546.

61. Fritz Stern, *Gold and Iron: Bismarck, Bleichröder, and the Building of the German Empire* (London: Allen & Unwin, 1977), p. 189.

62. Morgenstern, *International Financial Transactions*, p. 548.

63. L.S. Pressnell, 'The Sterling System and Financial Crises before 1914', in Kindleberger and Laffargue, eds, *Financial Crises*, pp. 148–64.

64. O.M.W. Sprague, *History of Crises under the National Banking System* (1910; reprint edn, New York: Augustus M. Kelley, 1968), p. 132.

65. C.P. Kindleberger, 'International Propagation of Financial Crises: the Experience of 1888–93', in idem, *Keynesianism vs. Monetarism and Other Essays in Financial History* (London: Allen & Unwin, 1985), pp. 226–39.

66. Franco Bonelli, 'The 1907 Financial Crisis in Italy: a Peculiar Case of the Lender of Last Resort in Action', in Kindleberger and Laffargue, eds, *Financial Crises*, pp. 51–65.

67. Franco Bonelli, *La crisi del 1907: una tappa dello sviluppo industriale in Italia* (Turin: Fondazione Luigi Einaudi, 1971), pp. 31–2.

68. Ibid., p. 34.

69. Ibid., pp. 42–3.

70. Frank Vanderlip, 'The Panic as a World Phenomenon', *Annals of the American Academy of Political and Social Science*, vol. 31 (January–June 1908), p. 303.

10 Policy Responses: Benign Neglect, Exhortation, and Bank Holidays

1. Ian Giddy, 'Regulation of Off-Balance Sheet Banking', in *The Search for Financial Stability: the Past Fifty Years* (San Francisco: Federal Reserve Bank of San Francisco, 1985), pp. 165–77.

2. Edward J. Kane, 'Competitive Financial Reregulation: an International Perspective', in R. Portes and A. Swoboda, eds, *Threats to International Financial Stability* (Cambridge: Cambridge University Press, 1987), pp. 111–45.

3. Stanley Zucker, *Ludwig Bamberger: German Liberal Politician and Social Critic, 1823–1899* (Pittsburgh: University of Pittsburgh Press, 1975), p. 78.

4. Thomas Joplin, *Case for Parliamentary Inquiry into the Circumstances of the Panic, in a Letter to Thomas Gisborne, Esq., M. P.* (London: James Ridgeway & Sons, n.d. [after 1832]), p. 10 (apropos the panic of 1825).

5. Sir John Clapham, *The Bank of England: a History* (Cambridge: Cambridge University Press, 1945), vol. 2, p. 236 (apropos the panic of 1847).

6. E. Victor Morgan, *The Theory and Practice of Central Banking, 1797–1913* (Cambridge: Cambridge University Press, 1943), p. 133.

7. The episode is noted in A. Andréadès, *History of the Bank of England* (London: P.S. King, 1909), p. 334. For an extended discussion, see Rudiger Dornbusch and Jacob Frenkel, 'The Gold Standard and the Bank of England in the Crisis of 1847', in Michael Bordo and Anna J. Schwartz, eds, *A Retrospective on the Classical Gold Standard, 1821–1931* (Chicago: University of Chicago Press, 1984), pp. 233–64.

8. Ministère des Finances et al., *Enquête sur les principes et les faits géneraux qui régissent la circulation monétaire et fiduciare* (Paris: Imprimerie impériale, 1867), vol. 1, p. 456.

9. 'The Revulsion of 1857 – Its Causes and Results', *New York Herald*, n.d., quoted in D. Morier Evans, *The History of the Commercial Crisis, 1857–1858, and the Stock Exchange Panic of 1859* (1859; reprint edn, New York: Augustus M. Kelley, 1969), p. 121.

10. Herbert Hoover, *The Memoirs of Herbert Hoover* (New York: Macmillan & Co., 1952), vol. 3, p. 30.

11. Murray N. Rothbard, *America's Great Depression*, 3rd edn (Kansas City: Sheed & Ward, 1975), p. 187.

12. Paul Johnson, *Modern Times: the World from the Twenties to the Eighties* (New York: Harper & Row, 1983), p. 244.

13. Ernst Baasch, *Holländische Wirtschaftsgeschichte* (Jena: Gustav Fischer, 1927), p. 238.

14. Joplin, *Parliamentary Inquiry*, pp. 14–15.

15. Great Britain, *Parliamentary Papers, Monetary Policy, Commercial Distress* (1857; Shannon: Irish University Press, 1969), vol. 1, pp. 427, 431.

16. D. Morier Evans, *The Commercial Crisis, 1847–1848*, 2nd edn (1849; reprint edn, New York: Augustus M. Kelley, 1969), p. 89n.

17. Evans, *Commercial Crisis*, p. 181.

18. Parliamentary Papers, *Commercial Distress*, p. xii.

19. Ibid., vol. 4, appendix 20; *Foreign Communications Relative to the Commercial Crisis of 1857*, Hamburg consular circular no. 76, November 23, 1857, pp. 435, 440, 441.

20. Henrietta M. Larson, *Jay Cooke, Private Banker* (Cambridge, Mass.: Harvard University Press, 1936), p. 80.

21. W.C.T. King, *History of the London Discount Market* (London: George Routledge & Sons, 1936), p. 243.

22. Theodore Dreiser, *The Financier* (1912), in *Trilogy of Desire* (New York: World Publishing, 1972), p. 491.

23. W. Jett Lauck, *The Causes of the Panic of 1893* (Boston: Houghton, Mifflin, 1907), p. 102.

24. *The Commercial and Financial Chronicle*, 16 May 1884, p. 589, quoted in O.M.W. Sprague, *History of Crises under the National Banking System* (1910; reprint edn, New York: Augustus M. Kelley, 1968), p. 112.

25. W.R. Brock, *Lord Liverpool and Liberal Toryism, 1820–1827* (Cambridge: Cambridge University Press, 1941), pp. 209–10 (cited by Clapham, *Bank of England*, vol. 2, p. 108).

26. Clapham, *Bank of England*, vol. 2, p. 332.

27. Stephan Skalweit, *Die Berliner Wirtschaftskrise von 1763 und ihre Hintergründe* (Stuttgart: Verlag W. Kohlhammer, 1937), pp. 49–73.

28. Arthur D. Gayer, W.W. Rostow, and Anna J. Schwartz, *The Growth and Fluctuation of the British Economy, 1790–1850* (Oxford: Oxford University Press, Clarendon Press, 1953), vol. 1, p. 272.

29. Larry T. Wimmer, 'The Gold Crisis of 1869: a Problem in Domestic Economic Policy and International Trade Theory' (PhD diss., University of Chicago, 1968), p. 79.

30. Andréadès, *Bank of England*, p. 137, citing Henry D. McLeod, *Theory and Practice of Banking*, 3rd edn (London: Longman Green, Reader & Dyer, 1879), p. 428.

31. John Carswell, *The South Sea Bubble* (London: Cresset Press, 1960), p. 184.

32. Andréadès, *Bank of England*, p. 151.

33. Alexander Dana Noyes, *The Market Place: Reminiscences of a Financial Editor* (Boston: Little, Brown, 1938), p. 333.

34. Sprague, *History of Crises*, p. 259.

35. Ibid.

36. Ibid., p. 181.

37. Max Wirth, *Geschichte der Handelskrisen*, 4th edn (1890; reprint edn, New York: Burt Franklin, 1968), p. 521.

38. Maurice Lévy-Leboyer, *Les banques européennes et l'industrialisation internationale dans la première moitié du XIX^e siècle* (Paris: Presses universitaires de France, 1964), p. 480, text and note 5.

39. George W. Van Vleck, *The Panic of 1857: an Analytical Study* (New York: Columbia University Press, 1943), p. 80.

40. Sprague, *History of Crises*, pp. 120, 182–3.

41. Ibid., pp. 75, 291–2.

42. Jacob H. Schiff, 'Relation of a Central Bank to the Elasticity of the Currency', *Annals of the American Academy of Political and Social Science*, vol. 31 (January–June 1908), p. 375. For a more contemporary account, see Jean Strouse 'The Brilliant Bailout', *New Yorker* (23 November 1998), pp. 62–77.

43. Myron T. Herrick, 'The Panic of 1907 and Some of Its Lessons', *Annals of the American Academy of Political and Social Science*, vol. 31 (January-June 1908), p. 309.

44. John Kenneth Galbraith, *The Great Crash, 1929*, 3rd edn (Boston: Houghton Mifflin, 1972), pp. 107–8.

45. For a discussion of the rescue committee (*Aufhilfsfonds*) in Vienna, see Eduard März, *Österreich Industrie- und Bankpolitik in der Zeit Franz Josephs I: Am Beispiel der k.k. priv. Österreichischen Credit-Anstalt für Handel und Gewerbe* (Vienna: Europa Verlag, 1968), pp. 177–82. März notes (p. 179) that the best account of the crisis is that of Josef Neuwirth, *Bank und Valuta in Österreich*, vol. 2, *Die Spekulationskrisis van 1873* (no publishing data given).

46. Lévy-Leboyer, *Banques européennes*, pp. 470–1.

47. Bertrand Gille, *La banque en France au XIX^e siècle* (Paris: Droz, 1970), p. 93.

48. Parliamentary Papers, *Commercial Distress*, vol. 4, appendix, consular dispatch from Hamburg, no. 75. p. 434.

49. Ibid., p. 435.

50. Hans Rosenberg, *Die Weltwirtschaftskrise von 1857–1859* (Stuttgart/Berlin: Verlag von W. Kohlhammer, 1934), p. 129.

51. Parliamentary Paper, *Commercial Distress*, vol. 4, appendix, consular dispatches from Hamburg, nos. 77, 80, 81, 82, 84, 86, pp. 435–7.

52. Albert E. Fr. Schäffle, 'Die Handelskrise von 1857 in Hamburg, mit besonderer Rücksicht auf das Bankwesen', in *Gesammelte Aufsätze* (Tübingen: H. Raupp'schen, 1885), vol. 2, pp. 44, 45, 52, 53.

53. Clapham, *Bank of England*, vol. 2, p. 156.

54. Ibid., p. 331. For more detailed accounts of the Baring crisis and rescue, see L.S. Pressnell, 'Gold Reserves, Banking Reserves and the Baring Crisis of 1890', in C.R. Whittlesey and J.S.G. Wilson, eds, *Essays in Money and Banking in Honour of R.S. Sayers* (Oxford: Clarendon Press, 1968), pp. 67–228; and idem, 'The Sterling System and Financial Crises before 1914', in C.P. Kindleberger and J.-P. Laffargue, eds, *Financial Crises: Theory, History and Policy* (Cambridge: Cambridge University Press, 1982), pp. 148–64.

55. Diary of John Biddulph Martin, in George Chandler, *Four Centuries of Banking* (London: B.J. Batsford, 1964), vol. 1, p. 330.

56. Ellis T. Powell, *The Evolution of the Money Market (1384–1915): an Historical and Analytical Study of the Rise and Development of Finance as a Central Coordinated Force* (1915: reprint edn, New York: Augustus M. Kelley, 1966), p. 528.

57. Ibid., p. 525.

58. 'New York Fed Assists Hedge Fund Bailout', *New York Times*, 24 September 1998, pp. A1, C11.

59. There are some purists or masochists who think that bank runs are an optimal response of depositors, and good for the system, with panic as a form of monitoring. 'If depositors believe that there are some underperforming banks, but cannot decide which ones may become insolvent, they may force out all the undesirable ones by a systemwide panic.' This is from Charles W. Calomoris and Gary Gorton, 'The Origins of Banking Panics: Models, Facts and Bank Regulation', in R. Glenn Hubbard, ed., *Financial Markets and Financial Crises* (Chicago: University of Chicago Press, 1991), pp. 120–1.

60. Hoover, *Memoirs*, vol. 3, pp. 211–12.

61. 'Bank Board Doubles Texas Cost Estimate', *New York Times*, 8 July 1988, sec. D.

62. 'Treasury Says Savings Aid Should Not Tax the Taxpayer', *New York Times*, 3 August 1988, sec. D.

63. Wirth, *Handelskrisen*, p. 100.

64. Andréadès, *Bank of England*, pp. 187–9; Clapham, *Bank of England*, vol. 1, pp. 263–5.

65. Gayer, Rostow, and Schwartz, *Growth and Fluctuation*, vol. 1, p. 34.

66. William Smart, *Economic Annals of the Nineteenth Century* (1911; reprint edn, New York: Augustus M. Kelley, 1964), vol. 1, pp. 267–8.

67. Ibid., p. 271.

68. Wirth, *Handelskrisen*, pp. 110–11.

11 The Domestic Lender of Last Resort

1. François Nicholas Mollien, *Mémoires d'un Ministre du Trésor Public, 1780–1815* (Paris: Fournier, 1845), vol. 2, pp. 298ff.

2. Ministère des Finances et al., *Enquête sur les principes et les faits généraux qui régissent la circulation monétaire et fiduciaire* (Paris: Imprimerie impériale, 1867), vol. 2, pp. 31–2.

3. Murray N. Rothbard, *America's Great Depression*, 3rd edn (Kansas City: Sheed & Ward, 1975), p. 167.

4. Herbert Spencer, from 'State Tampering with Money and Banks', in *Essays: Scientific, Political and Speculative* (London: Williams & Norgate, 1891), vol. 3, p. 354, quoted by Charles Lipson, *Standing Guard: Protecting Foreign Capital in the Nineteenth and Twentieth Centuries* (Berkeley: University of California Press, 1985), p. 45.

5. T.S. Ashton, *Economic Fluctuations in England, 1700–1800* (Oxford: Oxford University Press, Clarendon Press, 1959), p. 112.

6. Ibid., p. 111.

7. E. Victor Morgan, *The Theory and Practice of Central Banking, 1797–1913* (Cambridge: Cambridge University Press, 1943), p. 240.

8. Maurice Lévy-Leboyer, *Les banques européennes et l'industrialisation internationale dans la première moitié du XIXᵉ siècle* (Paris: Presses universitaires de France, 1964), p. 490.

9. See F.A. Hayek's introduction to Henry Thornton, *An Enquiry into the Nature and Effect of the Paper Credit of Great Britain* (1802; London: Allen & Unwin, 1939; reprint, London: Frank Cass, 1962), p. 38.

10. Thornton, *Paper Credit*, pp. 187–8.

11. *The Collected Works of Walter Bagehot*, ed. Norman St John Stevas (London: *The Economist*, 1978), vol. 11, p. 149.

12. Ibid., vol. 9, p. 267.

13. Thomas Joplin, *Case for Parliamentary Inquiry into the Circumstances of the Panic, in a letter to Thomas Gisborne, Esq., M. P.* (London: James Ridgeway & Sons, n.d. [after 1832]), p. 29.

14. Jacob Viner, *Studies in the Theory of International Trade* (New York: Harper & Bros., 1937), p. 233.

15. D.P. O'Brien, 'Overstone's Thought', in D.P. O'Brien, ed., *The Correspondence of Lord Overstone* (Cambridge: Cambridge University Press, 1971), vol. 1, p. 95. (A bodkin is a dagger, stiletto, ornamental hairpin, or a blunt needle with a large eye.)

16. Milton Friedman and Anna J. Schwartz, *A Monetary History of the United States, 1867–1960* (Princeton: Princeton University Press, 1963), pp. 418–19. The metaphor of an avalanche is also used by Alfred Marshall in stating that prompt action by the Bank of England in regard to the rate of discount often checks unreasonable expansions of credit, 'which might otherwise grow, *after the manner of a fall of snow on a steep mountain side*' and become an avalanche (Marshall's emphasis). See Marshall, *Money, Credit, and Commerce* (1923; reprint edn, New York: Augustus M. Kelly, 1965), pp. 258–9.

17. Walter Bagehot, *Lombard Street: a Description of the Money Market* (1873; reprint edn, London: John Murray, 1917), p. 160.

18. Elmer Wood, *English Theories of Central Banking Control, 1819–1858* (Cambridge, Mass.: Harvard University Press, 1939), p. 147.

19. J.H. Clapham, *The Bank of England: a History* (Cambridge: Cambridge University Press, 1945), vol. 2, p. 289.

20. Bagehot, *Lombard Street*, pp. 161–2.

21. Clapham, *Bank of England*, vol. 2, p. 108.

22. Marshall, *Money, Credit, and Commerce*, p. 307. In their concluding chapter on the Dutch Republic, Jan de Vries and Ad van der Woude note that in 1780 the Dutch international banking sector lacked but one critical feature: effective credit-creating institutions to maintain liquidity in time of crisis (*The First Modern Economy* [Cambridge: Cambridge University Press, 1997], p. 683).

23. Bertrand Gille, *La banque en France au XIXᵉ siècle: Recherches historiques* (Geneva: Librairie Droz, 1970), p. 32.

24. Bertrand Gille, *La Banque et le crédit en France de 1815 à 1848* (Paris: Presses universitaires de France, 1959), p. 367.

25. Michel Chevalier, *Lettres sur l'Amérique du Nord*, 3rd edn (Paris: Charles Gosselin, 1838), p. 37, note 1.

26. Ministère des Finances et al., *Enquête*, vol. 3, pp. 411–12.

27. Ibid., vol. 2, pp. 129–30.

28. Jean Bouvier, *Un siècle de banque française* (Paris: Hachette Littérature, 1973), pp. 83–4.

29. See Rondo Cameron, *France and the Economic Development of Europe, 1800–1914* (Princeton: Princeton University Press, 1961), pp. 191ff.

30. Maurice Lévy-Leboyer, *Histoire économique et sociale de la France depuis 1848* (Paris: Les Cours de Droit, Institut d'Etudes Politiques, 1951–52), p. 121.

31. Cameron, *France and Europe*, p. 117.

32. Alfred Pose, *La monnaie et ses insitutions* (Paris: Presses universitaires de France, 1942), p. 215.

33. Jean Bouvier, *Le krach de l'Union Générale, 1878–1885* (Paris: Presses universitaires de France, 1960), pp. 150, 152–53.

34. Esther Rogoff Taus, *Central Banking Functions of the United States Treasury, 1789–1941* (New York: Columbia University Press, 1943), pp. 22, 23, 29.

35. Ibid., pp. 39–131.

36. C.A.E. Goodhart, *The New York Money Market and the Finance of Trade, 1900–1913* (Cambridge, Mass.: Harvard University Press, 1969), p. 120.

37. George W. Van Vleck, *The Panic of 1857: an Analytical Study* (New York: Columbia University Press, 1943), p. 106.

38. O.M.W. Sprague, *History of Crises under the National Banking System* (1910; reprint edn, New York: Augustus M. Kelley, 1968).

39. Myron T. Herrick, 'The Panic of 1907 and Some of Its Lessons', *Annals of the American Academy of Political and Social Science*, vol. 31 (January–June 1908), p. 324.

40. Ridgely's essay in *Annals* was more narrowly entitled 'An Elastic Credit Currency as a Preventive for Panics', but the papers of the others, except for Seligman, were unqualified in their advocacy of central bank currency elasticity.

41. Note Bagehot's characterization of the character of the then members of the court of the Bank of England: 'A board of plain, sensible prosperous English merchants; and they have both done and left undone what such a board might be expected to do and not to do. Nobody could expect great attainments in economical science from such a board; laborious study is for the most part foreign to the habits of English merchants' (Walter Bagehot, *Lombard Street: a Description of the Money Market* [1873; reprint edn, London: John Murray, 1917], p. 166). Later he is more critical: 'Unluckily … directors of the Bank of England were neither acquainted with the right principles, nor were they protected by a judicious routine. They could not be expected them selves to discover such principles. The abstract thinking of the world is never to be expected from people in high places … No doubt when men's own fortunes are at stake, the insight of the trader does somehow anticipate the conclusions of the closet' (p. 169).

42. Wood, *English Central Banking Control*, pp. 169–70.

43. Franco Bonelli, *La crisi del 1907: una tappa dello sviluppo industriale in Italia* (Turin: Fondazione Luigi Einaudi, 1971), passim and esp. p. 165.

44. Leone Levi, *History of British Commerce, 1763–1870*, 2nd edn (London: John Murray, 1872), pp. 311–12.

45. Great Britain, *Parliamentary Papers, Monetary Policy, Commercial Distress* (1857; Shannon: Irish University Press, 1969), vol. 3, p. xxix.

46. *Collected Works of Walter Bagehot*, vol. 9, p. 147.
47. Ibid., vol. 11, pp. 149–50.
48. See C.P. Kindleberger, 'Rules vs. Men: Lessons from a Century of Monetary Policy', in Christoph Buchheim, Michael Hutter, and Harold James, eds, *Zerrissene Zwischenkriegsheit: Wirtschaftshistorische Beiträge: Knut Borchardt zum 65. Geburstag*, reprinted in C.P. Kindleberger, *The World Economy and National Finance in Historical Perspective* (Ann Arbor: University of Michigan Press, 1995), pp. 181–200.
49. Bouvier, *Le krach*, chap. 5.
50. Charles Wilson, *Anglo-Dutch Commerce and Finance in the Eighteenth Century* (Cambridge: Cambridge University Press, 1941), pp. 176–7.
51. Friedman and Schwartz, *Monetary History*, p. 309 and esp. pp. 309–10n; Peter Temin, *Did Monetary Forces Cause the Great Depression?* (New York: W.W. Norton, 1976).
52. Clapham, *Bank of England*, vol. 1, p. 261; vol. 2, p. 58.
53. Ibid., vol. 2, pp. 59–60.
54. Ibid., vol. 1, p. 249.
55. Ibid., vol. 2, pp. 82–4.
56. Ibid., p. 145.
57. Alain Plessis, *La Politique de la Banque de France de 1851 à 1870* (Geneva: Droz, 1985), pp. 89, 99, 107. The Banque also loaned on Paris bonds, but not on those of Marseilles or Bordeaux; ibid., pp. 105–6.
58. Bagehot, *Lombard Street*, p. 195.
59. Clapham, *Bank of England*, vol. 2, p. 59.
60. W.C.T. King, *History of the London Discount Market* (London: George Routledge & Sons, 1936), p. 36.
61. Clapham, *Bank of England*, vol. 2, pp. 206–7.
62. Ibid., vol. 1, p. 261.
63. D. Morier Evans, *The History of the Commercial Crisis, 1857–1858, and the Stock Exchange Panic of 1859* (1859; reprint edn, New York: Augustus M. Kelley, 1969), p. 80.
64. Lévy-Leboyer, *Banques européennes*, p. 559.
65. Ibid., p. 647.
66. Ibid., p. 492.
67. Evans, *Commercial Crisis*, pp. i–ii, vi–xviii.
68. A. Andréadès, *History of the Bank of England* (London: P.S. King, 1909), p. 266.
69. Clapham, *Bank of England*, vol. 2, p. 157. He refers to it later as 'the long drawn out W bank affair' (ibid., p. 337).
70. R.C.O. Matthews, *A Study in Trade-Cycle History: Economic Fluctuations in Great Britain, 1832–1842* (Cambridge: Cambridge University Press, 1954), p. 173.
71. Clapham, *Bank of England*, vol. 1, p. 245.
72. H.S. Foxwell, preface to Andréadès, *Bank of England*, p. xvii.
73. Clapham, *Bank of England*, vol. 1, p. 256.
74. See Friedman and Schwartz, *Monetary History*, pp. 339, 363–7.
75. Ibid., p. 395.
76. Ibid., p. 339.
77. See Milton Friedman, 'Rediscounting', in *A Program for Monetary Stability* (New York: Fordham University Press, 1960), pp. 35–6.
78. Paul Johnson, *Modern Times: the World from the Twenties to the Eighties* (New York: Harper & Row, 1983), p. 244.
79. Friedman and Schwartz, *Monetary History*, pp. 334–5.
80. Clapham, *Bank of England*, vol. 2, p. 102.

81. Evans, *Commercial Crisis*, p. 207.
82. Taus, *Central Banking*, pp. 55, 70.
83. Sprague, *History of Crises*, p. 256.

12 The International Lender of Last Resort

1. Niall Ferguson, *The House of Rothschild: Money's Prophets, 1798–1847* (New York: Viking, 1998), chap. 16, Table 16a on p. 461, shows the course of bond prices in Vienna, Paris, Rome, and London. Ferguson states that 'the London house was not the lender of last resort at all'; but Alphonse, Nathan's grandson, was sent to New York to turn Belmont away from preoccupation with the Mexican indemnity and focus on Europe (pp. 466–70).
2. Ron Chernow, *The Warburgs: the Twentieth-Century Odyssey of a Remarkable Jewish Family* (New York: Random House, 1993), p. 328.
3. R.D. Richards, 'The First Fifty Years of the Bank of England, 1694–1744', in *History of the Principal Public Banks*, compiled by J.G. van Dillen (The Hague: Martinus Nijhoff, 1934), p. 234.
4. Violet Barbour, *Capitalism and Amsterdam in the 17th Century* (1950; reprint edn, Ann Arbor: University of Michigan Press, 1966), p. 125.
5. Charles Wilson, *Anglo-Dutch Commerce and Finance in the Eighteenth Century* (Cambridge: Cambridge University Press, 1941), pp. 168–9.
6. Ibid., p. 176.
7. Sir John Clapham, *The Bank of England: a History* (Cambridge: Cambridge University Press, 1945), vol. 1, p. 249.
8. Alain Plessis, *La Politique de la Banque de France de 1850 à 1870* (Geneva: Droz, 1985), chap. 4, esp. pp. 241–5.
9. Knut Wicksell, *Lectures on Political Economy* (New York: Macmillan, 1935), vol. 2, pp. 37–8.
10. Jacob Viner, *Studies in the Theory of International Trade* (New York: Harper & Bros., 1937), p. 273; Clapham, *Bank of England*, vol. 2, p. 169.
11. Clapham, *Bank of England*, vol. 2, pp. 164–5.
12. Clément Juglar, *Des crises commerciales et leur retour périodiques en France, en Angleterre et aux Etats-Unis*, 2nd edn (1889; reprint edn, New York: Augustus M. Kelley, 1967), p. 417.
13. E. Victor Morgan, *The Theory and Practice of Central Banking, 1797–1913* (Cambridge: Cambridge University Press, 1943), p. 148.
14. Arthur D. Gayer, W.W. Rostow, and Anna J. Schwartz, *The Growth and Fluctuation of the British Economy, 1790–1850* (Oxford: Oxford University Press, Clarendon Press, 1953), vol. 1, p. 333.
15. British Parliamentary Papers, *Causes of Commercial Distress: Monetary Policy* (Shannon: Irish University Press, 1969), vol. 1, p. 153, question 2018.
16. Viner, *Studies in International Trade*, p. 273.
17. Clapham, *Bank of England*, vol. 2, p. 170.
18. Bertrand Gille, *La Banque et le crédit en France de 1818 à 1848* (Paris: Presses universitaires de France, 1959), p. 377.
19. Clapham, *Bank of England*, vol. 2, p. 229.
20. Helmut Böhme, *Frankfurt und Hamburg: Des Deutsches Reiches Silber und Goldloch und die Allerenglishte Stadt des Kontinents* (Frankfurt-am-Main: Europaïsche Verlagsanstalt, 1968), pp. 255–68.

21. Ibid., pp. 267–74.
22. Parliamentary Papers, *Commercial Distress*, vol. 4, appendix 20, consular dispatch no. 7, January 27, 1858, p. 441.
23. Ibid., dispatch no. 393 from Berlin, December 29, 1857, pp. 450–1.
24. Clapham, *Bank of England*, vol. 2, p. 234.
25. Morgan, *Theory and Practice of Central Banking*, p. 176.
26. Clapham, *Bank of England*, vol. 2, pp. 291–4.
27. Ibid., pp. 329–30, 344.
28. Quoted from Hartley Withers, *The Meaning of Money* (1909) in R.S. Sayers, *Bank of England Operations 1890–1914* (London: P.S. King, 1936), p. 111.
29. J.-L. Billoret, 'Système bancaire et dynamique économique dans un pays à monnaie stable: France, 1896–1914', thesis, Faculty of Law and Economic Science, Nancy, 1969, quoted in Jean Bouvier, *Un siècle de banque française* (Paris: Hachette Littérature, 1973), p. 240.
30. Hans Rosenberg, *Die Weltwirtschaftskrise von 1857–1859* (Stuttgart: Verlag von W. Kohlhammer, 1934), p. 38.
31. George W. Van Vleck, *The Panic of 1857: an Analytical Study* (New York: Columbia University Press, 1943), p. 42.
32. Billoret, 'Système bancaire', as quoted in Bouvier, *Un siècle de banque française*, p. 238.
33. Walter Bagehot, *Lombard Street: a Description of the Money Market* (1873; reprint edn, London: John Murray, 1917), pp. 32–4.
34. O.M.W. Sprague, *History of Crises under the National Banking System* (1910; reprint edn, New York: Augustus M. Kelley, 1968), pp. 248–85.
35. Franco Bonelli, *La crisi del 1907: una tappa dello sviluppo industriale in Italia* (Turin: Fondazione Luigi Einaudi, 1971), p. 42.
36. Oskar Morgenstern, *International Financial Transactions and Business Cycles* (Princeton: Princeton University Press, 1959), pp. 128–37.
37. See League of Nations, *The Course and Control of Inflation after World War I* (Princeton: League of Nations, 1945).
38. Paul E. Erdman, *The Crash of '79* (New York: Simon & Schuster, 1976).
39. Stephen A. Schuker, *The End of French Predominance in Europe, the Financial Crisis of 1924 and the Adoption of the Dawes Plan* (Chapel Hill: University of North Carolina Press, 1976), p. 67.
40. J.N. Jeanneney, 'De la Spéculation financière comme arme diplomatique: A propos de la première bataille de franc (november 1923–mars 1924)', *Relations internationales*, no. 13 (Spring 1978), pp. 9–15, is persuaded that the German government organized the bear raid on the franc. Schuker is skeptical: *The End of French Predominance*, p. 56. Jean-Claude Debeir, 'La crise du franc de 1924: Une exemple de la spéculation "internationale"', *Relations Internationales*, no. 13 (Spring 1978), p. 35, lays considerable blame for financing the attacks on American banks but states that the French did most of the speculation.
41. Schuker, *The End of French Predominance*, chap. 4.
42. Ibid., p. 111.
43. C.P. Kindleberger. *The World in Depression, 1929–1939*, 2nd edn (Berkeley: University of California Press, 1986), esp. chaps 7, 14.
44. R.G. Hawtrey, *The Art of Central Banking* (London: Longmans, Green, 1932), pp. 220–4; all italics are in the original. Hawtrey is prescient to a degree. Since the International Monetary Fund takes time to make credit decisions as a lender of last resort, and the Bank for International Settlements can act quickly, the latter has

operated in a number of recent financial crises by providing 'bridge loans', that is, short-term finance (though limited in amounts) to countries with immediate needs. These loans would then be repaid with funds from credits from the International Monetary Fund. See Harold James, *International Monetary Cooperation since Bretton Woods* (Washington, DC: International Monetary Fund, 1995; New York and Oxford: Oxford University Press, 1996), pp. 361–2, 389, 557, 562, 592.

45. Hawtrey, *The Art of Central Banking*, pp. 229–32; italics are in the original.
46. Susan Howson and Donald Winch, *The Economic Advisory Council, 1930–1939: a Study in Economic Advice during Depression and Recovery* (Cambridge: Cambridge University Press, 1977), pp. 188–9. The personnel of the various committees are given in ibid., appendix 1, pp. 354–70.
47. Ibid., pp. 272–81.
48. This discussion is largely based on Karl Erich Born, *Die deustche Bankenkrise, 1931: Finanzen und Politik* (Munich: R. Piper & Co. Verlag, 1967).
49. Ibid., p. 86.
50. Ibid., p. 83.
51. Norman to Harrison, cable, 3 July 1931, in Federal Reserve Bank of New York files.
52. Stephen V.O. Clarke, *Central Bank Cooperation, 1924–31* (New York: Federal Reserve Bank of New York, 1967), p. 44. This was after discussion by both sides, with the British and Americans saying no (Kindleberger, *World in Depression*, p. 153). The Germans for their part were debating a plan of Wilhelm Lautenbach of the Economics Ministry, which called for suspending service on foreign debt and expanding spending (ibid., p. 171, n. 5). The Lautenbach plan came up again after the British pound went off gold in September 1931. See Knut Borchardt and Hans Otto Schöltz, eds, *Wirtschaftspolitik in der Krise* (Baden-Baden: Nomos, 1931 [1991]). This is a transcript of the riveting debate in the Reichsbank among German officials and economists.
53. Clarke, *Central Bank Cooperation*, pp. 121, 148.
54. Howson and Winch, *Economic Advisory Council*, pp. 88–9.
55. Ibid., p. 162.
56. Kindleberger, *World in Depression*, pp. 168, 184.
57. D.E. Moggridge, 'Policy in the Crises of 1920 and 1929', in C.P. Kindleberger and J.-P. Laffargue, eds, *Financial Crises: Theory, History and Policy* (Cambridge: Cambridge University Press, 1982), pp. 171–87.
58. Stephen A. Schuker, *American 'Reparations' to Germany, 1919–33: Implications for the Third-World Debt Crisis*, Princeton Studies in International Finance no. 61 (Princeton: International Finance Section, Princeton University Press, 1988).
59. Richard N. Cooper, 'Economic Interdependence and Foreign Policy in the Seventies', *World Politics*, vol. 24 (January 1972), p. 167.
60. See Bank for International Settlements, *Annual Report* no. 63 to 31 March 1993, p. 196; and no. 64 to 31 March 1994. World turnover rose from $10 billion to $20 billion a day in 1973, to $80 billion in 1983, before soaring to over $1 trillion a day in 1995: *The Economist*, vol. 337, no. 7936 (14 October 1995), p. 10. For emerging markets and the European Monetary Union, see Bank for International Settlements, 68th *Annual Report* to 31 March 1998, p. 11, and chart p. 115.
61. Charles A. Coombs, *The Arena of International Finance* (New York: Wiley), pp. 77, 79, 81, 83, 195, 202, 292.
62. Coombs, *Arena*, p. 37.
63. Ibid., pp. 81, 85, 111, 121, 127, 134, 181, 185.
64. Susan Strange, 'International Monetary Relations', in Andrew Schonfield, ed., *International Economic Relations of the Western World, 1959–71* (Oxford: Oxford

University Press, for the Royal Institute of International Affairs, 1976), vol. 2, p. 136.

65. See Debeir, 'La crise du franc du 1924'.

66. Kathleen Burk and Alec Cairncross, *'Goodbye Great Britain': the 1976 IMF Crisis* (New Haven, Conn.: Yale University Press, 1992).

67. For background, see the paper on Mexico by Ernesto Zedillo in Donald R. Lessard and John Williamson, eds, *Capital Flight and Third World Debt* (Washington, DC: Institute for International Economics, 1987), pp. 174–85. The crisis itself is discussed at length in the *New York Times* issues from 21 to 29 December 1994.

68. 'President Clinton Sidesteps Congress Using Emergency Authority', *New York Times*, 1 February 1995, p. A1.

69. Bank for International Settlements, *Annual Report*, no. 68, p. 134.

70. Ibid.

71. Wynne Godley, 'Seven Unsustainable Processes', a special report of the Levy Institute, Annandale-on-Hudson, New York, 1999.

72. C.P. Kindleberger, *World Economic Primacy, 1500–1990* (New York: Oxford University Press, 1996), chaps 10, 12.

73. Robert O. Keohane, *After Hegemony: Cooperation and Discord in the World Political Economy* (Princeton: Princeton University Press, 1984).

Index